Reshaping Language and Cognition in Education Through AI

Lalu Nurul Yaqin
Universiti Brunei Darussalam, Brunei

Yabit bin Alas
Universiti Brunei Darussalam, Brunei

Akhmad Habibi
Jambi University, Indonesia

IGI Global
Scientific Publishing
Publishing Tomorrow's Research Today

Vice President of Editorial Melissa Wagner
Director of Acquisitions Mikaela Felty
Director of Book Development Jocelynn Hessler
Production Manager Mike Brehm
Cover Design Jose Rosado

Published in the United States of America by
IGI Global Scientific Publishing
701 East Chocolate Avenue
Hershey, PA, 17033, USA
Tel: 717-533-8845 | Fax: 717-533-7115
Website: https://www.igi-global.com E-mail: cust@igi-global.com

Library of Congress Cataloging-in-Publication Data

Names: Yaqin, Lalu Nurul, 1985- editor | Yabit Alas editor | Habibi, Akhmad, 1983- editor
Title: Reshaping language and cognition in education through AI / edited by Lalu Nurul Yaqin, Yabit bin Alas, Akhmad Habibi.
Description: Hershey, PA : IGI Global Scientific Publishing, 2026. | Includes bibliographical references and index. | Summary: "This book aims to critically explore the role of AI as a transformative force within the domains of linguistics and education"-- Provided by publisher.
Identifiers: LCCN 2025042318 (print) | LCCN 2025042319 (ebook) | ISBN 9798337343976 hardcover | ISBN 9798337343983 paperback | ISBN 9798337343990 ebook
Subjects: LCSH: Language and languages--Study and teaching--Technological innovations | Linguistics--Technological innovations | Cognition | Artificial intelligence--Educational applications | LCGFT: Essays
Classification: LCC P53.855 .R48 2026 (print) | LCC P53.855 (ebook)
LC record available at https://lccn.loc.gov/2025042318
LC ebook record available at https://lccn.loc.gov/2025042319

British Cataloguing in Publication Data
A Cataloguing in Publication record for this book is available from the British Library.

Table of Contents

Detailed Table of Contents

Artificial intelligence (AI) has rapidly reshaped language education, shifting the field from rule-based CALL systems to data-driven, adaptive, and generative environments. Advances in deep learning, natural language processing, and large language models (LLMs) have greatly expanded AI-mediated instruction, yet a comprehensive longitudinal synthesis of the field remains limited. This study fills that gap through a scientometric analysis of 4,390 publications (2016–2025) using CiteSpace to map co-citation patterns, keyword bursts, collaboration networks, and disciplinary trends. Results show a nearly thirty-fold growth in outputs, from 66 publications in 2016 to 1,841 in 2025. Six major research fronts emerge: intelligent tutoring, automated writing evaluation, speech technologies, adaptive learning and analytics, conversational agents, and neural machine translationeach with distinct temporal trajectories. Co-citation clusters indicate a progression from classical NLP and CALL toward LLM-driven ecosystems, with collaboration networks centered in East Asia and Anglophone regions.

Universities worldwide embrace internationalization strategies, many are requiring faculty to publish their research in English to improve global rankings. This shift presents challenges, especially for those unprepared to navigate new publishing norms. Many faculty members encounter complex procedural barriers that hinder their ability to publish in English. To address these challenges, some scholars with limited English proficiency, particularly from the Malay (OML) background, are beginning to use artificial intelligence (AI). A thematic survey was conducted

This chapter provides a critical reflection on the encounter between artificial intelligence and the linguistic turn in education as language-based AI systems reconfigure meaning-making, discourse, and epistemology in today's educational settings. Drawing on interdisciplinary perspectives from applied linguistics, cognitive science, and educational theory, the chapter examines the underlying epistemological issues of AI interventions such as large language models, intelligent tutoring systems, and natural language processing. The authors consider how AI challenges as well as accommodates the linguistic turn's dialogic and constructivist ideals by appropriating the key components of classroom interaction, learner role, and authorship. Particularly, the chapter investigates the implications of AI for multilingual education, low-resource language preservation, and linguistic diversity and inclusion in generally underrepresented contexts. The chapter also interrogates the ideological and ethical dimensions of algorithmic language, including bias, and linguistic imperialism.

The rapid rise of artificial intelligence (AI) and digital tools in education and work has transformed learning, communication, and decision-making. While AI enhances access and personalization, it also raises ethical issues related to privacy, bias, misinformation, and academic integrity. This chapter explores how educational psychology can foster ethical decision-making, digital responsibility, and moral reasoning in the AI era. Grounded in Piaget's and Kohlberg's developmental theories and Bandura's Social Learning Theory, it connects moral growth with reflective digital practice. Through case-based learning, empathy, and critical digital literacy, it equips educators to build ethical, responsible digital citizens.

This chapter investigates the convergence of neurocomputational linguistics and AI-augmented cognitive modeling as a transformative force in education. Moving beyond

Preface

INTRODUCTION

Bismillahirohmanirohim,

Greetings from the editors

The rapid integration of artificial intelligence (AI) into educational contexts has fundamentally reshaped how language is learned, taught, assessed, and theorised. What was once confined to rule-based computer-assisted language learning has evolved into complex, data-driven, and generative systems capable of mediating meaning, cognition, and interaction at scale. As editors of *Reshaping Language and Cognition in Education Through AI*, we were motivated by a growing recognition that AI is no longer merely a technological aid in education, but an epistemic force that reconfigures how knowledge, language, and learning are constructed.

This volume brings together interdisciplinary scholarship that situates AI within broader linguistic, cognitive, cultural, and ethical frameworks. Rather than treating AI as a neutral tool, the chapters interrogate its role in shaping learner agency, academic knowledge production, cultural representation, and linguistic equity. Contributions span diverse methodological traditions from scientometric analysis and mixed-methods research to philosophical inquiry and neurocomputational modelling, reflecting the multifaceted nature of AI's impact on education.

Importantly, this edited collection foregrounds voices and contexts that are often underrepresented in global AI and education discourse, including low-resource languages, localised cultural knowledge, and non-Anglophone academic communities. By doing so, the volume aims to contribute not only to theoretical advancement but also to more inclusive and responsible AI practices in language education worldwide.

The volume is organised into four chapters (each containing multiple contributions) that collectively trace the movement from macro-level patterns to micro-level practices and future-facing implications.

OVERVIEW OF CHAPTERS

Chapter 1 maps the global research landscape of AI in language education through a large-scale scientometric analysis, revealing key research fronts, collaboration patterns, and the field's transition toward large language model–driven ecosystems. It establishes a macro-level foundation for understanding how AI research trajectories have evolved over the past decade.

Chapter 2 shifts attention to academic knowledge production, examining how AI tools are being used by scholars from Malay language backgrounds to navigate English-dominant publication norms. The chapter highlights both the affordances and limitations of AI in addressing structural inequities in global academia.

Chapter 3 addresses the persistent digital marginalization of low-resource languages. Drawing on global case studies, it argues for linguistically just AI development grounded in participatory governance, multilingual design, and cultural sustainability.

Chapter 4 offers a culturally grounded evaluation of how major AI platforms interpret Bruneian proverbs. By comparing literal, metaphorical, and moral interpretations, the chapter underscores the risks of cultural flattening and the importance of community-informed AI systems.

Chapter 5 explores English teachers' cognitions regarding AI-supported autonomous reading. Through mixed-methods analysis, it reveals that the pedagogical impact of AI is contingent on teacher awareness, instructional design, and ethical sensitivity.

Chapter 6 introduces the concept of *thinking through the algorithm*, reframing AI from a productivity aid to a dialogic partner in meaning-making. Using empirical evidence from language learners, it emphasizes learner agency, pragmatics, and communicative effectiveness.

Chapter 7 engages critically with the linguistic turn in education, examining how AI-mediated language challenges traditional notions of discourse, authorship, and epistemology. It raises important questions about ideology, bias, and linguistic imperialism in AI systems.

Chapter 8 focuses on the cognitive and ethical dimensions of AI-generated language, drawing on educational psychology to argue for the development of moral reasoning, digital responsibility, and ethical literacy in AI-rich learning environments.

Chapter 9 explores the convergence of neurocomputational linguistics and AI-augmented cognitive modeling. It proposes biologically inspired and linguistically grounded approaches to designing adaptive learning environments that align with human cognition.

Chapter 10 concludes the volume with a philosophical exploration of human–AI cognitive entanglement. By proposing typologies of synthetic and human cognition, it challenges notions of seamless convergence and advocates for epistemic plurality and responsible co-evolution.

Bandar Seri Begawan, Brunei Darussalam

Lalu Nurul Yaqin
Universiti Brunei Darussalam, Brunei

Yabit Alas
Universiti Brunei Darussalam, Brunei

Akhmad Habibi
Jambi University, Indonesia

Acknowledgment

We extend our sincere gratitude to all contributors, reviewers, and colleagues whose expertise and commitment have helped shape this edited volume, *Reshaping Language and Cognition in Education Through AI*. We gratefully acknowledge the support of our colleagues at Universiti Brunei Darussalam and Universitas Negeri Jambi, whose encouragement and assistance were invaluable throughout the project.

We also thank the IGI Global Scientific Publishing Team for their professional guidance and continuous support across every stage of the process from the initial proposal and call for chapter proposals to manuscript development and production.

It is our hope that this book will serve as a meaningful resource for educators, researchers, policymakers, and other stakeholders with an interest in language, cognition, and AI in education.

Chapter 1
The Evolving Landscape of AI in Language Education:
Knowledge Structures, Research Trajectories, and Global Collaboration Patterns

Lalu Nurul Yaqin
Universiti Brunei Darussalam, Brunei

ABSTRACT

Artificial intelligence (AI) has rapidly reshaped language education, shifting the field from rule-based CALL systems to data-driven, adaptive, and generative environments. Advances in deep learning, natural language processing, and large language models (LLMs) have greatly expanded AI-mediated instruction, yet a comprehensive longitudinal synthesis of the field remains limited. This study fills that gap through a scientometric analysis of 4,390 publications (2016–2025) using CiteSpace to map co-citation patterns, keyword bursts, collaboration networks, and disciplinary trends. Results show a nearly thirty-fold growth in outputs, from 66 publications in 2016 to 1,841 in 2025. Six major research fronts emerge: intelligent tutoring, automated writing evaluation, speech technologies, adaptive learning and analytics, conversational agents, and neural machine translationeach with distinct temporal trajectories. Co-citation clusters indicate a progression from classical NLP and CALL toward LLM-driven ecosystems, with collaboration networks centered in East Asia and Anglophone regions.

DOI: 10.4018/979-8-3373-4397-6.ch001

1. INTRODUCTION

Artificial intelligence (AI) has experienced a remarkable and accelerating evolution over the past four decades, progressing from rule-based expert systems and symbolic approaches to highly sophisticated data-driven architectures grounded in machine learning, deep learning, and, most recently, large language models (LLMs). Early AI systems in education relied on predefined rules, handcrafted domain representations, and rigid problem-solving pathways, which limited their adaptability and responsiveness to learner diversity (Bilad et al., 2023). Although these systems could deliver structured feedback, they struggled to accommodate individual differences in linguistic development or interactional patterns. The emergence of statistical learning and neural network models marked a decisive shift, enabling computational systems to infer patterns from large-scale behavioural and linguistic data. Transformer-based architectures and large-scale pre-trained language models have further accelerated this transformation, empowering AI systems to generate extended coherent text, sustain multi-turn dialogic exchanges, and integrate multimodal inputs with unprecedented fluency and contextual sensitivity (Devlin et al., 2019; Radford et al., 2019; Xiao et al., 2025).

These technological advances intersect with the long-standing evolution of computer-assisted language learning (CALL), intelligent CALL (ICALL), and technology-enhanced language learning (TELL). Early CALL tools consisted largely of drill-and-practice grammar exercises, discrete-point vocabulary tasks, and text reconstruction activities with minimal adaptively (Inayah et al., 2024). ICALL systems later introduced richer domain models capable of diagnosing learner weaknesses and tailoring instruction, while TELL expanded the pedagogical landscape through multimedia input, networked communication, and mobile learning environments. Within this broader trajectory, AI has come to be understood not merely as another tool within CALL, but as a methodological paradigm that fundamentally reshapes how language input, interaction, feedback, and assessment are conceptualised, implemented, and personalised (Chen et al., 2022; Hwang et al. 2020; Zawacki-Richter et al., 2019).

Contemporary AI applications now permeate virtually all domains of language education. Intelligent tutoring systems offer adaptive grammar, vocabulary, and reading support; automated writing evaluation (AWE) systems and AI-driven feedback tools analyses lexical sophistication, syntactic complexity, cohesion, and linguistic accuracy, often supplemented by LLM-generated revision suggestions (Barrot, 2023; L. Feng, 2025; Marshall & Tsikouras, 2023). Speech technologies such as automated speech recognition and speech-scoring engines enable pronunciation practice, mispronunciation detection, prosody analysis, and real-time fluency monitoring (Habibi et al. 2025; Korzekwa, 2021; Ni & Yin, 2025; Yaqin, 2025).

Moreover, chatbots, conversational agents, and LLM-based dialogue systems facilitate interactive role-play, collaborative problem solving, and affective support in both formal and informal contexts (Huang & Wang, 2022; Jeon, 2024; B. Liang, 2025). Adaptive recommender systems increasingly personalise learning trajectories (Namaziandost, 2025). while learning analytics and educational data mining inform data-driven instructional design and policy-making (Chen et al. 2020).

Parallel advances in natural language processing (NLP), neural machine translation (NMT), and generative AI have further reconfigured the pedagogical landscape, raising complex questions about translanguaging, academic integrity, learner autonomy, and evolving norms of digital literacy (Syafiee & Yaqin, 2025; Yaqin, et al., 2025a). High-performing NMT systems such as Google Translate and DeepL now play a central role in learners' writing and reading practices, prompting debates about their pedagogical legitimacy and ethical constraints (M. Deng, 2022; Jolley & Maimone, 2025; Tsai, 2022). Since the public release of ChatGPT in late 2022, generative AI and LLM-supported learning have received substantial scholarly attention, particularly in relation to writing development, conversational practice, and informal digital learning (Foung & Kohnke, 2025; Lahby, 2024; Yan, 2023; Yaqin, et al., 2025b). Scientometric evidence underscores the scale and speed of this transformation. Annual publications on AI-supported language education increased nearly thirtyfold between 2016 and 2025, rising from 66 to 1,841 documents. This expansion occurred across three phases: an initial period dominated by rule-based CALL and early machine-learning applications (2016–2018); a deep-learning-driven phase characterised by rapid adoption of neural architectures for speech and text processing (2019–2021); and a post-LLM surge (2022-2025) defined by generative AI, chatbots, and LLM-mediated instruction, assessment, and feedback (R. Deng et al., 2025; Erdem & Coşgun, 2025; Ding et al, 2025) Co-citation networks reveal a strong core of CALL and EdTech journals linked to applied linguistics, speech technology, AI, psychology, and cognitive science, illustrating an increasingly interdisciplinary research ecosystem.

Despite these advances, key challenges persist. The rapid diversification of topics complicates efforts to track emerging trends such as affective computing, multimodal feedback, and AI-mediated informal learning. The coexistence of foundational NLP studies, empirical classroom research, and AI literacy work renders it challenging for narrative reviews to delineate the field's intellectual structure (Devlin et al., 2019; Hwang et al. 2020; Zawacki-Richter et al., 2019). Moreover, uneven adoption of deep learning and LLMs across subfields results in asynchronous development patterns that elude traditional systematic synthesis.

Although numerous reviews have examined AI in education and digital language learning (Chen et al., 2022; Hwang et al. 2020; Zawacki-Richter et al., 2019), a dedicated scientometric mapping of AI in language education remains lacking.

Existing syntheses rarely integrate co-citation, co-occurrence, and burst-detection analyses within a unified framework that can capture strategic themes, intellectual turning points, and evolving knowledge flows.

To address this gap, the present study conducts a large-scale scientometric analysis of Web of Science publications from 2016-2025 using CiteSpace. Its primary aim is to map the evolution of research themes through co-citation, co-occurrence, and temporal-burst analyses. Secondary objectives include identifying influential authors, institutions, countries, and journals; visualising collaboration networks; tracing the development of thematic clusters such as AWE, speech technology, adaptive learning, and LLM-mediated tutoring; and outlining emerging research fronts and future directions for AI-driven language education.

2. METHODS

2.1 Data Collection

This scientometric review employed a systematic data collection procedure grounded in established principles of knowledge-domain visualisation (Chen, 2004, 2006). Two major bibliographic databases, Web of Science (WoS) Core Collection and Scopus, were selected because they provide extensive coverage of high-quality peer-reviewed publications in applied linguistics, educational technology, computer science, and related interdisciplinary fields, and because they are fully compatible with advanced scientometric tools such as CiteSpace (Yaqin, et al., 2025c). All records were retrieved on 17 November 2025. The decision to use both WoS and Scopus ensured broad disciplinary coverage and maximised the retrieval of publications related to artificial intelligence in language education. WoS was chosen for its curated indexing and its detailed cited-reference data, while Scopus provided extensive international coverage and complementary indexing in computer science and engineering, thereby reducing database bias in the final dataset.

A comprehensive search query was devised to capture the full breadth of research intersecting AI and language education. The final search string was:

```
   ("artificial intelligence" OR "AI" OR "machine learning"
OR "deep learning" OR "natural language processing" OR "NLP")
AND ("language education" OR "language learning" OR "language
teaching" OR "second language" OR "L2 learning" OR "foreign
language education")
```

This query was applied to titles, abstracts, author keywords, and indexed keywords across both databases.

2.2 Inclusion, Exclusion and Data Cleaning

Document types: Articles, early-access articles, proceedings papers, reviews, editorials, and book reviews. The initial retrieval yielded 3,580 articles, 200 early-access articles, 2 article proceedings papers, 5 book reviews, 8 early-access book reviews, 25 editorial materials, 473 proceedings papers, 88 reviews, and 9 early-access reviews (Figure 1). Topical focus: Studies explicitly addressing AI-based or AI-enhanced tools, systems, models, or pedagogical applications related to language teaching, learning, or assessment. Studies focusing on AI in general education, excluding those with a language component. Research on speech recognition, NLP, or translation systems lacks explicit pedagogical or educational integration. Non-scholarly records, such as corrections or incomplete metadata entries. Deduplication, during which all duplicate and invalid entries were removed. Manual validation of a stratified random subset to ensure topical relevance and metadata completeness. The final cleaned dataset comprised 4,390 unique records, including 2,296 Scopus and 2,094 WoS entries. These records represented 9,158 authors, 1,310 journals, 9,727 institutions, and 505 countries/regions.

2.3 Data Analysis

All scientometric analyses were performed using CiteSpace (Version 6.3 R.3), a widely used knowledge-domain visualisation tool designed to detect emerging trends, co-citation structures, and temporal patterns in scholarly communication (Chen, 2004, 2006). CiteSpace's analytical functions enabled the identification of intellectual turning points, the tracing of the evolution of thematic clusters, and the visualisation of collaborative networks across authors, institutions, and countries. The analysis followed standard CiteSpace workflows. A time-slicing configuration of 2016-2025 with five-year intervals was used to support mid-range trend detection. Separate runs were conducted for each node type: author keywords and Keywords Plus for thematic mapping; cited references to uncover the intellectual base and turning-point studies; cited journals to identify influential publication venues; and countries, institutions, and authors to generate collaboration networks.

This multi-layered approach provided a holistic view of the structural and temporal dynamics of the field. To maintain network interpretability while preserving essential relational structures, CiteSpace's top-N or percentile-based thresholds were applied for each time slice. Pruning strategies, including Pathfinder, Minimum Spanning Tree (MST), and Pruning Sliced Networks, were implemented to reduce noise and highlight significant connections, ensuring that the visualisations remained analytically meaningful. Multiple network metrics were employed to assess structural properties and identify key elements. Betweenness centrality revealed nodes that served as bridges across clusters, such as pivotal authors, institutions, or references. Modularity Q measured the extent to which the network was divided into distinct clusters. Silhouette values assessed cluster cohesion and thematic clarity.

Citation burstness detected rapid increases in attention to specific keywords or references, signalling emerging research fronts or intellectual turning points. CiteSpace's clustering algorithms, including log-likelihood ratio (LLR), term frequency–inverse document frequency (TF–IDF), and Mutual Information, were employed to extract and label thematic clusters. Interpretation of cluster labels was refined through manual inspection of representative references, recurring keywords, and citation trajectories. Prominent cluster themes included automated writing evaluation, speech recognition in CALL, adaptive vocabulary learning, and LLM-based writing assistance, representing both established and developing areas of inquiry.

Figure 1. Flow Chart of Scientometric Analysis

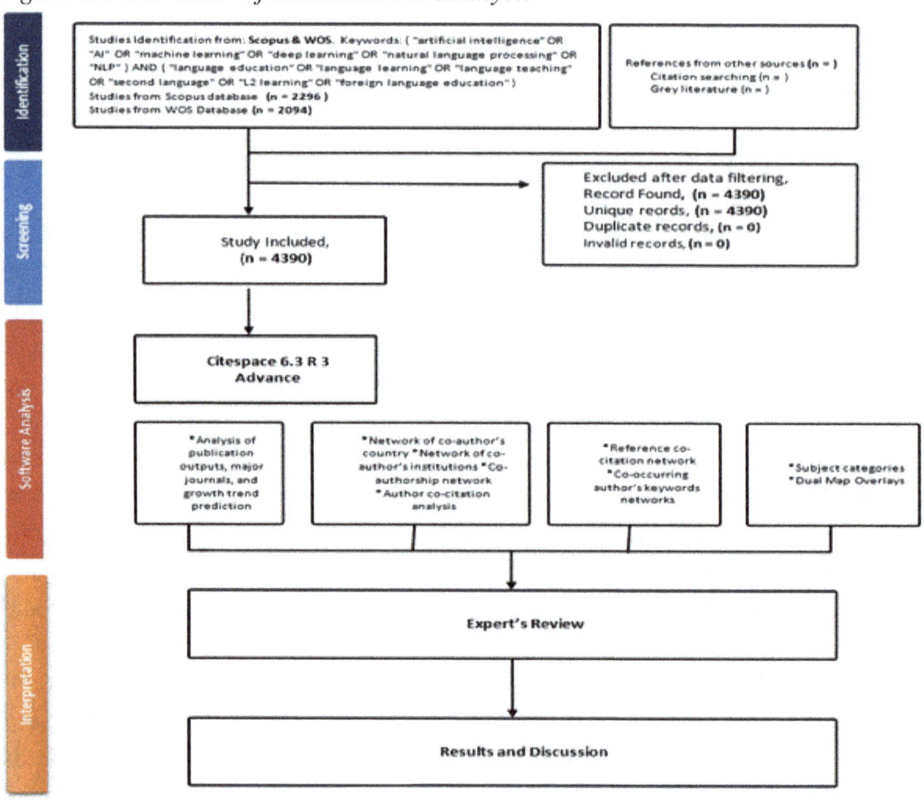

3. RESULTS

3.1 Publication Outputs and Journals (2016–2025)

3.1.1 Annual Publication Trends

Figure 2 presents the annual number of publications on AI-supported language education between 2016 and 2025. The corpus expands from 66 documents in 2016 to 1,841 in 2025, representing an almost thirtyfold increase over the decade. The trajectory is not linear but characterised by three distinct phases. The initial phase (2016–2018) is characterised by relatively modest yet steady growth, with a rise from 66 to 97 publications. During this period, AI applications in language education primarily focused on early computer-assisted language learning (CALL), automated writing evaluation, and rule-based or traditional machine learning techniques. Representative work on digital language learning and brain behaviour links began to appear at this time in outlets such as Bilingualism: Language and Cognition (Li & Ding 2025).

The acceleration phase (2019-2021) shows gradual expansion as deep-learning techniques and end-to-end speech and text processing become more widely adopted. Publication counts rise from 113 (2019) to 174 (2021). Studies on mispronunciation detection using deep tone models (Li & Jiang, 2019), neural natural language generation (Erdem & Coşgun, 2025), and large-scale surveys of digital language learning technologies (Li, 2025) exemplify the increasing methodological sophistication and the growing presence of AI in mainstream language-education venues. The post-LLM surge (2022–2025) has been explosive. Outputs almost double from 174 in 2021 to 266 in 2022, then rise to 473 in 2023, 1,139 in 2024, and 1,841 in 2025. This dramatic expansion aligns with the widespread diffusion of generative AI and large language models (LLMs), particularly after the release of ChatGPT at the end of 2022. Systematic reviews and meta-analyses on ChatGPT and generative AI in education (e.g., R. Deng et al., 2025; Lin et al., 2024) and empirical work on LLM-mediated instruction, assessment, and affective outcomes (Belda-Medina & Kokošková, 2024) contribute substantially to this late-period spike. The steep upward curve in Figure 1 thus reflects both the rapid diffusion of LLM tools into educational practice and the swift response of the research community.

Figure 2. Annual publication trends in AI-supported language education (2016–2025).

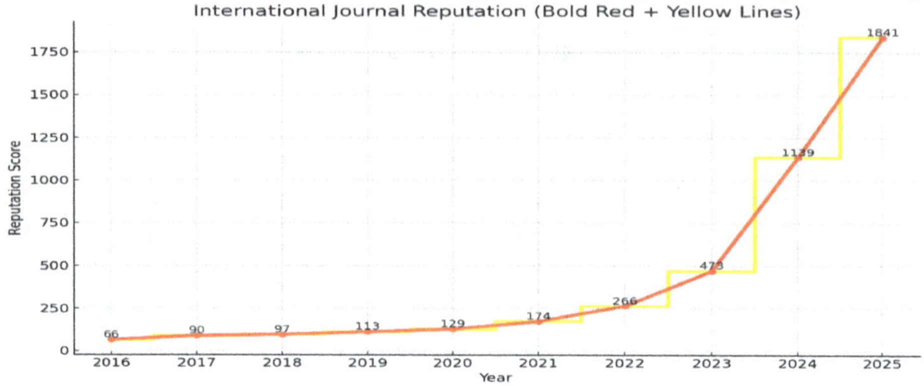

3.1.2 Core Publication Venues and Journal Influence

Figure 3 visualises the cited-journal co-citation network for the same period. Nodes represent journals and node size corresponds to citation frequency within the corpus, while coloured regions indicate major journal clusters. The map reveals a highly central cluster of specialised CALL and educational-technology journals surrounded by applied linguistics, general AI, and speech-technology outlets, along with a set of broad, high-impact multidisciplinary journals.

Within the CALL–education cluster, Computer Assisted Language Learning, Education and Information Technologies, System, Computers & Education, and Interactive Learning Environments form the most prominent hubs, each receiving more than 500 citations in the network. These venues host key empirical and review studies on AI-enhanced instruction, feedback, and learning outcomes for example, R. Deng et al., (2025) meta-analysis of ChatGPT's impact on student learning in Computers & Education, and several systematic reviews of AI chatbots and social robots for language learning in Education and Information Technologies and System (Lin & Liu 2024; Wang, 2024). Collectively, these journals constitute the core publication infrastructure for AI-supported language education.

A second set of influential journals is located in the applied linguistics and language education cluster, including TESOL Quarterly, Language Learning, Applied Linguistics, Assessing Writing, and the Journal of Second Language Writing. These outlets contribute theoretical grounding and methodological rigour to AI studies, for instance, work on computational modelling of bilingual language learning (Li & Xu, 2023), L2 writing analytics and automated evaluation (Eguchi & Kyle, 2024; Qin, 2025), and investigations of L2 development under LLM-mediated conditions

(Cong, 2024). Their strong co-citation links with CALL and educational-technology journals indicate that AI research is increasingly integrated into mainstream second-language acquisition (SLA) and assessment debates.

The general AI and computer science cluster encompasses high-impact venues, including the Journal of Artificial Intelligence Research, Neural Information Processing Systems, Advances in Neural Information Processing Systems, arXiv, and Lecture Notes in Computer Science. These journals contribute foundational techniques for neural natural language generation (Erdem Coşgun, 2025), cross-modal representation learning (Zhang et al., 2021), and speech self-supervision from raw audio (Dunbar et al., 2017), which are frequently cited by language-education researchers adapting state-of-the-art models to pedagogical contexts. Their presence in Figure 2 underscores the extent to which AI in language education is technically anchored in advances from the broader AI community.

A further cluster is formed by speech and signal-processing journals, including Speech Communication, Interspeech Proceedings, IEEE/ACM Transactions on Audio, Speech, and Language Processing, and Journal of the Acoustical Society of America. These outlets are significant sources for research on automatic speaking assessment, mispronunciation detection, and AI-driven speech evaluation systems (Li & Jiang, 2019). Their tight integration with CALL and educational journals reflects the growing importance of speech-based AI tools in oral language teaching and testing.

Multidisciplinary and high-impact science journals, such as PLOS ONE, Nature, Science, Cognition, and Psychological Bulletin, occupy key positions at the periphery of the central cluster. These journals are typically cited for their foundational work on human cognition, psycholinguistics, and AI evaluation, which provides a conceptual framework for studies of learner emotions, motivation, and goal orientation in AI-enhanced environments (Derakhshan, 2025; Wilcox et al., 2025).

Figure 3. Co-citation network of the most influential journals in AI language education (2016-2025).

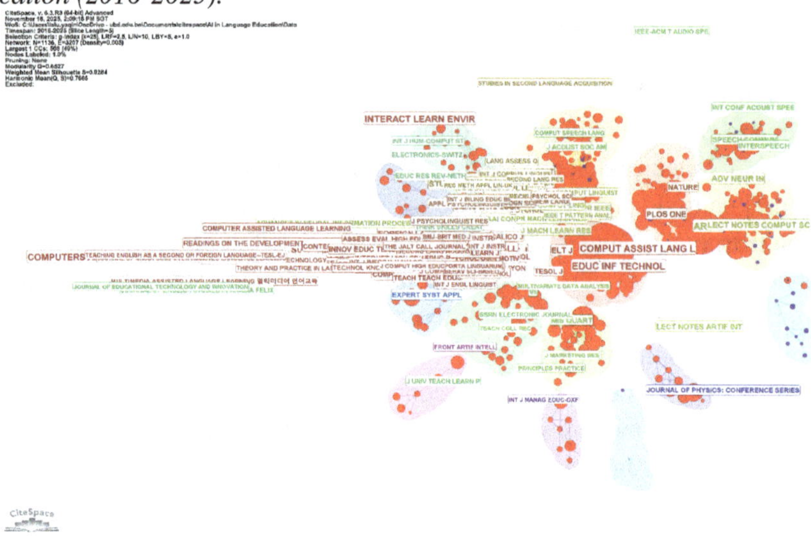

3.1.3 Disciplinary Spread of Publication Outlets

The distribution of journals across clusters highlights the multidisciplinary nature of AI in language education. On the citing side, many empirical and review articles are published in education and educational-technology outlets (e.g., Computers & Education, Education and Information Technologies, Review of Educational Research), where the primary focus is on learning outcomes, instructional design, and technology acceptance (X. Deng et al., 2025b; Lin & Yu, 2025). At the same time, a substantial number of studies appear in language-education and applied-linguistics journals such as Language Teaching Research, Arab World English Journal, and RELC Journal, especially those examining teachers' and learners' perceptions of ChatGPT, AI-mediated feedback, and translanguaging practices (Kohnke et al., 2023; Saeed Ahmad Mohammed & Mahmoud Ibrahim Alian, 2023; Waluyo & Rouaghe, 2025).

Parallel to these education-oriented outlets, AI-heavy venues in computer science and engineering publish work at the interface of language learning and core AI technologies, including speech recognition, neural text generation, and multimodal interaction (Cong, 2024). Bibliometric and scientometric analyses of AI-enhanced language learning are increasingly published in journals specialising in information science and research evaluation, such as Scientometrics and the Journal of Informetrics (Xue, 2024).

Taken together, the annual growth curve and the journal co-citation structure show that AI-supported language education has evolved from a relatively niche subfield of CALL into a rapidly expanding, highly interdisciplinary research domain. The field is anchored in specialist language education and educational technology journals. However, it draws heavily on advances in computer science, signal processing, and cognitive science, and increasingly publishes in broader social-science and multidisciplinary outlets.

3.2 Co-Occurring Keywords (Hotspots and Emerging Trends)

The co-occurring author keywords reveal how the intellectual focus of AI in language education has shifted from linguistically oriented, rule-based CALL to data-intensive, generative and affect-aware systems (Figures 4 and 5). The CiteSpace keyword network for 2016-2025 comprises multiple well-defined clusters with high silhouette values, indicating coherent thematic groupings. Across these clusters, artificial intelligence, language learning, students, teaching and natural language processing are the most frequently cited and central nodes, underscoring their foundational role in the domain.

Figure 4. Timeline visualisation of co-occurring author keywords networks (2016–2025)

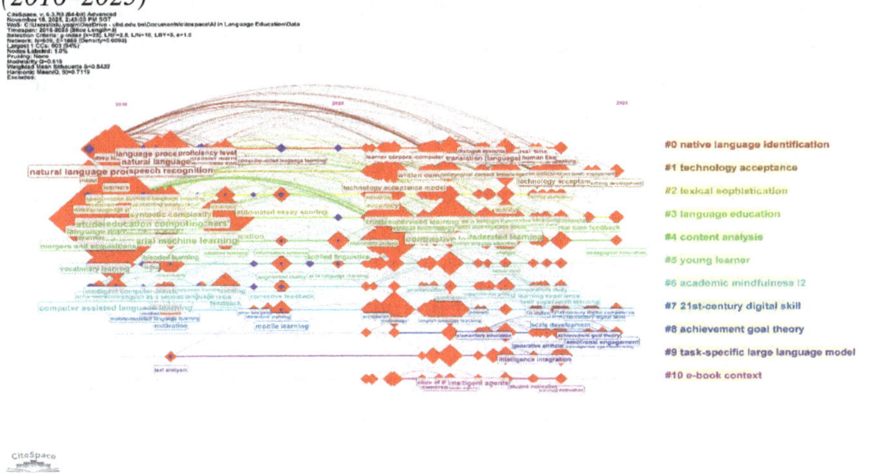

Figure 5. Timeline visualisation of co-occurring author keywords networks (2020–2025)

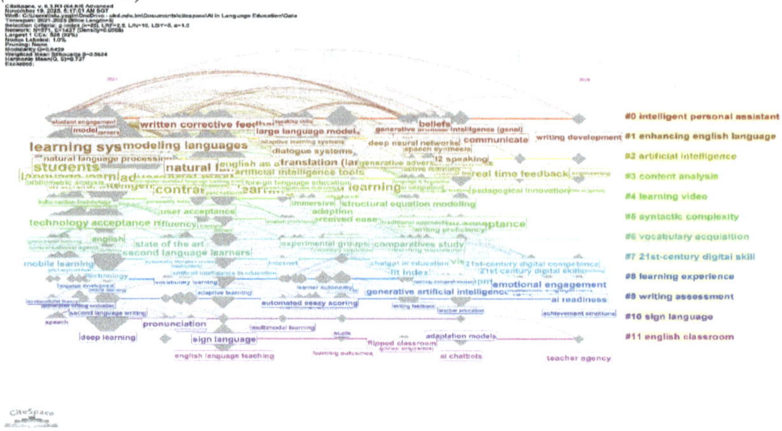

3.2.1 Core Thematic Clusters

The co-occurring keyword analysis reveals several major thematic clusters that collectively chart the evolution of AI-supported language education. One of the most prominent themes concerns intelligent tutoring systems (ITS) and adaptive scaffolding. These topics dominate clusters labelled artificial intelligence and enhancing English language, where high-centrality nodes such as teaching, learning systems, and computer-aided instruction connect multiple subfields. Recent studies on multimodal tutoring systems and dialogue-based intelligent tutors (Ding, 2024) illustrate how deep learning and contrastive learning models are increasingly used to analyse learner behaviour and provide personalised instructional pathways in English for academic purposes and discipline-specific contexts. Their prominence underscores the growing importance of AI-driven tutoring architectures as a foundational component within the broader research landscape.

A second major theme is automated writing evaluation (AWE) and AI-mediated feedback. This cluster is anchored in terms associated with syntactic complexity and writing assessment. Early contributions emphasised corpus-informed scoring, rule-based analysis, and constructs such as writing assessment, second-language writing, and corpus-based second-language writing, as well as corpus-based analysis (Appendix in Table 1). Recent studies, however, reflect a shift toward generative and interactive assessment paradigms. Citing articles now examine chatbot-assisted essay evaluation and LLM-supported peer review (Guo & Xia, 2025; Lan et al., 2025; Moorhouse et al., 2025), corresponding with keyword expansions to include task-specific large language models and academic writing. This development marks the field's transition from static evaluation toward dynamic, LLM-mediated formative feedback cycles.

Speech recognition, pronunciation training, and speaking assessment represent another major cluster, evident across the intelligent personal assistant, vocabulary acquisition, and sign language networks. Early burst terms (e.g., speech synthesis, frequency) indicate foundational work in acoustic modelling and text-to-speech systems (Appendix, Table 1). From 2021 onward, strong bursts in recognition and speech (see Appendix in Table 2) align with intensified research on mobile-assisted pronunciation training (Ni & Yin, 2025), chatbot-supported oral practice (Jeon et al. 2023), and automatic sign language recognition and assessment (Khandaqji et al., 2025). Deep learning serves as a central enabling technology, particularly in accent conversion and sign language processing, highlighting how speech-related AI applications have expanded both in technical sophistication and educational relevance.

Another rapidly expanding domain is adaptive learning, personalisation, and learning analytics. Keywords such as personalised learning, learning models, and performance-based assessment appear consistently in the 2016-2020 burst list (Appendix in Table 1), indicating early moves toward precision education models. Between 2021 and 2025, this trend intensifies as precision education, big data, online learning, mobile learning, and e-learning become high-burst terms (Appendix in Table 2). Recent work has applied learning analytics to track LLM-mediated writing development (Du, 2025) and examined psychological variables, such as autonomy, mindfulness, and emotion regulation, within AI-rich learning environments (Huang & Wang, 2025; Namaziandost, 2025). The high degree and centrality of learning systems affirm their foundational role in today's data-driven, adaptive instructional ecosystems.

The cluster on chatbots, conversational agents, and LLM-based dialogue systems constitutes one of the most dynamic research fronts. Keywords such as chatbot, AI chatbot, conversational agents, and intelligent personal assistants are closely tied to pedagogical concerns, including engagement, collaboration, and learning outcomes. Representative studies highlight applications ranging from speech-recognition chatbots (Jeon et al., 2023) and generative AI-mediated informal English learning (Liang, 2025) to flipped classrooms supported by chatbots (Ling & Jan, 2025) and collaborative note-taking through conversational AI (Chen, 2024). These findings suggest that LLM-based dialogue systems are now mainstream tools for interaction, feedback, and affective support across diverse learner populations.

The neural machine translation (NMT) and AI-mediated translanguaging practices form a dispersed yet increasingly visible cluster. The burst term translation (languages) (2023-2025) and NMT's prominence in clusters on English enhancement and syntactic complexity reflect growing interest in classroom-oriented translation technologies. Research spans immersive sign-language translation through 3D avatars (Gamage et al., 2023), automated identification of machine-translation use in L2 writing (Jolley & Jolley, 2025), and pedagogical orchestration of NMT and peer feedback in task-based learning. Collectively, these studies highlight issues of translanguaging, academic integrity, and shifting norms regarding the acceptable use of machine translation in multilingual education.

3.3 Co-Cited References (Key Clusters, Turning Points, Citation Bursts)

3.3.1 Major Co-Citation Clusters and the Intellectual Base

The co-citation network for AI in language education comprises 14 well-defined clusters, characterised by high silhouette values (0.79-1.00), which indicate clear thematic coherence across the corpus. As illustrated in Figures 6 and 7, and in the Appendix (Table 3), these clusters collectively map the intellectual progression of the field from foundational CALL and AI-in-education frameworks to recent developments in generative AI, LLM-mediated tools, and AI-supported informal learning. Each cluster encapsulates a specific segment of the knowledge base, forming a layered architecture of conceptual and methodological influences.

Clusters #9 ("artificial intelligence") and #16 ("leveraging artificial intelligence") represent the foundational layer of the network. These clusters consolidate seminal frameworks synthesising AI in education and CALL, including major reviews by Zawacki-Richter et al., (2019). Hwang et al., (2020) and Chen, (2025). Their prominence reflects their ongoing role in shaping theoretical and methodological orientations in AI-supported language learning. Frequently co-cited across subsequent work, these references anchor later syntheses examining chatbots, generative AI, and AI-mediated informal learning, forming the conceptual backbone for emerging research strands.

Methodological and evaluative concerns constitute the second major cluster. Cluster #6 ("automated writing evaluation") gathers foundational studies on AWE systems, grammar-checking technologies, and AI-driven written corrective feedback. Highly cited works by Barrot, (2023), Koltovskaia, (2020), This cluster is closely tied to reviews of AI-mediated feedback (Shi, 2023) and second-language writing (D. Feng, 2025), underscoring how advances in NLP and writing analytics have influenced pedagogical applications and assessment frameworks in language education.

Cluster #12 ("neural machine translation") complements the AWE cluster by focusing on AI-mediated translation and reading support. Representative works by Jolley & Maimone, (2025), Lee et al., (2023), Tsai, (2022) investigate how neural machine translation (NMT) tools such as DeepL and Google Translate reshape writing processes, accuracy, and the use of strategies. Together with Cluster #13 ("writing anxiety complexity fluency"), these studies form the intellectual basis for discussions on AI-supported writing, translanguaging, academic integrity, and evolving norms around machine translation in multilingual classrooms.

Conversational agents and generative AI form a third major domain. Cluster #0 ("using chatbot") focuses on speech-recognition chatbots and text-based dialog systems for language learning, anchored by pivotal studies from Jeon (2024), Liang et al. (2023). These contributions analyse motivational, interactional, and proficiency-related outcomes associated with conversational AI, and they are widely cited in systematic reviews on language-learning chatbots (e.g., Jeon et al. 2023; Li & Ding, 2025; Wu & Yu, 2024). Cluster #1 ("using ChatGPT") captures the surge of post-2022 research on LLM-based writing, feedback, and literacy development. Influential studies include Barrot (2024), Foung et al. (2023), Kasneci et al. (2023), and Yan (2023), which form the empirical core for emerging meta-analyses of generative AI in education (Deng et al, 2025; Wang, 2025).

Clusters #2 ("student engagement") and #3 ("emotional experience") add an affective dimension to the landscape, focusing on learners' motivation, enjoyment, anxiety, and socio-emotional responses in AI-mediated environments. Co-cited works by Derakhshan & Ghiasvand (2024), Fathi & Rahimi (2024), Wang (2024) document how AI tools shape self-concept, emotional engagement, and identity development in L2 learning. These clusters frequently intersect with chatbot and generative AI clusters, signaling a convergence of technological and psychological research agendas.

The Cluster #10 ("AI-mediated informal digital learning") encapsulates research on AI-mediated informal digital learning of English (AI-IDLE). Led by a series of studies by Liu and colleagues, this cluster examines how learners engage with AI tools outside formal classrooms, exploring the links among AI use, motivation, enjoyment, autonomy, and the ideal L2 self. Its strong ties to clusters on engagement, emotion, and generative AI indicate that informal learning environments are increasingly shaped by emergent AI technologies.

Figure 6. Co-citation reference network of AI in language education (2016–2025).

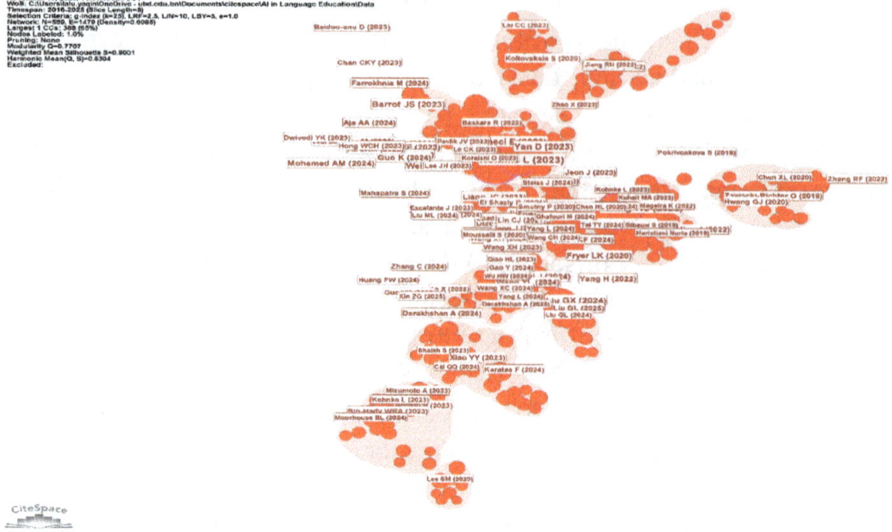

Figure 7. Co-citation reference network with 14 clusters of AI in language education (2016–2025).

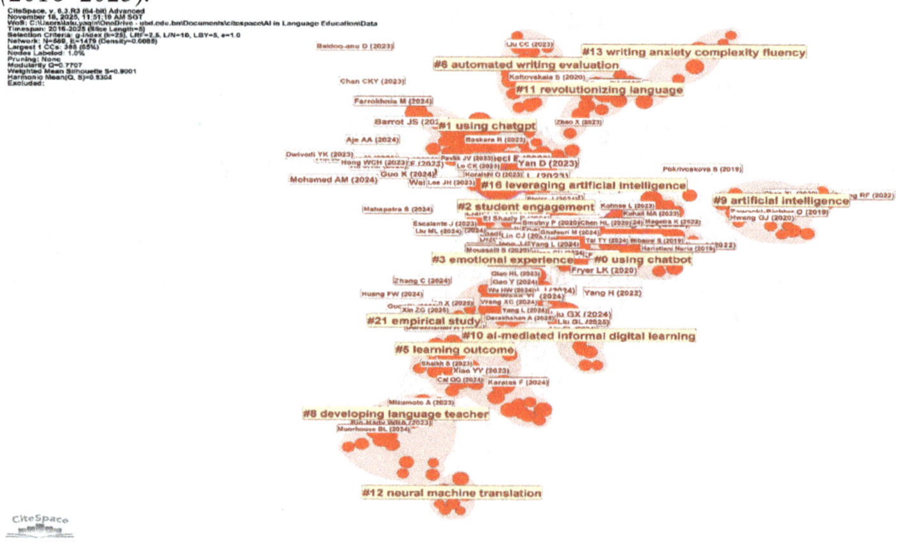

3.3.2 Turning-Point References and Bridging Roles

Betweenness centrality values highlight references that serve as bridges across otherwise separate clusters. The most central of these is Huang & Wang (2022) (centrality = 0.15), whose work on chatbot-mediated speaking forms a conceptual link between CALL-design research, student engagement, and affective experience. Similarly, Kohnke et al. (2023) and Yan (2023) exhibit high centrality values (0.15 and 0.14, respectively), indicating their bridging influence between studies on LLM-mediated writing, AI-in-education reviews (Zawacki-Richter et al, 2019; Chen et al., 2022). In methodological domains, Tsai (2022) and Shi (2023) emerge as pivotal references (centrality = 0.08), connecting learning analytics, structural-equation modelling, AI-mediated feedback, and teacher practices. Additional bridging roles are played by Fathi & Rahimi (2024) in linking affective studies with everyday AI use, as well as Jeon (2024), who consolidate conceptual and empirical work on educational chatbots. Collectively, these references function as intellectual "turning points" (Chen, 2004), enabling cross-cluster diffusion of theories, methods, and technologies.

3.4 Country and Institution Networks

This section examines how research on AI-supported language education is geographically distributed and how collaborations among countries and institutions shape the development of the field (Figure 8). Country- and institution-level co-authorship networks were generated in CiteSpace based on the address information of the citing articles, and were analysed in terms of citation counts, citation bursts, degree, betweenness centrality, and sigma.

Figure 8. Network of co-authors' countries in AI language education (2016-2025).

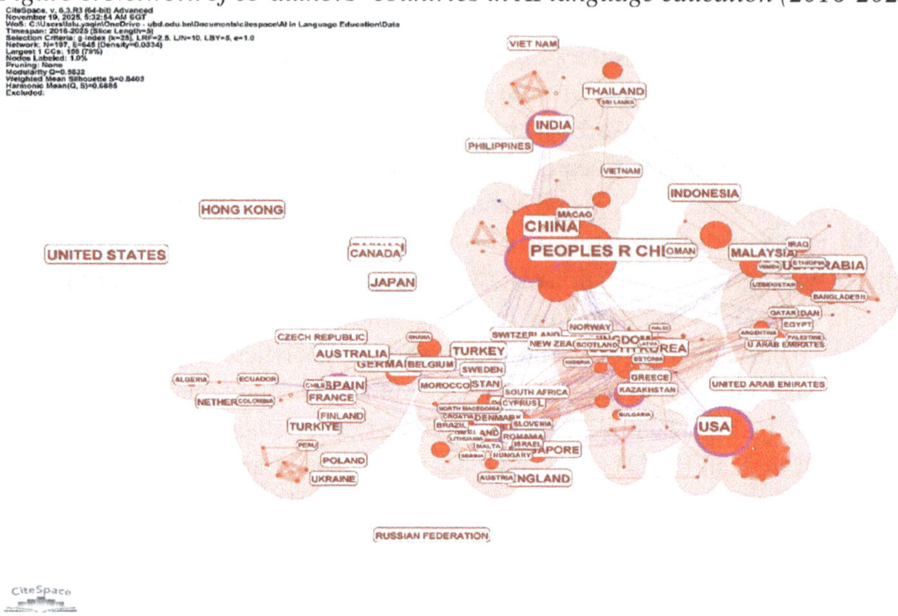

3.4.1 Country Collaboration Patterns

The country collaboration network reveals seven distinct clusters (Clusters #0–#6), each aligned with thematic areas such as affective support, teacher support, multi-agent systems, AI-mediated informal digital learning, automated scoring, 21st-century digital skills, and writing skill. As illustrated in Figure 6, global participation in AI-supported language education is markedly uneven. Mainland China, represented by the People's Republic of China and China, forms the largest and most influential hub (719 and 486 citations), followed by the USA/United States (336 and 314 citations). Other East Asian economies, including Japan, South Korea, Hong Kong, and Taiwan, also hold central positions, alongside emerging contributors such as Saudi Arabia and India. Together, these regions demonstrate that the Asia–Pacific zone is currently the most active global centre for AI-enhanced language-education research.

The thematic distribution of clusters underscores region-specific research orientations. Cluster #3, associated with AI-mediated informal digital learning, is dominated by the People's Republic of China, the United States, Japan, and Hong Kong, with a focus on investigations into AI-IDLE and chatbot-supported out-of-class learning. Cluster #1, focusing on teacher support, is concentrated in the Gulf and Southeast Asian regions, with Saudi Arabia, Malaysia, Indonesia, Oman, and The United Arab

Emirates contributing substantial work on AI-assisted examination systems, digital teacher-support platforms, and sustainability-aligned English-language education. Cluster #0 (affective support) highlights South Korea, the United Kingdom/England, Turkey, and Germany, reflecting strong engagement in studies examining anxiety, motivation, and socio-emotional dimensions of AI-enhanced learning. Cluster #5, comprising Spain, Iran, Türkiye, Ecuador, and Algeria, aligns with research on 21st-century digital skills and higher-education applications. Cluster #6 signals an emerging South and Southeast Asian hub, with India, Thailand, Vietnam, Sri Lanka, and Mongolia, focusing on speech-recognition tools and writing-related AI applications.

Network indicators provide additional insight into structural influence. Degree centrality identifies the USA as the most connected node (degree = 30), followed by Italy and the Netherlands (degree = 28) and Slovenia (degree = 25), revealing their roles as key collaboration brokers. Betweenness centrality further highlights the USA (0.27) and the United States (0.20) as the principal mediators linking East Asian, European, and Middle Eastern research networks. Other countries with notable mediating roles include SPAIN (0.16), India (0.13), England (0.12), Qatar (0.10), Peru (0.10), France (0.08), the United Kingdom (0.08), and Malaysia (0.08). Sigma values reinforce these findings, with the United States ($\sigma = 5.63$) and the USA ($\sigma = 3.59$) demonstrating both structural prominence and strong temporal dynamics.

Citation-burst analysis (Appendix in Table 4) reveals shifting geographical contributions. Italy exhibits the strongest early burst (strength = 9.88, 2016–2020), followed closely by Germany (9.68, 2016–2025) and the United States (9.54, 2016–2020). Subsequent bursts in the Netherlands, Sweden, Japan, Belgium, and Qatar indicate diversification across Europe and the Gulf. More recent bursts in Singapore, Ireland, Egypt, Hungary, and South Africa suggest the need for expanding engagement from both established innovation hubs and emerging research systems in the Global South. Collectively, these findings depict a polycentric yet hierarchical global structure, with East Asia and Anglophone regions at the core and broadening participation from diverse world regions.

3.4.2 Institutional Collaboration Patterns

The institutional co-authorship network comprises 12 clusters, reflecting diverse research foci, including L2 writing cohesion, AI-mediated informal digital learning, creativity and self-reflection, speech evaluation systems, lexical sophistication and automated indices, feedback literacy, Saudi Arabia–based applications, AI-supported language learning, and writing instruction (Figure 9). Consistent with country-level patterns, the network reveals a dense cluster of East Asian and Hong Kong institutions closely connected to Anglophone and European universities, forming an extensive transregional research infrastructure. Citation indicators identify several institutions as highly influential. The Chinese University of Hong Kong ranks first (83 citations), followed by The Education University of Hong Kong (45), EdUHK (43), National Taiwan Normal University (37), The Hong Kong Polytechnic University (35), and Xi'an Jiaotong–Liverpool University (35). These universities anchor major research programmes on AI-IDLE, learner engagement, writing development, and AI-based speech evaluation. Western institutions, including the University of Cambridge (34 citations) and Beijing Language and Culture University (32), make foundational contributions to lexical sophistication research, automated complexity indices, and corpus-driven writing analytics.

Degree centrality measures illuminate patterns of collaboration. A group of predominantly public universities—including the University of Kansas (degree = 33), Utah State University (32), Missouri State University (31), Texas A&M University System (29), University of Washington (27), Utah System of Higher Education (27), University System of Georgia (26), Arizona State University–Tempe (26), NTNU (26), and Ghent University (26) forms the structural backbone of the network. These institutions host leading research groups in CALL, AI-enhanced education, and learning analytics, frequently collaborating across North America, Europe, and East Asia.

Betweenness centrality identifies several brokerage institutions: Northeastern University (0.16), University of Auckland (0.12), and University of Washington (0.09) serve as key connectors across regional clusters. Outside Anglophone systems, institutions such as National Taichung University of Education, Chongqing University, King Faisal University, and Macao Polytechnic University serve as important bridges, linking East Asia, the Middle East, and broader international networks. Sigma values corroborate the structural importance of these institutions and their engagement in emerging high-momentum research areas.

Burst detection at the institutional level reveals shifting research centres, as shown in Appendix Table 5. Early bursts occur at longstanding CALL and AWE hubs, including the University of Cambridge (strength = 11.37), Georgia State University, the University System of Georgia, and the University of Hawai'i at Mānoa. Later bursts at Peking University, National Central University, Beijing Normal University, and Kyushu University signal the rise of East Asian centres specialising in NLP, speech recognition, and AI-based assessment. More recent bursts at Lingnan University, the University of Oregon, Tohoku University, Tokyo Gakugei University, NTNU, Sookmyung Women's University, Korea University, Temple University, and Athabasca University reflect the rapid proliferation of generative AI research and AI-supported teacher education across Asia, North America, and Europe.

Taken together, the country- and institution-level networks demonstrate that AI-supported language education is shaped by interconnected innovation hubs concentrated in East Asia, Anglophone regions, and selected European and Middle Eastern systems. These hubs exhibit dense collaboration patterns, high network centrality, and intense temporal bursts, indicating that advancements in LLMs, learning analytics, and AI-based assessment will continue to emerge from these global centres, with increasing contributions from emerging research communities in Latin America, Africa, and the wider Global South.

Figure 9. Network of co-authors' institutions of AI in language education (2016–2025).

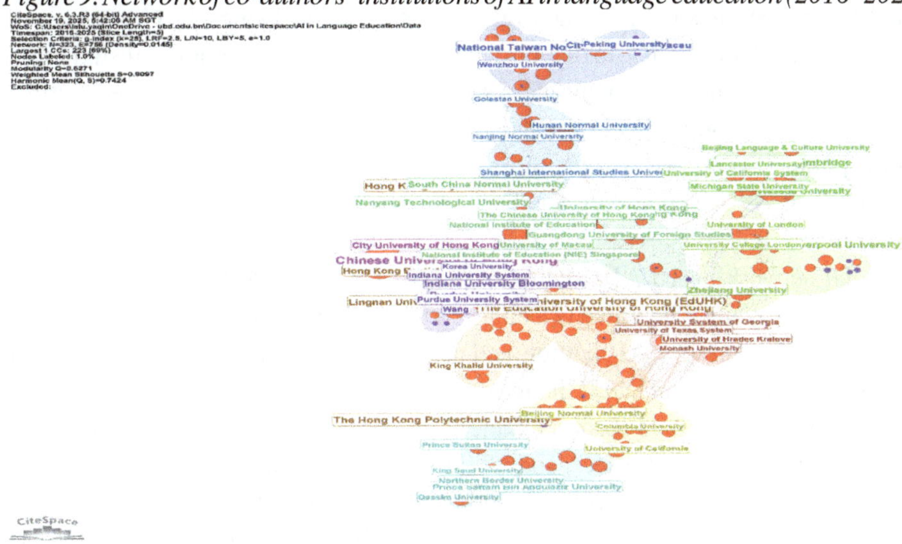

3.5 Author Collaboration and Impact

3.5.1 Productive and Highly Cited Authors

The author co-authorship network for 2016-2025 reveals a compact yet clearly delineated research community contributing to AI-supported language education (Figure 8). Two major clusters dominate the landscape, both centred on artificial-intelligence-driven speech evaluation and broader AI applications in education. Cluster #0 ("artificial intelligence speech evaluation system") comprises 15 authors and exhibits a silhouette value of 1.00, indicating exceptionally high thematic cohesion. This cluster is anchored by influential scientometric and review works, Chen's bibliometric mappings of AI, particularly in education and technology-enhanced learning (Chen et al, 2022) and Huang's syntheses of AI applications in language learning (Huang & Wang, 2022). Collectively, these studies provide essential conceptual and methodological infrastructure for empirical research on AI-enhanced instruction.

Within Cluster #0, Zou Di emerges as the most cited author (41 citations). Zou's contributions span AI-enhanced speaking practice, intelligent evaluation of oral proficiency, and AI-mediated informal digital learning, establishing his work as a reference point for speech-based CALL. Kohnke Lucas (24 citations) and Xie Haoran (23 citations) occupy similarly central roles, particularly through their research on learner engagement with AI tools, generative AI for language instruction, and mixed-method studies of informal digital learning. Jeon Jaeho (17 citations) and Chen Xieling (15 citations) further enrich this cluster through empirical studies on AI-mediated speaking assessment and systematic analyses of AI-enhanced education.

A second key group, Cluster #5, also labelled "artificial intelligence speech evaluation system",—comprises six authors whose work heavily emphasises empirical evaluations of AI-driven speech feedback systems. This cluster is led by Zou Bin, whose research investigates EFL learners' perceptions of automated speech evaluation and social network-based AI speaking practice (Zou et al, 2023). Supporting this line of inquiry, Wang Chenghao (18 citations) and Du Yiran (11 citations) examine conversational generative AI chatbots and personalised AI-based speaking tools. At the same time, Crosthwaite and Zhang contribute assessment-focused perspectives. Collectively, these authors form a coherent research front that bridges automated speech evaluation, learner affect, and willingness to communicate.

Figure 10 visualises the most influential authors across the dataset. Beyond the prominence of Zou Di and Zou Bin, Kyle Kristopher (20 citations) stands out for contributions to lexical sophistication and syntactic complexity, which underpin many automated writing evaluation (AWE) systems. The influence of Kyle's work demonstrates the foundational role of corpus-linguistic and NLP-informed measures in shaping contemporary AI-driven writing support. Additional highly cited authors include Wang Chenghao, Jeon Jaeho, Wang Chaoran (15 citations), Chen Xieling, and Lee Seongyong (14 citations), who represent a cluster of researchers advancing AI-supported speaking, listening, and classroom-based experimentation.

3.5.2 Co-Authorship Clusters and Core Research Groups

The co-authorship network exhibits a small-world structure, comprising dense local clusters connected by a handful of key bridging authors (Figure 8). Zou Di, with the highest degree centrality (degree = 13), functions as the central hub, collaborating extensively with Kohnke, Xie, Moorhouse, and Lee. This Hong Kong-centred group has produced influential work on AI chatbots, automated speech evaluation tools, and generative-AI-enhanced language learning, frequently employing survey-based, mixed-method, and experimental designs.

A separate set of authors forms a cluster around AWE and writing analytics. Here, the contributions of Kyle and Crossley—particularly in lexical sophistication and syntactic complexity are frequently referenced, as indicated by the cited-author cluster on "lexical sophistication." These patterns demonstrate the foundational role of applied linguistics, corpus linguistics, and psycholinguistics in contemporary AI-driven writing analytics. Speech-based CALL constitutes another distinct research group. Authors such as Evanini, Mouri, and Zhang Jinsong, who exhibit intense citation bursts between 2017 and 2020, contribute core advances in speech recognition, prosody modelling, and automatic pronunciation assessment. Their technical innovations form the basis for more pedagogically oriented studies by later authors such as Zou and Wang Chenghao.

Multidisciplinary collaborations also feature prominently. For example, Chen's bibliometric and topic-modelling studies draw from information science and computer science while remaining grounded in educational technology. Similarly, Huang & Wang as well as Godwin-Jones, widely cited in clusters focused on AI-mediated informal digital learning, integrate perspectives from psychology, learning sciences, and applied linguistics to examine AI bots, metaverse environments, and ambient intelligence.

3.5.3 Citation Bursts and Evolving Author Impact

Citation-burst analysis (Appendix in Table 6) identifies two significant waves of influence. The first wave (2016–2020) is dominated by foundational CALL and AWE researchers, including Kyle (burst 5.68), Crossley (4.99), Evanini (3.43), Meurers (3.41), and Zhang Jinsong (2.69). These authors provided essential methods for automated scoring, learner modelling, and speech processing, forming the technical basis of subsequent AI-enhanced systems. A second, more recent wave (2021–2025) reflects the explosive growth of generative AI, AI-mediated informal learning, and automated speech evaluation. Chen Xieling (5.83) and Xie Haoran (5.50) show the strongest bursts, fuelled by large-scale bibliometric syntheses and empirical studies of AI-supported engagement. Cheng Gary (3.62) and Cheng Gary often collaborate, representing a sustained interest in conversational agents and multimodal analytics. Emerging authors such as Kou Xiaojing and Pack Austin (3.04 each) mark diversification toward LLM-supported writing, prompt engineering, and adaptive learning.

Several rising scholars specialise in affective and socio-critical dimensions of AI-enhanced learning. Wen Yun and Yang Hyejin contribute to research on enjoyment, anxiety, and motivation in AI-mediated contexts, while Al-ghezi and Getman investigate AI literacy, ethics, and teacher attitudes. Their intense bursts indicate that affective, ethical, and critical perspectives are becoming central to the field.

3.5.4 Influential Cited Authors and Knowledge Bases

The cited-author network (Figure 11) illuminates the theoretical and methodological foundations of AI-supported language education. Clusters labelled "systematic review," "AI-mediated informal digital learning," and "using ChatGPT" are dominated by scholars such as Jeon, Huang W.J., Hwang, and Wang, whose work on AI-mediated learning, motivation, and technology acceptance informs much of the empirical research. In parallel, the "lexical sophistication" cluster highlights the enduring impact of Crossley, Chapelle, Lu, and Kyle on the measurement of L2 writing complexity, while the "corrective feedback" cluster, grounded in the work of Barrot, Ellis, Warschauer, and Hyland, provides foundational perspectives on feedback, CALL, and writing pedagogy. Together, these analyses indicate that AI-supported language education is driven by a relatively compact but highly interconnected global author community. A small number of productive hubs, primarily located in East Asia, North America, and Europe, shape the field through influential collaborations, while emerging scholars in generative AI and informal digital learning signal a rapid diversification that will likely reshape the research landscape in the years ahead.

Figure 10. Co-authorship network (2016–2025).

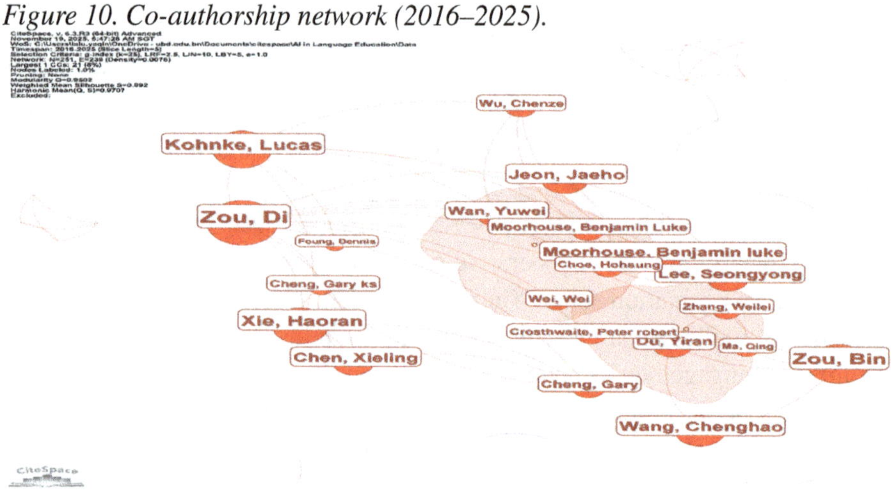

Figure 11. Top authors by citation (2016–2025).

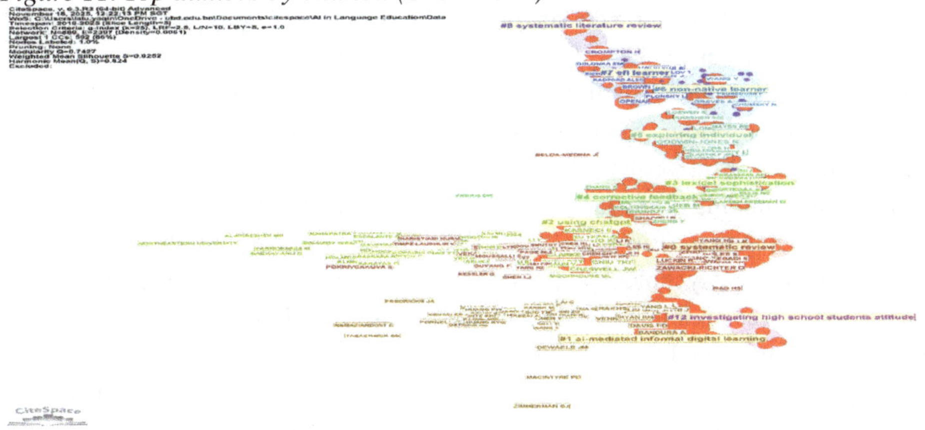

3.6 Subject Categories (2016-2025)

Analysis of Web of Science (WoS) subject categories provides a comprehensive view of the disciplinary configuration of AI-supported language education. Because every WoS-indexed publication receives one or more subject assignments, all 4,390 documents in the corpus were included in this analysis, spanning education, linguistics, computer science, engineering, psychology, and multiple branches of the natural and social sciences. This broad distribution highlights the inherently interdisciplinary nature of the field. Burst-detection analysis identified 25 subject categories with significant citation surges between 2016 and 2025 (Appendix in Table 7), visualised in Figure 12 as temporal bursts marking periods of intensified disciplinary influence.

The overall landscape indicates that Education & Educational Research (865 citations) is the dominant category, followed by Linguistics (488 citations), confirming that the field remains firmly rooted in educational and language sciences. On the computational side, Computer Science, Artificial Intelligence (330 citations) and Computer Science, Interdisciplinary Applications (290 citations) form the core technical foundations, complemented by Language & Linguistics (287 citations). Network indicators further underscore the prominence of these domains: Computer Science, Artificial Intelligence has the highest degree (28) and betweenness centrality (0.25), with similarly central roles played by Computer Science, Interdisciplinary Applications (degree 25), Engineering, Electrical & Electronic (degree 22; centrality 0.18), Education & Educational Research (degree 20; centrality 0.22), and Linguistics (degree 20; centrality 0.18). Collectively, these categories serve as disciplinary hubs that bridge technical fields such as audio engineering, neuroscience, and sustainability research with language learning and pedagogical scholarship.

CiteSpace clustering identifies eight macro-clusters that further reflect this integrative disciplinary structure. Clusters labelled “large language model” and “learning model” bridge AI and computer science with education and linguistics, incorporating work on mispronunciation detection (Korzekwa, 2021), gesture-based sign language training (Tan et al, 2022), and family-oriented generative AI for spatial language development (Liu et al, 2025). Additional clusters highlight intersections with cognitive and psychological domains, including bilingualism and motivational orientations (MacWhinney, 2025), as well as sustainability-driven approaches to AI-mediated language learning (Wen et al., 2025; Yang, 2025).

Early bursts, occurring primarily between 2016 and 2019, were mainly in technology-driven categories. Computer Science, Interdisciplinary Applications (2016-2018; burst strength 4.50), Computer Science, Theory & Methods (2017-2019; 10.77), Computer Science, Artificial Intelligence (2018-2022; 18.19), and Engineering, Electrical & Electronic (2018–2019; 11.17) reflect foundational research applying speech recognition, acoustic modelling, and signal processing to pronunciation training and speech-based CALL. Studies such as (2021) focus on mispronunciation detection, and Tan’s (2022) gesture-recognition systems exemplify how early developments were grounded in engineering and computer science, with language education serving as the applied context.

Between 2019 and 2021, the disciplinary scope expanded toward linguistics, neuroscience, and information systems. Bursts in Audiology & Speech-Language Pathology (2019-2023; 4.43), Computer Science, Cybernetics (2019-2020; 2.60), and Imaging Science & Photographic Technology (2019-2021; 1.30) reveal increasing integration of multimodal and speech-pathology perspectives. Concurrent bursts in Computer Science, Information Systems (2020-2021; 7.53) and Neurosciences (2020-2022; 5.16) reflect a growing interest in learning analytics, neurocognitive modelling, and data-intensive educational platforms. The emergence of Literary Theory & Criticism, as well as Physics, Applied as burst categories, highlights an interdisciplinary expansion into multimodal composition and physics-informed neural modelling.

From 2021 to 2025, diversification accelerated in line with the development of generative AI and LLM-focused innovations. Technology-oriented categories, including Computer Science, Software Engineering (2021-2023; 4.69), Acoustics (2021–2023; 1.92), Mathematics (2021–2023; 1.31), and Computer Science, Hardware & Architecture (2018–2023; 3.53), show sustained bursts, reflecting growing work on LLM deployment, multimodal infrastructures, and scalable model design. Simultaneously, educational and social-science categories rose in prominence. Education, Scientific Disciplines (2023–2025; 7.63) and Education & Educational

Research became increasingly central, as studies examined generative-AI-supported instruction and assessment.

Ethical and sustainability concerns also gained momentum, reflected in bursts for Green & Sustainable Science & Technology (2022–2023; 1.93) and clusters exploring sustainable AI-mediated language learning (Yang, 2025; Wen et al, 2025). Concurrently, categories in philosophy and logic addressed epistemic and critical-thinking issues arising from LLM-enhanced EFL instruction. Emerging bursts in Engineering, Biomedical and Literature (2023-2025) further demonstrate the application of AI language models in health communication, patient literacy, and literary analysis, illustrating the increasingly expansive disciplinary reach of AI-supported language education.

Figure 12. Burst detection of Web of Science subject categories of AI in language education (2016–2025).

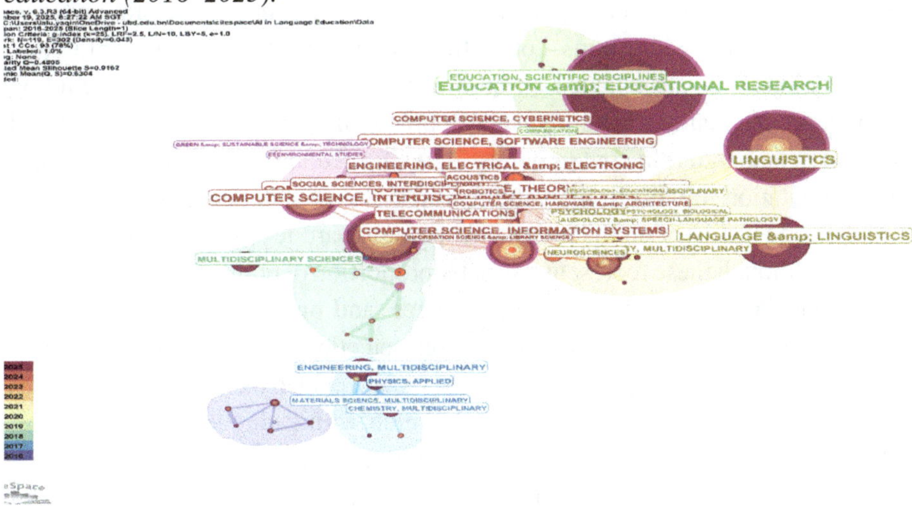

and their referenced sources on the right. Curved trajectories indicate citation paths, revealing cross-disciplinary knowledge flows.

The citing map indicates that most AI-in-language-education outputs cluster within the psychology, education, and health regions. This concentration reflects the field's strong grounding in educational technology, applied linguistics, and learning sciences, where empirical studies on classroom practice, learner affect, teacher cognition, and instructional design predominate. These citing nodes primarily connect to the psychology–education–social cluster on the cited side, highlighting the foundational influence of educational psychology, sociocognitive theories, and second language acquisition research on the pedagogical and affective dimensions of AI-supported learning.

A second prominent citing region is systems, computing, and computer science research situated at the intersection of CALL and computational sciences. Studies on automated writing evaluation, speech-recognition-driven CALL, and learning-analytics dashboards originate largely from this region. The dominant citation flows from this cluster to the systems–computing–computer and molecular–biology–genetics regions indicate reliance on core computer-science literature, including machine learning, neural modelling, speech processing, and large language model architectures that underpin contemporary AI-enhanced language-learning tools.

Smaller citation streams from mathematics, systems, mathematical and physics, materials, and chemistry on the citing side point toward technical domains on the cited side. Although these represent a smaller proportion of publications, they signal contributions from signal processing, acoustics, and optimisation research areas that support innovations in mispronunciation detection, acoustic modelling, and uncertainty estimation for speech assessment systems. Such technical contributions serve as bridges linking foundational AI research with pedagogical applications. On the cited map, beyond the major education- and computing-oriented clusters, two additional knowledge bases emerge. First, citation flows toward the health–nursing–medicine region reflect growing use of research on anxiety, self-efficacy, and well-being to conceptualise affective experiences in AI-mediated language learning. Second, linkages to the economics–economic–political region suggest a selective integration of scholarship on technology acceptance, digital inequality, and governance frameworks, particularly in studies addressing ethical use, access, and policy considerations surrounding AI tools.

Temporal patterns in the dual-map overlay further illustrate the field's evolution. Early citation activity (2016-2018) centres almost exclusively on psychology, education, and social domains, consistent with early CALL research that treated AI as a supplementary instructional technology. During 2019-2021, citation flows increasingly incorporate systems-computing and computer regions, paralleling the uptake of deep learning, automated feedback, and speech-based assessment systems.

From 2022 onward, citation paths diversify substantially, connecting education and linguistics journals more closely with computer science and engineering clusters in response to the rapid adoption of generative AI and large language models. Concurrently, new research arcs in molecular biology, genetics, and health-related fields indicate an emerging interest in neurocognitive approaches, multimodal analytics, and ethical debates concerning learner well-being in AI-driven environments.

Figure 13. Dual-map overlay of AI in language education (2016–2025).

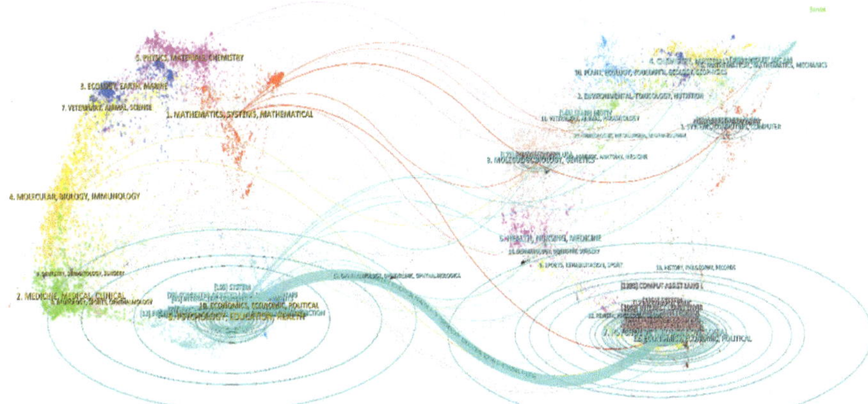

4. DISCUSSION

4.1 Main Findings Summary

This scientometric review offers a comprehensive and longitudinal synthesis of how artificial intelligence in language education has evolved over the past decade, demonstrating a clear transition from a specialised extension of computer-assisted language learning (CALL) to an expansive, interdisciplinary research domain shaped by deep learning and large language models (LLMs). The publication trajectory reveals three major developmental phases. The formative phase (2016-2018) is characterised by modest yet steady growth, mainly driven by rule-based CALL systems, early automated writing evaluation tools, and traditional machine-learning approaches to linguistic analysis. The transitional phase (2019-2021) aligns with the widespread integration of deep-learning architectures into natural language and speech processing, which brought substantial advances in mispronunciation detection, neural text generation, and discourse processing (Cruz, 2020; Erdem Coşgun,

2025). This period marks the mainstreaming of AI methodologies in language education. The transformative phase (2022-2025) coincides with the rapid diffusion of generative AI and LLMs, catalysed by the release of ChatGPT, leading to an unprecedented rise in research output, reaching 1,841 publications in 2025. This surge reflects escalating scholarly interest in LLM-mediated instruction, feedback, assessment, and learner affect, as documented in systematic reviews and empirical studies (Belda-Medina & Kokošková, 2024; Deng et al., 2024; Ding et al, 2025; Lin, 2024; Wang, 2025; Wei et al., 2023).

Co-citation network analysis further elucidates the field's intellectual foundations. A tightly interconnected core of CALL and educational technology journals, such as Computer Assisted Language Learning, Education and Information Technologies, System, and Computers & Education, anchors empirical work on AI-enhanced pedagogy, feedback systems, and social robots (X Deng, 2025; Lin, 2024; Wang, 2025). These are complemented by applied linguistics journals (e.g., TESOL Quarterly, Language Learning, Applied Linguistics, Assessing Writing), which provide conceptual grounding for research on computational modelling, writing analytics, and LLM-supported development (Cong, 2024; Eguchi & Kyle, 2024; Qin, 2025). Interdisciplinary linkages to journals in computer science and speech technology including Journal of Artificial Intelligence Research, NeurIPS, Speech Communication, and IEEE/ACM Transactions on Audio, Speech, and Language Processing demonstrate that progress in language education increasingly depends on technical advancements in neural architectures, multimodal representation learning, and automated speech evaluation (Dunbar et al. 2017; Erdem Coşgun, 2025; Li & Jiang, 2019; Liu et al. 2025; Wang, 2018).

Thematic analyses identify several enduring research clusters: intelligent tutoring systems, automated writing evaluation (AWE), speech recognition and pronunciation training, adaptive and analytics-driven learning environments, conversational agents and chatbots, and neural machine translation with translanguaging practices. These clusters follow a chronological progression from early corpus-based and linguistic descriptors to advanced constructs aligned with deep learning, NLP, and LLM-driven learning, particularly after 2021, when terms such as deep learning, machine learning, and mobile learning show intense citation bursts.

Collaboration patterns indicate that the field is organised around major geographic and institutional hubs. East Asian regions, particularly mainland China, Hong Kong, Taiwan, Japan, and South Korea, along with Anglophone countries such as the United States, the United Kingdom, Australia, Canada, and New Zealand, form the core of the global network. Institutions in Hong Kong and Taiwan serve as particularly dense collaboration nexuses connected to leading research centers in North America and Europe. Author-level networks reveal cohesive and influential teams, including those led by Zou, Kohnke, Xie, and Chen, who contribute extensively to AI-mediated speaking evaluation and informal digital learning. Foundational work by Kyle, Crossley, and colleagues continues to shape research on lexical sophistication and AWE. Subject-category and dual-map overlay analyses reveal that AI-supported language education remains rooted in Education and Linguistics, but increasingly draws on Computer Science, Psychology, Neuroscience, Health Sciences, and Sustainability Studies. These interdisciplinary linkages highlight the field's increasing engagement with the cognitive, social, and ethical aspects of AI-enhanced learning environments.

5. CONCLUSION

This study responds to the growing need to synthesise an increasingly expansive and fragmented body of research on artificial intelligence in language education. This field has accelerated rapidly with the integration of deep-learning technologies, speech recognisers, and large language models (LLMs). As these tools become embedded in diverse learning ecologies, researchers and practitioners face mounting challenges in tracking emerging themes, identifying foundational knowledge bases, and understanding the field's evolving interdisciplinary structure. Through a comprehensive scientometric analysis of publications from 2016 to 2025, this study systematically maps the conceptual, methodological, and collaborative configurations that define contemporary AI in language education.

The findings reveal a decisive shift from rule-based and corpus-analytic CALL toward generative, multimodal, and analytics-driven ecosystems. Major research fronts include automated writing evaluation, speech technologies, adaptive and data-informed learning environments, conversational agents, and AI-mediated informal digital learning. Dense collaboration networks drive the field centred in East Asia and Anglophone regions. At the same time, dual-map overlays and subject-category bursts demonstrate increasing convergence with computer science, neuroscience, psychology, and sustainability studies. Looking ahead, the study underscores the importance of ongoing critical inquiry into AI literacy, data governance, equity, and the ethical implications of LLM-mediated learning. Ensuring the responsible and inclusive implementation of AI technologies will be essential for supporting high-quality language education worldwide.

REFERENCES

Barrot, J. S. (2023). Using ChatGPT for second language writing: Pitfalls and potentials. *Assessing Writing*, *57*, 100745.

Barrot, J. S. (2024). ChatGPT as a language learning tool: An emerging technology report. *Technology. Knowledge and Learning*, *29*(2), 1151–1156.

Belda-Medina, J., & Kokošková, V. (2024). ChatGPT for language learning: Assessing teacher candidates' skills and perceptions using the Technology Acceptance Model (TAM). *Innovation in Language Learning and Teaching*, ●●●, 1–16.

Bilad, M. R., Yaqin, L. N., & Zubaidah, S. (2023). Recent progress in the use of artificial intelligence tools in education. *Jurnal Penelitian dan Pengkajian Ilmu Pendidikan: e-Saintika, 7*(3), 279-314.

Chen, C. (2004). Searching for intellectual turning points: Progressive knowledge domain visualization. *Proceedings of the National Academy of Sciences of the United States of America*, *101*(suppl_1), 5303–5310.

Chen, C. (2006). CiteSpace II: Detecting and visualizing emerging trends and transient patterns in scientific literature. *Journal of the American Society for Information Science and Technology*, *57*(3), 359–377.

Chen, H., Han, J., & Wright, D. (2020). An investigation of lecturers' teaching through English medium of instruction, A case of higher education in China. *Sustainability*, *12*(10), 4046.

Chen, J. (2025). Modeling Chinese second language learners' motivation, engagement, and resilience in AI-enhanced contexts: A self-determination theory. *Learning and Motivation*, *92*, 102199.

Chen, X., Zou, D., Xie, H., Cheng, G., & Liu, C. (2022). Two decades of artificial intelligence in education. *Journal of Educational Technology & Society*, *25*(1), 28–47.

Chen, Y. (2024). Chinese primary school students improve English through VR-based drama education games. *Addressing Global* Challenges-*Exploring Socio-Cultural Dynamics and Sustainable Solutions in a Changing World*, 295-301.

Cruz, S. D. D. (2020). Saving the face: Language courtesy strategies among students. *Asian EFL Journal*, *27*(32), 39–60.

Deng, J., Liu, C., & Zhou, Z. (2024). *AI-Based English Vowel Formant Recognition and Pronunciation Correction Platform*. IEEE.

Deng, M. (2022). Application of Artificial Intelligence Technology in College Students' Foreign Language Learning. *2022 IEEE 5th Eurasian Conference on Educational Innovation (ECEI)*, 302-304.

Deng, R., Jiang, M., Yu, X., Lu, Y., & Liu, S. (2024). Does ChatGPT enhance student learning? A systematic review and meta-analysis of experimental studies. *Computers & Education*, •••, 105224.

Deng, X., Xu, Y., Hussin, S., & Sulaiman, N. A. (2025). Use of AI Tools in EFL Writing Instruction: A Case Study of Chinese Vocational College Instructors. In *Forum for Linguistic Studies* (Vol. 7, No. 5, pp. 114-129).

Derakhshan, A. (2025). EFL students' perceptions about the role of generative artificial intelligence (GAI)-mediated instruction in their emotional engagement and goal orientation: A motivational climate theory (MCT) perspective in focus. *Learning and Motivation*, *90*, 102114.

Derakhshan, A., & Ghiasvand, F. (2024). Is ChatGPT an evil or an angel for second language education and research? A phenomenographic study of research-active EFL teachers' perceptions. *International Journal of Applied Linguistics*, *34*(4), 1246–1264.

Devlin, J., Chang, M. W., Lee, K., & Toutanova, K. (2019, June). Bert: Pre-training of deep bidirectional transformers for language understanding. In *Proceedings of the 2019 conference of the North American chapter of the association for computational linguistics: human language technologies, volume 1 (long and short papers)* (pp. 4171-4186).

Ding, D., & Yusof, M. B. (2025). Investigating the role of AI-powered conversation bots in enhancing L2 speaking skills and reducing speaking anxiety: A mixed methods study. *Humanities & Social Sciences Communications*, *12*(1), 1–16.

Ding, Y. (2024, January). Development of Intelligent Tutors Based on Dialogue Systems. In *2024 Asia-Pacific Conference on Software Engineering, Social Network Analysis and Intelligent Computing (SSAIC)* (pp. 927-930). IEEE.

Du, Y., & Liu, Y. (2025). How generative AI affordance drives users' continuous usage intention: The mediation role of self-expansion and self-extension. *Behaviour & Information Technology*, •••, 1–16.

Dunbar, E., Cao, X. N., Benjumea, J., Karadayi, J., Bernard, M., Besacier, L., . . . Dupoux, E. (2017, December). The zero resource speech challenge 2017. In *2017 IEEE Automatic Speech Recognition and Understanding Workshop (ASRU)* (pp. 323-330). IEEE.

Eguchi, M., & Kyle, K. (2024). Building custom NLP tools to annotate discourse-functional features for second language writing research: A tutorial. *Research Methods in Applied Linguistics*, *3*(3), 100153.

Erdem Coşgun, G. (2025). Artificial intelligence literacy in assessment: Empowering pre-service teachers to design effective exam questions for language learning. *British Educational Research Journal.*

Fathi, J., & Rahimi, M. (2024). Utilising artificial intelligence-enhanced writing mediation to develop academic writing skills in EFL learners: A qualitative study. *Computer Assisted Language Learning*, •••, 1–40. DOI: 10.1080/09588221.2024.2374772

Feng, D. (2025). A Critical Bibliometric Review of Technology-Enhanced Vocabulary Learning: Uncovering a Decade of Tensions and Trends. *SAGE Open*, *15*(4), 21582440251393196.

Feng, L. (2025). Investigating the effects of artificial intelligence-assisted language learning strategies on cognitive load and learning outcomes: A comparative study. *Journal of Educational Computing Research*, *62*(8), 1741–1774.

Foung, D., & Kohnke, L. (2025). Generating synthetic data for CALL research with GenAI: A proof-of-concept study. *Research Methods in Applied Linguistics*, *4*(3), 100248.

Foung, D., Kohnke, L., & Chen, J. (2023). Predicting student success with self-regulated behaviors: A seven-year data analytics study on a Hong Kong University English Course. In *Data analytics and adaptive learning* (pp. 92–109). Routledge.

Gamage, B. G. J., Paranagama, R. D., Ranaweera, R. M. S. H., Dilshan, A. V. R., Weerathunga, I., & Kasthurirathna, D. (2023, December). Sinhala sign language translation through immersive 3D avatars and adaptive learning. In *2023 5th International Conference on Advancements in Computing (ICAC)* (pp. 1-6). IEEE.

Guo, S., & Xia, M. (2026). Exploring Learners' Perceptions of AI-mediated informal digital Multilingual Learning: A Q-Methodology Approach. *International Journal of Applied Linguistics.* Advance online publication. DOI: 10.1111/ijal.12815

Habibi, A., Attar, R. W., Atan, N. A., Said, M. N. H. B. M., Sofwan, M., Yaqin, L. N., & Al-Adwan, A. S. (2025). Technology use during teaching practicum: Beliefs/knowledge in urban and suburban schools. *Asia Pacific Journal of Education*, •••, 1–26. DOI: 10.1080/02188791.2025.2521112

Huang, G., & Wang, Z. (2022). The application of virtual reality technology in the coordination and interaction of regional economy and culture in the sustainable development of ecotourism. *Mathematical Problems in Engineering*, *2022*(1), 9847749.

Huang, S., & Wang, W. (2025, March). Foreign Language Learners' Perceptions of Chatbot-Assisted Speech-Giving Tasks. In *2025 14th International Conference on Educational and Information Technology (ICEIT)* (pp. 79-83). IEEE.

Hwang, W. Y., Manabe, K., Cai, D. J., & Ma, Z. H. (2020). Collaborative kinesthetic English learning with recognition technology. *Journal of Educational Computing Research, 58*(5), 946–977.

Inayah, N., Afra, A., Azmar, C. N., & Yaqin, L. N. (2024). A linguistic analysis of grammatical cohesion devices in discussion sections of undergraduate theses. *International Journal of Language Studies, 18*(4), 149–166. DOI: 10.5281/zenodo.13956443

Jeon, J. (2024). Exploring AI chatbot affordances in the EFL classroom: Young learners' experiences and perspectives. *Computer Assisted Language Learning, 37*(1-2), 1–26.

Jeon, J., Lee, S., & Choe, H. (2023). Beyond ChatGPT: A conceptual framework and systematic review of speech-recognition chatbots for language learning. *Computers & Education, 206*, 104898.

Jolley, J., & Maimone, L. (2025). Machine translation and language teaching and learning. *Language Teaching, 58*(2), 173–191.

Kasneci, E., Seßler, K., Küchemann, S., Bannert, M., Dementieva, D., Fischer, F., & Kasneci, G. (2023). ChatGPT for good? On opportunities and challenges of large language models for education. *Learning and Individual Differences, 103*, 102274.

Khandaqji, F., Ashqar, H. I., & Atawnih, A. (2025, July). A Survey of Using Artificial Intelligence (AI) in Sign Language for Deaf and Hard-of-Hearing Students. In *2025 International Conference on Smart Learning Courses (SCME)* (pp. 172-177). IEEE.

Kohnke, L., Moorhouse, B. L., & Zou, D. (2023). Exploring generative artificial intelligence preparedness among university language instructors: A case study. *Computers and Education: Artificial Intelligence, 5*, 100156.

Koltovskaia, S. (2020). Student engagement with automated written corrective feedback (AWCF) provided by Grammarly: A multiple case study. *Assessing Writing, 44*, 100450.

Korzekwa, D., Lorenzo-Trueba, J., Zaporowski, S., Calamaro, S., Drugman, T., & Kostek, B. (2021, June). Mispronunciation detection in non-native (L2) English with uncertainty modeling. In *ICASSP 2021-2021 IEEE international* conference *on acoustics, speech and signal processing (ICASSP)* (pp. 7738-7742). IEEE.

Lahby, M. (Ed.). (2024). *Empowering Digital Education with ChatGPT: From Theoretical to Practical Applications*. CRC Press.

Lan, G., Li, Y., Yang, J., & He, X. (2025). Investigating a customized generative AI chatbot for automated essay scoring in a disciplinary writing task. *Assessing Writing*, *66*, 100959.

Lee, J. H., Shin, D., & Noh, W. (2023). Artificial intelligence-based content generator technology for young English-as-a-foreign-language learners' reading enjoyment. *RELC Journal*, *54*(2), 508–516.

Li, P., & Xu, Q. (2023). Computational modeling of bilingual language learning: Current models and future directions. *Language Learning*, *73*(S2), 17–64.

Li, T., & Ding, Y. (2025). Exploring the GenAI Literacy of Chinese University Students in EFL Learning. [IJCALLT]. *International Journal of Computer-Assisted Language Learning and Teaching*, *15*(1), 1–24.

Li, W., & Jiang, W. (2019). Requests made by Australian learners of Chinese as a foreign language. *Journal of Language Teaching and Research*, *10*(1), 23–34.

Li, Y. (2025, June). An Enhanced Efficient English Learning System based on Artificial Intelligence Driven B/S Architectures. In *2025 International Conference on Intelligent Computing and Knowledge Extraction (ICICKE)* (pp. 1-6). IEEE.

Liang, B. (2024, December). Artificial Intelligence in Language Education: CiteSpace-based Visualisation and Analysis. In *Proceedings of the 2024 2nd International Conference on Information Education and Artificial Intelligence* (pp. 721-725).

Liang, J. C., Hwang, G. J., Chen, M. R. A., & Darmawansah, D. (2023). Roles and research foci of artificial intelligence in language education: An integrated bibliographic analysis and systematic review approach. *Interactive Learning Environments*, *31*(7), 4270–4296.

Lin, H., & Liu, X. (2025). InVideo AI: Creating instructional audiovisual input for English learning. *RELC Journal*, *56*(3), 803–809.

Lin, L., Zhou, D., Wang, J., & Wang, Y. (2024). A systematic review of big data driven education evaluation. *SAGE Open*, *14*(2), 21582440241242180.

Lin, S., & Crosthwaite, P. (2024). The grass is not always greener: Teacher vs. GPT-assisted written corrective feedback. *System*, *127*, 103529.

Lin, Y., & Yu, Z. (2025). Learner perceptions of artificial intelligence-generated pedagogical agents in language learning videos: Embodiment effects on technology acceptance. *International Journal of Human-Computer Interaction, 41*(2), 1606–1627.

Ling, Y., & Jan, J. M. (2025). Voices from the Flip: Teacher Perspectives on Integrating AI Chatbots in Flipped English Classrooms. *Education Sciences, 15*(9), 1219.

Liu, Z., Zhang, W., & Yang, P. (2025). Can AI chatbots effectively improve EFL learners' learning effects? A meta-analysis of empirical research from 2022–2024. *Computer Assisted Language Learning*, •••, 1–27.

MacWhinney, B. (2025). Understanding Language Through TalkBank. *Current Directions in Psychological Science, 34*(2), 75–81.

Marshall, D. J., & Tsikouras, B. (2023). Clay-shielded estuarine gastropods are better protected against environmental acidification than unshielded individuals. *The Science of the Total Environment, 866*, 161367.

Moorhouse, B. L., Ho, T. Y., Wu, C., & Wan, Y. Pre-service Language Teachers' Task-specific Large Language Model Prompting Practices. *RELC Journal*. Advance online publication. DOI: 10.1177/00336882251313701

Namaziandost, E. (2025). Integrating flipped learning in AI-enhanced language learning: Mapping the effects on metacognitive awareness, writing development, and foreign language learning boredom. *Computers and Education: Artificial Intelligence*, 100446.

Ni, X., & Yin, H. (2025, May). The Impact of Mobile-Assisted Pronunciation Training on Intelligibility and Comprehensibility in EFL Learners. In *2025 5th International Conference on Artificial Intelligence and Education (ICAIE)* (pp. 839-844). IEEE.

Qin, Y. (2025, April). Key Factors Influencing the Deployment of Generative Large Models in Language Teaching. In *2025 7th International Conference on Computer Science and Technologies in Education (CSTE)* (pp. 592-595). IEEE.

Radford, A., Wu, J., Child, R., Luan, D., Amodei, D., & Sutskever, I. (2019). Language models are unsupervised multitask learners. *OpenAI Blog, 1*(8), 1–24.

Saeed Ahmed Mohamed, S., & Mahmoud Ibrahim Alian, E. (2023). *Students' attitudes toward using chatbot in EFL learning. Arab World English Journal* (Vol. 14). AWEJ.

Shi, L. (2023). Enriching early years English language education with translanguaging. In *Innovative practices in early English language education* (pp. 81–99). Springer International Publishing.

Syafiee, A. S. E. E. A., & Yaqin, L. N. (2025). A comparative analysis of Google Translate and ChatGPT in translating Borneo Bulletin News into standard Malay. *Journal of Research on English and Language Learning (J-REaLL), 6*(1), 115–126.

Tan, J. C. A., Chan, W. P., Robinson, N. L., Kulić, D., & Croft, E. A. (2022, August). Impacts of Teaching towards Training Gesture Recognizers for Human-Robot Interaction. In *2022 31st IEEE International Conference on Robot and Human Interactive Communication (RO-MAN)* (pp. 618-623). IEEE.

Tsai, S. C. (2022). Chinese students' perceptions of using Google Translate as a translingual CALL tool in EFL writing. *Computer Assisted Language Learning, 35*(5-6), 1250–1272.

Waluyo, B., & Rouaghe, F. (2025). Beyond teacher-led approaches: Student-initiated translanguaging with artificial intelligence tools in foreign language acquisition. *SAGE Open, 15*(3), 21582440251362998.

Wang, H. (2024). *The design and evaluation of intelligent computer-assisted language learning tools for beginner learners of Mandarin Chinese*. Trinity College.

Wang, X. (2025, June). Design and Development of Smart Learning System for College English based on Intelligent Cloud Platform. In *2025 International Conference on Intelligent Computing and Knowledge Extraction (ICICKE)* (pp. 1-8). IEEE.

Wei, L. Z., Vijayasingam, V., & Kaur, P. (2023, December). Empowering Local Communities through an Intelligent Digital Platform for Rural Tourism in Malaysia. In *2023 IEEE 21st Student Conference on Research and* Development *(SCOReD)* (pp. 455-462). IEEE.

Wen, Y., Zhao, X., Li, X., & Zang, Y. (2025). Attitude Mining Toward Generative Artificial Intelligence in Education: The Challenges and Responses for Sustainable Development in Education. *Sustainability, 17*(3), 1127.

Wilcox, E. G., Hu, M. Y., Mueller, A., Warstadt, A., Choshen, L., Zhuang, C., & Linzen, T. (2025). Bigger is not always better: The importance of human-scale language modeling for psycholinguistics. *Journal of Memory and Language, 144*, 104650.

Wu, T., & Yu, Z. (2024). Bibliometric and systematic analysis of artificial intelligence chatbots' use for language education. *Journal of University Teaching & Learning Practice, 21*(6), 174–198.

Xiao, F., Zhu, S., & Xin, W. (2025). Exploring the landscape of generative AI (ChatGPT)-powered writing instruction in English as a foreign language education: A scoping review. *ECNU Review of Education*, 20965311241310881.

Xue, Y. (2024). Towards automated writing evaluation: A comprehensive review with bibliometric, scientometric, and meta-analytic approaches. *Education and Information Technologies*, *29*(15), 19553–19594.

Yan, D. (2023). Impact of ChatGPT on learners in a L2 writing practicum: An exploratory investigation. *Education and Information Technologies*, *28*(11), 13943–13967.

Yang, X. (2025, May). Generative AI Driven College English Multimodal Teaching Based on BERT Optimization: An Empirical Study on Interactive Participation and Learning Outcomes in Digital Twin Classrooms. In *2025 International Conference on Artificial Intelligence and Digital Ethics (ICAIDE)* (pp. 513-516). IEEE.

Yaqin, L. N. (2025). Emerging research in AI-assisted language learning: A systematic literature review. *Multi-Industry* Digitalization *and Technological Governance in the AI Era*, 29–50.

Yaqin, L. N., Bilad, M. R., Yusof, B., Hassan, H., Damit, A. R., & Chuchu, I. F. (2025a). Mapping research evolution in higher education: A scientometric analysis of Brunei Darussalam (1986–2024). *Discover Sustainability*, *6*(1). Advance online publication. DOI: 10.1007/s43621-025-00917-3

Yaqin, L. N., Hassan, H., & Yusof, B. (2025b). Performance And Accuracy Of Chatgpt In Generating Malay Academic Texts: A Comparative Study With Expert Corrections. *LLT Journal: A Journal on Language and Language* Teaching, *28*(1), 495–517.

Yaqin, L. N., McLellan, J., & Najib, N. (2025c). Shifting paradigms in translanguaging: A scientometric analysis of research trends and trajectories. *Studies in English Language and Education*, *12*(2), 684–702. https://doi.org/https://doi.org/10.24815/siele.v12i2.40243

Zawacki-Richter, O., Marín, V. I., Bond, M., & Gouverneur, F. (2019). Systematic review of research on artificial intelligence applications in higher education–where are the educators? *International Journal of Educational Technology in Higher Education*, *16*(1), 1–27.

Zhang, C., Song, J., Zhu, X., Zhu, L., & Zhang, S. (2021). Hcmsl: Hybrid cross-modal similarity learning for cross-modal retrieval. [TOMM]. *ACM Transactions on Multimedia Computing Communications and Applications*, *17*(1s), 1–22.

APPENDIX

Table 1. Top 25 keywords with the strongest citation bursts, 2016–2020

Keywords	Year	Strength	Begin	End	2016 - 2020
linguistics	2016	2.36	**2016**	2020	
semantics	2016	1.57	**2016**	2020	
learning algorithms	2016	1.16	**2016**	2020	
end-to-end learning	2018	1.12	**2018**	2020	
frequency	2016	1.1	**2016**	2020	
learner corpus	2016	1.1	**2016**	2020	
index	2016	1.08	**2016**	2020	
writing assessment	2016	1.08	**2016**	2020	
supervised machine learning	2016	1.08	**2016**	2020	
second language writing	2016	1.08	**2016**	2020	
personalized learning	2018	0.84	**2018**	2020	
courseworks	2018	0.84	**2018**	2020	
unsupervised learning	2018	0.84	**2018**	2020	
arabic language	2018	0.84	**2018**	2020	
speech synthesis	2018	0.84	**2018**	2020	
portfolio assessment	2018	0.84	**2018**	2020	
capstone projects	2018	0.84	**2018**	2020	
graduate level	2018	0.84	**2018**	2020	
esl learners	2018	0.84	**2018**	2020	
performance based assessment	2018	0.84	**2018**	2020	
corpus based	2018	0.84	**2018**	2020	
theory and practice	2018	0.84	**2018**	2020	
learning models	2017	0.83	**2017**	2020	
syntactic complexity	2017	0.83	**2017**	2020	

Table 2. Top 25 keywords with the strongest citation bursts, 2021–2025

Keywords	Year	Strength	Begin	End	2021 - 2025
deep learning	2021	20.65	**2021**	2025	
natural language processing	2021	15.04	**2021**	2025	
machine learning	2021	13.82	**2021**	2025	
computational linguistics	2021	7.57	**2021**	2025	
human-computer interaction	2022	5.53	**2022**	2025	
computer-assisted language learning	2021	5.22	**2021**	2025	
second language learning	2021	5.1	**2021**	2025	
mobile learning	2021	4.71	**2021**	2025	
e learning	2021	4.66	**2021**	2025	
recognition	2021	4.16	**2021**	2025	
big data	2021	3.91	**2021**	2025	
computer aided instruction	2021	3.85	**2021**	2025	
second language	2021	3.4	**2021**	2025	
online learning	2021	3.37	**2021**	2025	
human computer interaction	2021	3.29	**2021**	2025	
intelligent tutoring system	2021	3.16	**2021**	2025	
object detection	2022	3.16	**2022**	2025	
intelligent personal assistants	2022	2.96	**2022**	2025	
precision education	2021	2.89	**2021**	2025	
acquisition	2021	2.82	**2021**	2025	
text processing	2021	2.8	**2021**	2025	
classification (of information)	2021	2.8	**2021**	2025	
translation (languages)	2023	2.78	**2023**	2025	
mobile-assisted language learning	2021	2.73	**2021**	2025	
speech	2021	2.73	**2021**	2025	

Table 3. Top 25 co-cited references with the strongest citation bursts in AI-in-language-education research (2016–2025).

References	Year	Strength	Begin	End	2016 - 2025
Devlin J, 2019, 2019 CONFERENCE OF THE NORTH AMERICAN CHAPTER OF THE ASSOCIATION FOR COMPUTATIONAL LINGUISTICS: HUMAN LANGUAGE TECHNOLOGIES (NAACL HLT 2019), VOL. 1, P4171	2019	20.96	**2019**	2025	
Vaswani A, 2017, ADV NEUR IN, V30, P0	2017	15.05	**2017**	2025	
Kyle K, 2015, TESOL QUART, V49, P757, DOI 10.1002/tesq.194, DOI	2015	6.68	**2016**	2020	
Radford Alec, 2019, LANGUAGE MODELS ARE UNSUPERVISED MULTITASK LEARNERS, V1, P9, DOI 10.4018/978-1-5225-9348-5.CH006, DOI	2019	6.49	**2019**	2025	
Ng HweeTou, 2014, PROCEEDINGS ...... LEARNING: SHARED TASK, V0, PP1, DOI	2014	4.72	**2016**	2020	
Mikolov T, 2013, ADVANCES IN NEURAL INFORMATION PROCESSING SYSTEMS, V0, P3111	2013	4.09	**2016**	2020	
Manning CD, 2014, PROCEEDINGS ...... YSTEM DEMONSTRATIONS, V0, PP55, DOI	2014	4.05	**2016**	2020	
Zawacki-Richter O, 2019, INT J EDUC TECHNOL H, V16, P0, DOI 10.1186/s41239-019-0171-0, DOI	2019	13.95	**2021**	2025	
Fryer LK, 2019, COMPUT HUM BEHAV, V93, P279, DOI 10.1016/j.chb.2018.12.023, DOI	2019	10.77	**2021**	2025	
Pokrivcakova S, 2019, J LANG CULT EDUC, V7, P135, DOI 10.2478/jolace-2019-0025, DOI	2019	9.81	**2021**	2025	
Bibauw S, 2019, COMPUT ASSIST LANG L, V32, P827, DOI 10.1080/09588221.2018.1535508, DOI	2019	8.22	**2021**	2025	
Haristiani Nuria, 2019, JOURNAL OF PHYSICS: CONFERENCE SERIES, V1387, P0, DOI 10.1088/1742-6596/1387/1/012020, 10.1088/1742-6596/1387/1/012020, DOI	2019	7.91	**2021**	2025	
Fryer LK, 2020, LANG LEARN TECHNOL, V24, P8	2020	7.51	**2021**	2025	
Devlin J, 2019, ARXIV, V0, P0	2019	6.01	**2021**	2025	
Na-Young Kim, 2019, MULTIMEDIA-ASSISTED LANGUAGE LEARNING 멀티미디어 언어교육, V22, P32	2019	5.69	**2021**	2025	
Chen XL, 2020, COMPUTERS AND EDUCATION ARTIFICIAL INTELLIGENCE, V1, P100002, DOI 10.1016/j.caeai.2020.100002, 10.1016/j.caeai.2020.100002, 10.1016/J.CAEAI.2020.100002, 10.1016/j.caeai.2020.100002 10.1016/j.caeai.2020.100002, DOI	2020	5.62	**2021**	2025	
Na-Young Kim, 2019, JOURNAL OF DIGITAL CONVERGENCE, V0, P0	2019	4.74	**2021**	2025	

continued on following page

Table 3. Continued

References	Year	Strength	Begin	End	2016 - 2025
Bao MH, 2019, ARAB WORLD ENGL J, V0, PP28, DOI 10.24093/awej/call5.3, DOI	2019	4.74	**2021**	2025	
Brown TB, 2020, ADV NEUR IN, V33, P0	2020	4.5	**2021**	2025	
Ayedoun E, 2019, INT J ARTIF INTELL E, V29, P29, DOI 10.1007/s40593-018-0171-6, DOI	2019	4.42	**2021**	2025	
Wolf T, 2020, PROCEEDINGS OF THE 2020 CONFERENCE ON EMPIRICAL METHODS IN NATURAL LANGUAGE PROCESSING: SYSTEM DEMONSTRATIONS, V0, P38	2020	4.33	**2021**	2025	
Smutny P, 2020, COMPUT EDUC, V151, P0, DOI 10.1016/j.compedu.2020.103862, DOI	2020	4.24	**2021**	2025	
van den Berghe R, 2019, REV EDUC RES, V89, P259, DOI 10.3102/0034654318821286, DOI	2019	4.11	**2021**	2025	
Zhang RF, 2022, COMPUT ASSIST LANG L, V35, P696, DOI 10.1080/09588221.2020.1744666, DOI	2022	4.08	**2022**	2025	
Loewen S, 2019, RECALL, V31, P293, DOI 10.1017/S0958344019000065, DOI	2019	3.87	**2021**	2025	

Table 4. Top 25 countries with the strongest citation bursts in AI-supported language education (2016–2025).

Countries	Year	Strength	Begin	End	2016 - 2025
Italy	2016	9.88	**2016**	2020	
Germany	2016	9.68	**2016**	2025	
United States	2016	9.54	**2016**	2020	
Netherlands	2016	5.79	**2016**	2025	
USA	2016	5.36	**2016**	2025	
Sweden	2016	4.78	**2016**	2025	
Japan	2016	4.62	**2016**	2020	
Belgium	2017	4.33	**2017**	2020	
Qatar	2016	4.28	**2016**	2020	
Finland	2019	4.24	**2019**	2020	
Denmark	2017	3.88	**2017**	2025	
South Korea	2016	3.72	**2016**	2025	
England	2016	3.65	**2016**	2020	
Morocco	2016	3.49	**2016**	2020	
Canada	2016	3.48	**2016**	2020	

continued on following page

Table 4. Continued

Countries	Year	Strength	Begin	End	2016 - 2025
United Kingdom	2016	3.42	**2016**	2020	
Portugal	2016	3.28	**2016**	2025	
Ireland	2019	2.8	**2019**	2025	
Romania	2016	2.54	**2016**	2020	
Egypt	2020	2.46	**2020**	2025	
France	2016	2.38	**2016**	2025	
Greece	2016	2.22	**2016**	2020	
Singapore	2017	5.23	**2021**	2025	
Hungary	2022	2.48	**2022**	2025	
South Africa	2022	2.42	**2022**	2025	

Table 5. Top 25 institutions with the strongest citation bursts in AI-supported language education (2016–2025).

Institutions	Year	Strength	Begin	End	2016 - 2025
University of Cambridge	2016	11.37	**2016**	2025	
Georgia State University	2016	8.2	**2016**	2020	
University System of Georgia	2016	5.38	**2016**	2020	
University of Hawai'i at Mānoa	2016	5.1	**2016**	2020	
Peking University	2018	4.81	**2018**	2025	
Eberhard Karls University of Tubingen	2016	4.62	**2016**	2025	
University of Hawaii Manoa	2016	3.92	**2016**	2025	
University of Hawaii System	2016	3.92	**2016**	2025	
Kyushu University	2016	3.43	**2016**	2020	
Eberhard Karls Universität Tübingen	2016	3.22	**2016**	2025	
University of Piraeus	2018	3.16	**2018**	2020	
National Central University	2018	3.15	**2018**	2025	
Arizona State University	2016	2.85	**2016**	2025	
Arizona State University-Tempe	2016	2.72	**2016**	2025	
Beijing Normal University	2016	2.67	**2016**	2020	
Evanini	2017	2.54	**2017**	2020	
Lingnan University	2021	4.04	**2021**	2025	
University of Oregon	2023	3.24	**2023**	2025	
Tohoku University	2022	3.07	**2022**	2025	
Tokyo Gakugei University	2021	2.99	**2021**	2025	

continued on following page

Table 5. Continued

Institutions	Year	Strength	Begin	End	2016 - 2025
Norwegian University of Science & Technology (NTNU)	2022	2.89	**2022**	2025	
Sookmyung Women's University	2021	2.84	**2021**	2025	
Korea University	2021	2.71	**2021**	2025	
Temple University	2023	2.59	**2023**	2025	
Athabasca University	2022	2.56	**2022**	2025	

Table 6. Top 25 authors with the strongest citation bursts (2016–2025)

Authors	Year	Strength	Begin	End	2016 - 2025
Kyle, Kristopher	2016	5.68	**2016**	2020	
Crossley, Scott Andrew	2016	4.99	**2016**	2020	
Evanini, Keelan	2017	3.43	**2017**	2020	
Meurers, Detmar	2016	3.41	**2016**	2025	
Mouri, Kousuke	2019	3.12	**2019**	2020	
Troussas, Christos	2018	2.87	**2018**	2020	
Zhang, Jinsong	2017	2.69	**2017**	2020	
Hasnine, Mohammad Nehal	2019	2.49	**2019**	2020	
Crossley, Scott A	2016	2.37	**2016**	2020	
Drozdal, Jaimie	2018	2.29	**2018**	2020	
Chen, Xieling	2021	5.83	**2021**	2025	
Xie, Haoran	2021	5.5	**2021**	2025	
Cheng, Gary	2021	3.62	**2021**	2025	
Chen, Xiaobin	2021	3.26	**2021**	2025	
Kou, Xiaojing	2023	3.04	**2023**	2025	
Pack, Austin	2023	3.04	**2023**	2025	
Ehara, Yo	2021	2.97	**2021**	2025	
Zou, Di	2021	2.95	**2021**	2025	
Cheng, Gary ks	2021	2.65	**2021**	2025	
Chang, Maiga	2022	2.6	**2022**	2025	
Al-ghezi, Ragheb	2022	2.6	**2022**	2025	
Wen, Yun	2021	2.48	**2021**	2025	
Shin, Dongkawang	2021	2.43	**2021**	2025	
Getman, Yaroslav	2022	2.42	**2022**	2025	
Yang, Hyejin	2021	2.36	**2021**	2025	

Table 7. Top 25 Subject Categories with the Strongest Citation Bursts

Subject Categories	Year	Strength	Begin	End	2016 - 2025
Computer science, interdisciplinary applications	2016	4.5	**2016**	2018	
Computer science, theory & methods	2016	10.77	**2017**	2019	
Computer science, artificial intelligence	2016	18.19	**2018**	2022	
Engineering, electrical & electronic	2016	11.17	**2018**	2019	
Computer science, hardware & architecture	2018	3.53	**2018**	2023	
Automation & control systems	2018	2.72	**2018**	2020	
Audiology & speech-language pathology	2019	4.43	**2019**	2023	
Computer science, cybernetics	2017	2.6	**2019**	2020	
Imaging science & photographic technology	2019	1.3	**2019**	2021	
Literary theory & criticism	2019	1.13	**2019**	2021	
Computer science, information systems	2016	7.53	**2020**	2021	
Neurosciences	2016	5.16	**2020**	2022	
Mathematical & computational biology	2020	3.21	**2020**	2022	
Biochemistry & molecular biology	2020	1.19	**2020**	2021	
Physics, applied	2019	1.17	**2020**	2021	
Computer science, software engineering	2016	4.69	**2021**	2023	
Acoustics	2016	1.92	**2021**	2023	
Mathematics	2021	1.31	**2021**	2023	
Mathematics, interdisciplinary applications	2017	2.05	**2022**	2023	
Green & sustainable science & technology	2019	1.93	**2022**	2023	
Psychology, multidisciplinary	2017	1.04	**2022**	2023	
Genetics & heredity	2022	0.98	**2022**	2023	
Education, scientific disciplines	2016	7.63	**2023**	2025	
Engineering, biomedical	2021	1.01	**2023**	2025	
Literature	2023	1.01	**2023**	2025	

Chapter 2
AI Bridges the Publication Output Gap Among Scholars Skilled in the Malay Language

Yabit Alas
https://orcid.org/0000-0003-3643-361X
Universiti Brunei Darussalam, Brunei

Siti Ifwah Fauzani Chuchu
https://orcid.org/0009-0002-1822-7767
Universiti Brunei Darussalam, Brunei

Ashrol Rahimy Damit
https://orcid.org/0000-0002-1860-4211
Universiti Brunei Darussalam, Brunei

ABSTRACT

Universities worldwide embrace internationalization strategies, many are requiring faculty to publish their research in English to improve global rankings. This shift presents challenges, especially for those unprepared to navigate new publishing norms. Many faculty members encounter complex procedural barriers that hinder their ability to publish in English. To address these challenges, some scholars with limited English proficiency, particularly from the Malay (OML) background, are beginning to use artificial intelligence (AI). A thematic survey was conducted among a targeted group at Universiti Brunei Darussalam (UBD), including both proficient and struggling English speakers. The research aims to understand the challenges OML scholars face in meeting the university's internationalization goals.

DOI: 10.4018/979-8-3373-4397-6.ch002

Preliminary findings show that OML scholars are more likely to publish in Malay. Although many are willing to explore AI tools for manuscript enhancement, their English publications often do not meet academic standards and frequently rely on co-authorship with more proficient colleagues.

INTRODUCTION

Being a small nation situated in Southeast Asia, Brunei recognizes the critical importance of investing significantly in education to enhance the relevance of its human capital in an ever-evolving global landscape. This commitment to education is not merely a response to current demands but a proactive strategy aimed at preparing its citizens for future challenges and opportunities. To this end, Universiti Brunei Darussalam (UBD), the premier institution of higher learning in the country, has undergone a profound transformation since 2009 (Ishamina & Deterding, 2017; Hamdan, 2017). This transformation has involved embarking on various initiatives meticulously designed to enhance its relevance in a dynamic international context. For instance, UBD has introduced new academic programs that align with global trends, such as digital technology and sustainable development, thereby ensuring that its curriculum remains pertinent and forward-thinking.

Moreover, the university is actively cultivating employable graduates by integrating practical experiences and internships into its academic framework. This hands-on approach not only equips students with valuable skills but also enhances their employability in a competitive job market. Most critically, UBD is advancing its endeavors toward the internationalization of the university, fostering partnerships with renowned institutions worldwide and attracting a diverse student body. Such initiatives not only enrich the educational experience but also promote cultural exchange and collaboration, positioning Brunei as a significant player in the global academic arena. Through these comprehensive efforts, Brunei is striving to create a robust educational ecosystem that prepares its citizens to thrive in an interconnected world.

Since its inception in 1985, UBD has operated in both Malay and English mediums. The Faculty of Arts and Social Sciences (FASS) exclusively administers its programs in the Malay language, whereas the other faculties—specifically, the Faculty of Education—employ both Malay and English mediums, contingent upon the specific programs. In contrast, the Faculty of Management Studies and Public Policy, along with the Faculty of Sciences (FOS), offers its courses exclusively in English. Over the years, FASS has progressively transitioned from a predominantly Malay medium faculty to a more preeminent English medium faculty, with only the Malay Language and Linguistics and Malay Literature courses continuing to be taught in Malay. Following the comprehensive transformation initiated under the internationalization initiatives that commenced in 2009, the university has officially proclaimed itself an English-medium institution. Consequently, it is anticipated that both teaching and learning within the university will be conducted in English. Fortunately, this announcement poses minimal challenges for the academic staff, particularly the Bruneian faculty members, as the majority of the younger educators are products of the Bilingual Education System established in 1984 (Jones, 2015).

At the same time, there exists a cohort of Bruneian academic staff members who were educated within the preceding system, which comprised either English Medium or Malay Medium institutions. A considerable number of these educators emerged from the Malay Medium schools, where they instructed various subjects in the Malay language. Concurrently, there exists a considerable contingent of academic staff members within the Bilingual Education framework who demonstrate limited proficiency in English, who are allocated to instruct subjects exclusively in Malay. Consequently, for these individuals, the institutional transformation presents formidable challenges, particularly regarding the internationalization initiatives that predominantly necessitate academic staff to engage in research and publication in English.

Even though there were no formal mandates from the university requiring publication in English and no discrimination regarding promotion, many academic staff members personally felt an intrinsic pressure to align with mainstream practices prevalent in the global academic community. This inclination is largely due to the fact that Bruneian academic staff who lack proficiency in English constitute less than 10% of the overall academic population, making them a minority within the institution. This situation creates a subtle yet pervasive environment that encourages conformity to English-language publication norms. The majority of academic staff, whether international or Bruneian, possess language skills that are either native or nearly native in English, which further amplifies the expectation to publish in English. For example, many researchers believe that publishing in high-impact English-language journals is essential for gaining recognition and advancing their careers. Consequently, this pressure not only influences individual choices regarding language in publication but also shapes the overall academic culture at the university, reinforcing the dominance of English as the preferred medium for scholarly communication.

In the past, academic staff with proficiency in the Malay language (hereinafter referred to as OML) who were actively engaged in research typically published their findings in local or regional journals. These publications were particularly common in Malaysia and Indonesia, where researchers sought to share their work with a more localized audience. Many of these publications, however, unfortunately lacked international recognition. As a result, they are not listed under reputable indexed journals such as Scopus or Web of Science, which are critical for gaining visibility and credibility in the global academic community. This situation has posed challenges for OML researchers who aspire to make an impact beyond their immediate geographical confines.

However, with the advent of internationalization initiatives, particularly in the context of higher education and research collaborations, such practices are no longer considered the norm. The global academic landscape is evolving, and peers, alongside institutional pressures, instigate a significant transition in their preferences. Researchers are increasingly encouraged to shift their focus from local journals written in the Malay language to renowned international journals, which primarily publish in English. This shift not only enhances the reach and influence of their work but also aligns with the broader goals of academic excellence and international collaboration. Consequently, OML researchers are now exploring avenues to publish in high-impact journals, engaging in rigorous peer review processes, and adapting their research narratives to meet the expectations of a global audience.

Due to mounting pressures, the OML engaged the services of translation and editorial professionals who typically charge higher fees. Beyond financial considerations, many OML members felt uncomfortable submitting their articles for various reasons, including linguistic inadequacies and a perception that their work fell short of international standards. However, with the advancement of technologies, particularly artificial intelligence (AI), the sense of insecurity is no longer a concern. The OML can engage the services of AI discreetly within their own comfort zone and at their convenience.

AI has significantly impacted many aspects of society in the 21st century. From healthcare to transportation, AI technologies have revolutionized the way we live and work. For example, in healthcare, AI-powered systems can analyze medical images to detect diseases at an early stage, leading to better treatment outcomes. In transportation, self-driving cars use AI algorithms to navigate roads and ensure passenger safety. Communications, particularly while traveling, have become increasingly effortless due to AI-powered translation tools. Language is no longer an obstacle when navigating a foreign country, as Google Translate and other similar applications can seamlessly convert the local language. Indeed, engaging in shopping or dining at restaurants can immerse you in the local culture, as certain translation applications adeptly interpret menus in various languages.

Moreover, AI has transformed the way businesses operate. Companies now use AI for data analysis, customer service, and even predictive maintenance. For instance, e-commerce platforms utilize AI to recommend products to customers based on their browsing history and preferences (Egbuhuzor et al., 2025; Cherian et al., 2025; Thenmozhi, 2025). This personalized approach has led to increased sales and customer satisfaction. In the field of education, AI is being used to create personalized learning experiences for students. Adaptive learning platforms use AI algorithms to tailor educational content to individual needs and learning styles (Taşkın, 2025; Naseer et al., 2024; Gligorea et al., 2023). This has proven effective in improving student performance and engagement in the classroom.

Overall, AI in the 21st century has opened a world of possibilities, making tasks more efficient and improving decision-making processes (Rashid & Kausik, 2024; Shrivastava, 2024). As AI continues to advance, it is essential for society to adapt and embrace these technologies to stay competitive in a rapidly evolving digital landscape. This study aims to illuminate the diverse challenges faced by the OML during their research endeavors while also investigating their adaptability to the dynamic environment, especially regarding the utilization of AI.

RESEARCH AIMS AND OBJECTIVES

This research endeavors to elucidate the impact of the university's internationalization initiatives on the OML and to explore how artificial intelligence can be leveraged to mitigate the publication disparity among the OML. To achieve this, the research aims to:

1. Examine the dynamics between research-active and non-research-active individuals within the OML.
2. Assess the significance of language in this context.
3. Investigate OML initiatives aimed at bridging the gap.
4. Determine whether OML engages with AI to enhance its scholarly output.

LITERATURE REVIEW

1. Internationalization in higher education

Internationalization in higher education is a pivotal focus for universities worldwide. In today's interconnected and globalized society, the significance of internationalizing higher education cannot be overstated (Gacel-Ávila, 2005; Mittelmeier et al., 2024). It plays a crucial role in enhancing the visibility and world rankings of universities (De Wit, 2020; Shakirova & Smolnikova, 2017; Izharuddin et al., 2023). For instance, when universities establish strong partnerships with international institutions, it not only enriches the academic experience for students but also opens opportunities for collaborative research projects that can lead to groundbreaking discoveries. Moreover, internationalization allows for the exchange of diverse perspectives and ideas, fostering a more inclusive and culturally rich learning environment (Bakay, 2023; Liang, 2024). Additionally, international collaborations attract top talent from around the world, leading to a more diverse faculty and student body.

Being an international university, there exists a necessity for a robust presence of international faculty and students. This necessity is not merely a matter of diversity; it plays a crucial role in enhancing the educational experience and fostering a rich cultural exchange within the academic environment. In fact, this is one of the criteria assessed in Quacquarelli Symonds (QS) World University Rankings evaluations, which are widely recognized as a benchmark for assessing the quality and reputation of higher education institutions worldwide. Institutions that boast a substantial representation of international faculty are perceived as being of higher caliber, as these faculty members bring varied perspectives, teaching methodologies, and research insights that contribute to a more comprehensive learning atmosphere. Moreover, the presence of international students signifies the demand of a globalized market, reflecting the university's ability to attract talent from various corners of the globe. This not only enriches the student body but also prepares all students to operate in an increasingly interconnected world, where cross-cultural competencies are essential for success in many professional fields. Such a dynamic environment ultimately enhances the university's reputation, making it more appealing to prospective students and faculty alike.

One of the crucial tools for promoting internationalization is publication in reputable journals. Most reputable journals require publications in English. Publication in reputable journals plays a significant role in fostering internationalization efforts within academic and research communities. For instance, when researchers submit their work to well-known journals, it not only increases the visibility of their findings but also enhances the credibility and impact of their research on a global scale. Moreover, publishing in English is essential for reaching a wider audience and facilitating knowledge exchange among scholars from diverse linguistic backgrounds.

Furthermore, internationalization enhances the reputation of universities on a global scale (Rosyidah et al., 2020). When universities engage in partnerships with prestigious international institutions, it elevates their status and attracts top-tier faculty and students. This, in turn, contributes to higher rankings and increased funding opportunities for research and innovation. Overall, internationalization is not just a trend but a strategic imperative for universities looking to excel in today's competitive academic landscape. By embracing internationalization, universities can broaden their horizons, enrich their academic programs, and establish themselves as global leaders in education.

In the present day, internationalization in higher education encompasses a wide range of activities, including student and staff mobility, joint research projects, collaborations with international institutions, and the establishment of branch campuses abroad (De Wit, 2020; Mittelmeier et al., 2020; Mittelmeier et al., 2024). Universities now place a strong emphasis on diversifying their student body and faculty, offering international programs, and creating a multicultural learning environment (Li & Xue, 2022). Internationalization has not only enriched the academic experience for students and faculty but has also contributed to the overall reputation and competitiveness of universities. Institutions that actively promote internationalization are more likely to attract top talent, increase research output, and enhance their global visibility.

In the realm of academia, reputable journals serve as a platform for scholars to disseminate their research findings, engage in scholarly discussions, and contribute to the advancement of knowledge in their respective fields. By adhering to the publication requirements set by these journals, researchers demonstrate their commitment to upholding rigorous academic standards and promoting excellence in scholarly communication.

Research output serves as a crucial indicator of the internationalization of academic institutions. When we talk about internationalization, we refer to the process of integrating an international or global perspective into the core functions of an institution (Abramo et al., 2011; Kwiek, 2020). In this context, research output plays a significant role in showcasing the institution's engagement with global academic communities and its contribution to the advancement of knowledge on a global scale (Queirós et al., 2023; Nast et al., 2025). For example, universities that aim for internationalization often prioritize research collaborations with scholars from different countries. These collaborations result in joint research projects, co-authored publications, and shared resources, all of which contribute to the institution's research output. By engaging in such collaborations, universities not only enhance their visibility in the global academic landscape but also enrich the quality and diversity of their research endeavors.

Moreover, research output can also be measured through metrics such as publication citations, journal rankings, and research funding (Walters, 2017; Heyard & Hottenrott, 2021). Institutions with a strong international presence tend to have higher citation counts, which reflect the impact and influence of their research on a global scale. Additionally, securing research funding from international sources further demonstrates the institution's ability to attract resources and support for its research activities. Therefore, research output serves as a tangible and quantifiable indicator of an institution's internationalization efforts. By expanding their research collaborations, enhancing publication quality, and securing international funding, academic institutions can strengthen their global presence and make meaningful contributions to the advancement of knowledge worldwide.

2. English as an Instrument for Internationalization.

Most universities around the globe increasingly regard the adoption of English as the most effective and expedient pathway to achieving internationalization within their academic programs. This trend is particularly evident in the growing emphasis on publishing research findings in English, which underscores the preeminence of the language within the academic realm. English has undeniably evolved into the lingua franca of academia, serving as a vital tool that facilitates effective communication and collaboration among researchers from diverse nations and cultural backgrounds (Kurusu et al., 2024; Yildiz, 2021; Galloway & Mckinley, 2021). For instance, international conferences and collaborative research projects often require participants to present their work in English, thus reinforcing its status as the dominant language of scholarly discourse.

Consequently, the mandate for English publications in esteemed journals not only guarantees a superior level of readability and accessibility for a global audience but also promotes the cross-border dissemination of research findings. This requirement ensures that significant scholarly contributions are shared widely, enabling researchers from various parts of the world to engage with one another's work. In turn, this fosters a collaborative environment where ideas can be exchanged freely, leading to advancements in knowledge and innovation across disciplines. Ultimately, the prioritization of English in academic publishing serves as a crucial mechanism for enhancing the visibility and impact of research on a global scale, thereby supporting the overarching goals of international cooperation and academic excellence.

It is hardly surprising that English has emerged as a global lingua franca, increasingly dominating our interconnected world in various aspects of daily life and professional environments. With over 1.5 billion speakers worldwide, English not only stands as the most widely spoken language but also serves as a pivotal medium in diverse domains such as business, academia, and technology. The influence of English can be seen in international communication, where it acts as a common ground for people from different linguistic backgrounds to interact and exchange ideas (Sofyan, 2021; Wy & Kurniawan, 2024; Pratiwi, 2021). For instance, in multinational corporations, English is often used as the primary language for meetings and negotiations to ensure effective communication among employees from diverse cultural backgrounds. This is especially evident in companies that operate across various countries, where English serves as the unifying language that facilitates collaboration and understanding among teams that may otherwise struggle to communicate effectively due to language barriers.

Moreover, the dominance of English in the digital realm has further solidified its status as a global language, impacting not just personal interactions but also shaping global trends and information dissemination. From websites to social media platforms like Twitter, Facebook, and Instagram, a significant amount of online content is in English, making it essential for individuals to have at least a basic understanding of the language to navigate the digital landscape. This trend is particularly critical in the context of e-commerce, where English often serves as the default language for product descriptions, customer service interactions, and user interfaces, thereby influencing purchasing decisions and consumer behavior on a global scale. As a result, proficiency in English has become increasingly important for individuals seeking to enhance their career prospects and engage meaningfully in the global economy. In summary, the pervasive reach of English continues to reshape communication and interaction across various sectors, reinforcing its role as a fundamental tool for connectivity in our modern world.

In academia, it is widely recognized that many scholarly journals and research papers are published predominantly in English. This reality underscores the importance of English proficiency for researchers and academics alike, as it is essential for them to engage with the existing literature, share their findings, and contribute effectively to their respective fields (Kawakibi & Indrawan, 2024; Fadhil et al., 2025). For instance, a researcher aiming to publish their work in a high-impact journal must be able to write clearly and persuasively in English, as this is often the primary language of communication within the international academic community. Furthermore, English proficiency is frequently a prerequisite for students who wish to study abroad or pursue higher education in English-speaking countries, where the medium of instruction is typically English. This requirement can significantly influence a student's ability to access global academic resources and participate in diverse educational environments, highlighting the necessity of language skills in achieving academic success and fostering international collaboration. Thus, the ability to communicate effectively in English is not only beneficial but essential for thriving in today’s interconnected academic landscape.

Overall, the widespread use of English as a global language highlights the importance of language learning and cross-cultural communication in today's interconnected world. In various fields such as business, education, and technology, English has emerged as a common medium that facilitates international dialogue and collaboration. For instance, multinational companies often adopt English as their corporate language to ensure clear communication among employees from diverse linguistic backgrounds. Moreover, academic institutions around the globe increasingly offer courses in English, enabling students from different countries to engage with one another and share knowledge more effectively.

As the world continues to become more interconnected through advancements in technology and travel, the role of English as a global language is likely to grow even more prominent. This trend emphasizes the need for individuals to adapt and embrace linguistic diversity, not only by learning English but also by appreciating and respecting other languages and cultures. By doing so, individuals can enhance their ability to connect with others on a deeper level, fostering mutual understanding and cooperation in an increasingly globalized society. Ultimately, the ability to communicate across cultures will be a crucial skill for navigating the complexities of our modern world.

However, for a nation whose language is not among the predominant tongues globally, such as those spoken by major world powers like English, Mandarin, or Spanish, publishing in its native language may not be favored by its institutions. This lack of preference can lead to a significant portion of its scholarly articles often failing to gain acceptance in tier-one or tier-two journals, which are highly regarded in the academic community. These scholars often find themselves in a dilemma, torn between fulfilling university requirements, which may prioritize publications in more widely recognized languages, and the desire to publish in their own language, thus risking their recognition on a broader scale.

For instance, a researcher in a smaller nation may produce valuable insights in their field, yet struggle to have their work acknowledged simply because it is written in their native tongue. The repercussions of such unacknowledged publications are substantial, impacting their prospects for promotion, tenure, and international acknowledgment. This situation not only affects individual scholars but also reflects poorly on the university as a whole. For the university, this poses a risk of losing its opportunity to be ranked among the top institutions globally, as rankings often consider research output and visibility. Consequently, the challenge of balancing local language publication with the demands of the global academic landscape continues to be a pressing issue, demanding attention and innovative solutions from both scholars and academic institutions.

3. Artificial Intelligence (AI)

Artificial Intelligence (AI) refers to the simulation of human intelligence processes by machines, particularly computer systems, as noted by Haenlein and Kaplan in 2019, along with insights from Deng in 2018. This technological domain encompasses a wide array of technologies, including but not limited to machine learning, natural language processing, and computer vision. Each of these components plays a crucial role in enabling machines to perform tasks that typically require human intelligence. For instance, machine learning allows systems to learn from data and improve over time, while natural language processing enables machines to understand and respond to human language in a meaningful way.

One compelling example of AI in action is the development of autonomous vehicles. In these systems, sophisticated AI algorithms empower cars to perceive their surroundings through sensors and cameras, allowing them to make informed decisions without the need for human intervention. This technology not only enhances convenience for users but also has the potential to significantly reduce traffic accidents caused by human error. Moreover, AI is making substantial strides in the field of healthcare, where it is utilized for medical image analysis. By leveraging advanced algorithms, AI assists doctors in diagnosing diseases with greater accuracy and efficiency, ultimately improving patient outcomes and streamlining the diagnostic process. The integration of AI in these various sectors highlights its transformative impact on society, as it continues to evolve and reshape the way we live and work.

AI has the potential to revolutionize various industries by automating a wide array of tasks, significantly improving operational efficiency, and enabling the development of entirely new capabilities that were previously unimaginable (Dwivedi et al., 2019). For instance, in the finance sector, sophisticated AI algorithms can analyze vast amounts of data at lightning speed, allowing them to detect fraudulent activities with remarkable accuracy and make informed investment decisions based on complex market trends. This capability not only helps in minimizing financial losses but also enhances overall market stability.

Moreover, in the realm of customer service, chatbots powered by advanced AI technologies can provide instant support to users, addressing their queries and concerns in real-time. This level of responsiveness greatly enhances the customer experience, as clients can receive assistance at any hour of the day without the need for human intervention. Furthermore, these AI-driven systems can learn from interactions, continuously improving their responses and becoming more adept at understanding customer needs. As such, the integration of AI into various sectors is not merely a trend but a transformative movement that holds the promise of reshaping how businesses operate and interact with their clients.

Despite its benefits, AI also raises ethical concerns regarding privacy, bias, and job displacement (Ko, 2025). For example, AI algorithms trained on biased data may perpetuate discrimination. It is crucial for organizations and policymakers to address these ethical issues and ensure that AI is developed and deployed responsibly.

AI is a powerful technology with the potential to transform various sectors. By understanding its capabilities and limitations, we can harness the benefits of AI while mitigating its risks. As AI continues to advance, it is essential to prioritize ethical considerations and ensure that it serves the greater good of society. AI also has significantly impacted various aspects of society in the 21st century. From healthcare to transportation, AI technologies have revolutionized the way we live and work. For example, in healthcare, AI-powered systems can analyze medical images to detect diseases at an early stage, leading to better treatment outcomes (Jiang et al., 2017). In transportation, self-driving cars use AI algorithms to navigate roads and ensure passenger safety (Rathore et al., 2024).

Moreover, AI has fundamentally transformed the operational paradigms of businesses. Organizations now harness AI for sophisticated data analysis, enhanced customer service, and even predictive maintenance. For instance, e-commerce platforms deploy AI to curate product recommendations tailored to customers' browsing histories and preferences (Tulasi & Ahamed, 2025). This personalized methodology has resulted in heightened sales and augmented customer satisfaction. In the educational sector, AI is being utilized to craft bespoke learning experiences for students. Adaptive learning platforms leverage AI algorithms to customize educational content according to individual needs and learning styles (Taşkın, 2025). This approach has proven to be effective in bolstering student performance and engagement within the classroom (Mahafdah et al., 2024). Another illustration is the adaptive learning platforms that facilitate the analysis of student performance, thereby personalizing teaching materials to meet their unique requirements. This level of customization guarantees that students receive focused support and guidance, ultimately culminating in enhanced academic outcomes.

Overall, AI in the 21st century has unveiled a plethora of possibilities, streamlining tasks and refining decision-making processes. As AI technology continues to evolve, it is imperative for society to adapt and embrace these innovations to maintain competitiveness in an increasingly dynamic digital landscape. AI in education has revolutionized the methodologies by which students learn and educators instruct. It is being seamlessly integrated into educational frameworks to elevate personalized learning experiences. For example, adaptive learning platforms employ AI algorithms to scrutinize student performance and customize teaching materials to suit their individual needs. This level of tailored instruction ensures that students receive concentrated support and guidance, ultimately leading to enhanced academic achievement.

In addition, AI-powered virtual tutors are becoming increasingly popular in online education. These virtual tutors can provide immediate feedback, answer questions, and guide students through challenging concepts. For instance, a virtual tutor in a language learning app can correct pronunciation errors and suggest practice exercises to improve fluency. This interactive and responsive form of learning motivates students to engage with the material actively.

Furthermore, AI is transforming the assessment process in education. Automated grading systems can quickly evaluate student responses to quizzes and assignments, providing instant feedback to both students and teachers (Dimari et al., 2024). This immediate feedback loop allows students to track their progress in real time and address any misconceptions promptly. Moreover, AI can analyze large datasets to identify patterns in student performance and suggest interventions to support struggling learners.

AI in education is not about replacing teachers but empowering them with valuable tools to enhance the learning experience. By leveraging AI technologies, educators can personalize instruction, provide individualized support, and streamline administrative tasks. As AI continues to advance, its role in education will likely expand, offering new opportunities to improve student outcomes and educational practices.

In the realm of academic writing, artificial intelligence is fundamentally transforming research by dismantling linguistic barriers and facilitating efficient data analysis (Owan et al., 2023). AI-driven tools significantly enhance the quality of academic writing by meticulously checking grammar and detecting plagiarism, thereby conserving time and elevating the overall caliber of scholarly papers (Malik et al., 2023; Bing & Leong, 2025). Furthermore, AI aids in the generation of innovative ideas, the organization of complex thoughts, and the structuring of coherent narratives. By streamlining various processes, augmenting quality, and promoting global collaboration in the composition of academic papers, AI emerges as an invaluable asset in the scholarly landscape.

Being an international university, it is imperative for all academic staff to produce high-caliber publications, preferably in English. This requirement ensures that the university's research output reaches a global audience and makes a significant contribution to the advancement of knowledge across diverse fields. For instance, a professor in the Department of Economics might publish a comprehensive study on the ramifications of globalization on emerging markets, offering invaluable insights for policymakers and businesses operating within those regions.

Furthermore, publishing in English enhances the visibility and impact of research, as English is predominantly recognized as the lingua franca of academia (Stockemer & Wigginton, 2019). By disseminating their work in English, scholars can engage with a wider audience, foster collaborations with international researchers, and attract funding for prospective research endeavors. This practice not only benefits the individual academics but also elevates the university's reputation and standing on the global stage. Moreover, the production of quality publications necessitates that academic staff adhere to stringent standards of research integrity and scholarly writing. They must ensure that their work is original, meticulously researched, and makes a meaningful contribution to the existing corpus of knowledge. This process entails conducting exhaustive literature reviews, devising robust methodologies, and articulating findings in a clear and coherent manner.

Consequently, it is essential for academic staff at an international university to prioritize the generation of high-quality publications in English. By doing so, they not only fulfill their professional obligations but also further the university's mission of promoting excellence in research and scholarship.

METHODOLOGY

This research began with the recognition that a substantial proportion of academic personnel at UBD are struggling with the formidable task of composing scholarly papers in English. To tackle this issue, a questionnaire was meticulously disseminated online to a specifically targeted cohort. As of 2024, UBD data indicates that the institution employs 342 academic staff members, comprising 226 Bruneians and 116 expatriates. Among the 226 Bruneian faculty, 26 individuals were identified as OML, representing 11.50 percent or 7.6 percent of the total academic staff at the university.

However, out of the 26 targeted OML, only 17 respondents completed the questionnaire, culminating in a response rate of 67.38 percent. This response percentage is regarded as commendable, particularly when compared with the average rate of online surveys within education-related domains, which stands at 44% (Wu et al., 2022). Furthermore, it has been reported that online survey responses are notably lower, at 42.8%, in comparison to other survey formats. Nevertheless, employing a pre-contact strategy yielded a higher response rate of 54.6% compared to scenarios without pre-contact (Wu et al., 2022). The questionnaire was distributed directly to the participants via email. All individuals involved are well-acquainted with the research team, having collaborated as colleagues for several years at the same university. Nevertheless, even after a few days and the dispatch of a follow-up email, (Wu et. al. 2022) observed that the response rate remained disappointingly low at 37.7%. This stands in stark contrast to the significantly higher response rate of 79.0% achieved when reminders are communicated via telephone, as reported by Wu et al. (2022). The researchers did not initially contact the subjects via telephone, primarily to mitigate any discomfort associated with the feeling of coercion. Furthermore, it was explicitly stated in the preamble of the questionnaire that participation in the survey is entirely voluntary.

The questionnaire was thematically categorized into three distinct sections:

1. Participants' professional experiences at UBD.
2. Participants' publication histories.
3. Participants' readiness for AI.

The aforementioned categories, which are pivotal for elucidating the correlations among various informants, play a significant role in comprehending diverse aspects such as academic ranks, duration of employment, and publication output. For instance, when scrutinizing academic ranks, it is imperative to consider how they influence the overarching dynamics within an institution. Senior academic ranks frequently entail augmented responsibilities and leadership roles, which can substantially impact decision-making processes and research trajectories. With regard to the duration of employment, the length of tenure an academic has with the university can directly influence their level of expertise and research productivity. Academics who have been affiliated with the university for an extended period may exhibit superior research quality and publication records in comparison to their mid-career or early-career counterparts.

DISCUSSIONS

Universiti Brunei Darussalam was established in 1985 and remained primarily a teaching institution until 2019 (Muhammad & Petra, 2021). Consequently, it is unsurprising to encounter numerous academic staff members who are deemed senior, having joined the faculty prior to 2019, yet still demonstrate a deficiency in either the volume or caliber of their publications. This information is pivotal for discerning between those who are research-inactive due to personal choices or institutional "legacies" and those who are hindered in their publishing endeavors by linguistic barriers. So, the respondents' backgrounds are quite intriguing, as nearly half (47%) have been employed at UBD for a duration of 11 to 20 years, which we have classified in this study—and within the UBD context—as mid-career academics. Only 24% have been with the institution for less than 10 years, while a mere 29% are regarded as senior staff, having served UBD for over 21 years. It is noteworthy that prior to 2019, UBD cultivated its personnel from the tutor level, which corresponds to those who had recently graduated with their Bachelor degrees. Subsequently, they would dedicate their professional journey to pursuing their Master degree and, consequently, PhD degree while simultaneously fulfilling their roles as academics at UBD. However, there were also a few exceptional instances where staff were appointed as tutors with advanced academic qualifications, specifically holding either a Master degree or a PhD. During that period, the role of tutor served as an entry-level academic position for new faculty members, who would subsequently be promoted to lecturer upon the successful completion of their MA or PhD degrees. For those who already possessed an MA or PhD, promotion to lecturer would occur after three years of service. S0, this framework justifies our categorization of service duration into three distinct categories within UBD.

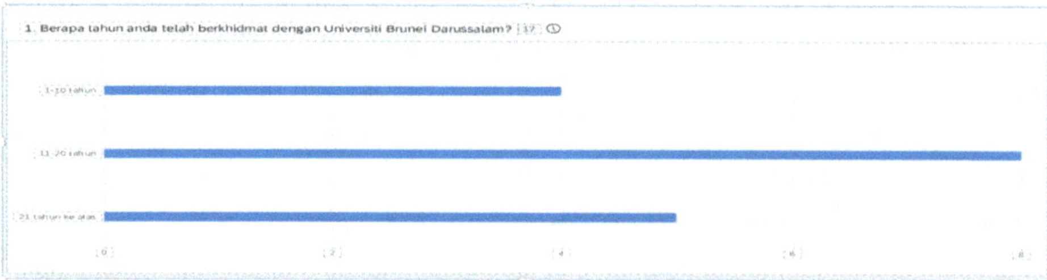

1. Berapa tahun anda telah berkhidmat dengan Universiti Brunei Darussalam? 17

Q2 - 1. Berapa tahun anda telah berkhidmat dengan Universiti Brunei Darussalam?	Count	Count
1-10 tahun	24%	4
11-20 tahun	47%	8
21 tahun ke atas	29%	5

The survey reveals that all 17 respondents, which constitutes a remarkable 100% of the participants, have successfully published their work in various formats, specifically in either academic journals or book formats. This accomplishment not only highlights their dedication to their field but also underscores their commitment to fulfilling their responsibilities as academicians. Among these responsibilities, the imperative of publication stands out as a fundamental aspect of academic life, serving both to advance knowledge and to share insights with the wider scholarly community. Each respondent exhibited considerable experience in the realm of publication, having disseminated their scholarly contributions across a wide spectrum of reputable journals and books. For instance, many of these individuals have contributed to prestigious journals that are well-regarded in their respective disciplines, thereby enhancing their professional credibility and visibility. Additionally, some have authored or co-authored books that delve into complex subjects, further showcasing their expertise and commitment to academic discourse. This level of engagement in publishing not only reflects their individual dedication to their research but also contributes to the broader academic landscape, enriching the dialogue within their fields.

However, their publication output revealed considerable variability; the majority (53%) had published between 11 and 30 articles, which indicates a moderate level of productivity among the faculty members. In contrast, a mere 18% had produced only 1 to 10 publications, suggesting that a small fraction of the faculty may be struggling with their research output or may have recently entered the academic field. Furthermore, 29% had surpassed the threshold of 30 articles, showcasing a group of highly productive individuals who likely have established themselves as prominent scholars within their respective fields. Considering that the majority have been employed at the university for a tenure of 11-20 years, it is reasonable to expect that their output should ideally reside within the range of 21-40 publications, with a minimum publication rate of 2 articles per annum for each faculty member. This expectation is not arbitrary; it aligns with the findings of Rostad & Aksnes (2015), who noted that scholars in the humanities published an average of 2.02 articles per year, while those in the social sciences published an average of 1.5 articles per year. These statistics imply that with a decade or more of experience, faculty members should be contributing significantly to the academic discourse. The respondents in this research predominantly belong to either the social sciences or humanities disciplines, which further emphasizes the importance of understanding publication trends within these areas, as they often influence the broader academic landscape and contribute to the intellectual community as a whole.

Although the modern educational landscape in Brunei traces its origins back to 1951, when two government English schools were inaugurated—one in Brunei Town and another in Kuala Belait—by 1966, a Malay-medium government school commenced operations (Yunos, 2013). This institution later transitioned to a Bilingual Education framework in 1984, evolving into the SPN-21 system in 2008. Consequently, it is anticipated that several staff members at UBD may be aligned with either of the two medium systems or the Bilingual Education model.

At the same time, as an international university, it is anticipated that the academic staff at UBD will disseminate their research not only in the Malay language but also in English, reflecting the global nature of academia today. This dual-language approach is essential for reaching a broader audience and enhancing the visibility of their work on an international scale. The survey results indicate that only three academic staff members, which constitutes 18% of the total, exclusively publish in Malay. In contrast, the majority, comprising fourteen academic staff members or 82%, actively engage in publishing their research in both Malay and English. This trend highlights the increasing importance of bilingualism in academia, as it allows researchers to bridge cultural and linguistic divides. However, the elevated figure of those publishing in both languages can be attributed to the burgeoning trend of co-authorship, where a staff member proficient in English often assumes the role of lead author. This collaborative approach not only enhances the quality of the research but also provides opportunities for less experienced researchers to learn from their more seasoned colleagues, thereby fostering a supportive academic environment.

Universiti Brunei Darussalam (UBD) is positioned at 367th globally in the 2026 Quacquarelli Symonds (QS) World University Rankings (UBD, 2025). As a member of the top 25% of universities worldwide, it is only natural for UBD's academic staff to seize opportunities to present their research on a global stage. This inclination is evidenced by an impressive 93% of faculty members expressing that they publish in English to ensure their articles are globally accessible. From the university's perspective, they perceive that they are fulfilling institutional mandates (93%) by publishing in English. However, their motivation is not rooted in linguistic proficiency; rather, it arises from necessity. This is underscored by survey results indicating that only 7% of respondents feel comfortable publishing in English.

Untuk penerbitan Bahasa Inggeris 14 ⓘ

Q7 - Untuk penerbitan Bahasa Inggeris	Count	Count
Jangkauan pembaca antarabangsa yang lebih luas	93%	13
Jangkaan institusi	93%	13
Pangkalan data dan pengindeksan yang lebih komprehensif	100%	14
Matlamat penyebaran penyelidikan antarabangsa	100%	14
Amalan standard bidang/disiplin	36%	5
Lebih selesa menerbitkan dalam bahasa Inggeris	7%	1

The decision to embrace the Malay language is deeply intertwined with notions of nationalism as well as the preservation of language and culture, achieving an impressive score of 94%. This percentage aligns with the fact that OML targets its publications for local consumption. This is evidenced by the significant 88% of respondents indicating that their articles are primarily aimed at a local readership. The majority of this target demographic hails from disciplines such as linguistics, cultural studies, literature, history, sociology, anthropology, and education, with their research predominantly centered on Brunei. Accordingly, the emphasis and demand for their publications are predominantly derived from the Bruneian populace and the neighboring regions. Having stated that, the utilization of the Malay language as a mechanism for local capacity building registers a comparatively modest score of 82%. This suggests that language does not serve as the predominant determinant influencing the capacity development of local academics. Notwithstanding their inclination to publish in the Malay language, the respondents demonstrate a notable confidence in their proficiency to publish in English, as evidenced by a mere 24% expressing a deficiency in this regard. In contrast, a rather low score of 59% emerges when respondents are queried about their comfort and fluency in publishing in Malay.

The OML, despite not being proficient in English writing, demonstrates a noticeable tendency to publish in English, albeit with limited productivity. This phenomenon is hardly astonishing, given the educational backgrounds of many UBD Bruneian academics, which are often rooted in English. For instance, a significant number of these scholars have received their higher education in institutions where English is the primary medium of instruction, such as various universities in Malaysia or the United Kingdom. This exposure inevitably shapes their academic pursuits and influences their publication choices. Furthermore, as academics, they typically engage with English in various contexts whenever necessary, whether it be through reading international journals, attending conferences, or collaborating with peers from different countries. Thus, although their command of English is relatively modest and they encounter challenges, such as difficulties in articulating complex ideas or adhering to specific academic writing conventions, they still possess a minimal capability to produce competent scholarly articles in English. This situation reflects a broader trend where non-native English speakers strive to contribute to global academic discourse, despite the inherent obstacles they face.

From the various categories delineated above, the OML publications are deemed adequate both in terms of output and international visibility. This adequacy is particularly notable when considering the diverse range of topics addressed within the publications, which cover significant areas of research and inquiry. However, as previously articulated, the elevated output in English can be attributed to collaborative publishing efforts with various colleagues, who bring a wealth of expertise and perspectives to the table. Analyzing the publication output reveals that the majority of OML's works transcend linguistic boundaries, thus reaching a broader audience and enhancing their impact on the global academic community. Indeed, "the legacies" are evolving in response to the contemporary university environment and its exigencies, adapting to new challenges such as funding cuts and shifts in academic priorities. In recognition of the current demands, this research investigates how OML is embracing technology, particularly artificial intelligence, to streamline research processes, improve accessibility, and foster innovative approaches to scholarship. By integrating AI tools, OML not only enhances the quality of its publications but also ensures that they remain relevant in an ever-changing educational landscape.

AI is undeniably more advantageous for users of the English language in comparison to those who communicate in Malay. This disparity can be attributed to several factors, including the vast availability of resources, tools, and applications developed primarily for English speakers. Consequently, for individuals proficient in English, the integration of AI into their research and the composition of academic papers undoubtedly simplifies their endeavors. For instance, AI-driven tools can assist with grammar checks, content generation, and data analysis, thereby allowing researchers to focus more on the critical aspects of their work. Conversely, for those lacking proficiency in English, the adoption of AI is regarded as a transformative aid, enabling them to bridge the language gap and access a wealth of knowledge that may have previously been beyond their reach. The feedback indicates that a commendable 71% of OML has embraced AI to enhance their scholarly work, utilizing it to refine their writing and improve the quality of their submissions. While AI is gradually making inroads within the Malay language, with some applications beginning to cater to Malay speakers, the remaining 29% of OML either exhibit a lack of research engagement or prefer to publish exclusively in Malay. This preference may stem from cultural factors or a desire to maintain the integrity and richness of the Malay language in academic discourse. As such, the journey towards greater AI integration in academic settings for Malay speakers remains ongoing, highlighting the need for continued development and support in this area.

OML, having embraced the potential of artificial intelligence, has employed a diverse array of applications, among which the foremost are ChatGPT, Gemini, Claude.ai, Scispace, QuillBot, Grammarly, Perplexity, and Co-Pilot. This extensive selection of AI tools signifies that OML has judiciously opted for applications that are conducive to its academic pursuits. Although most of these applications possess the capability to read and respond in the Malay language, it can be contended that OML primarily utilized them to enhance the quality of its English manuscripts. This assertion is substantiated by data indicating that 67% of respondents employed AI assistance for translation purposes. Furthermore, they harnessed AI to generate writing ideas and facilitate the creation of early drafts, accounting for 50% of OML's usage. Additionally, they capitalized on AI for the revision and editing of their papers. Notably, despite these various applications, OML utilized AI minimally for grammatical and stylistic enhancement, which constituted a mere 33% of the survey responses.

On the contrary, despite OML's familiarity with AI applications and the fact that over 50% have availed themselves of AI for various English language services, only 58% of their articles were accepted for publication. Conversely, 42% of OML members who possess knowledge of and have even employed AI have refrained from submitting their articles for publication. The data present a paradox wherein, on one hand, they have embraced AI to assist in composing their papers, yet on the other hand, they hesitate to pursue publication. This phenomenon may be attributed to their motivation and self-assurance, suggesting that mentorship could play a pivotal role in bolstering their confidence. One possibility is that their reluctance stems from a deficiency in self-assurance, arising from the perception that their papers are evaluated by AI, rather than by a human endowed with professional acumen. Having stated that, and considering the adoption of AI is relatively new, particularly among the OML, they have published between 1 and 8 articles. This number is considered justified, as publications in English are primarily aimed at fulfilling university requirements and garnering swift citations on platforms such as Google Scholar. Furthermore, they predominantly publish in Malay as is customary.

To address the potential challenges associated with OML responses in relation to the use of AI, particularly stemming from language barriers or technophobia, and to gain a better understanding of their perspectives on the ethical considerations surrounding AI utilization, we specifically asked whether they find the incorporation of AI in academic writing to be morally acceptable. It is worth noting that a significant portion of the OML participants, precisely 10 individuals, expressed their approval of using AI and deemed it ethical within the realm of academic paper composition. Furthermore, a substantial majority of 12 out of the total 17 OML respondents believe that the integration of AI into their writing processes does not compromise the integrity of their academic work.

This positive stance towards AI in academic writing was corroborated by the survey findings, which indicated that only 9 out of the OML participants perceive AI engagement as posing a potential threat to their writing perspectives. Nonetheless, a notable concern raised by 13 OML individuals revolves around the risk of AI leading to inadvertent plagiarism. Consequently, a considerable number of 13 participants advocated for the mandatory disclosure of AI usage in all their publications to ensure transparency and academic integrity. By acknowledging these perspectives and concerns, it becomes evident that the OML community is actively grappling with the implications of AI integration in academic writing, seeking to strike a balance between technological advancements and ethical considerations.

Thirteen respondents believed that by employing artificial intelligence, they could produce academic papers more rapidly while simultaneously enhancing the quality of their work, as affirmed by twelve participants. Their grammatical accuracy and stylistic expression improved significantly, according to eleven respondents. Furthermore, thirteen individuals contended that AI assisted in refining their ideas and writing structure. Additionally, fourteen respondents believed that engaging with AI made a substantial difference in their academic writing. In fact, fourteen participants believed that utilizing AI bolstered their productivity. Consequently, thirteen individuals were confident that leveraging AI would undoubtedly enable them to meet their Key Performance Indicators (KPIs). However, they remained somewhat apprehensive, as only twelve believed that their writing would be accepted by international journals. Simultaneously, twelve respondents felt that their papers would facilitate connections with international researchers.

On a positive note, thirteen respondents believed AI would become a standard tool for fulfilling academic writing requirements, particularly in competitive academic environments where the pressure to produce high-quality work is intense. This perspective reflects a growing trend in various educational institutions, where technology is increasingly integrated into the learning process. When asked whether those who do not employ AI would be left behind in this rapidly evolving landscape, only nine respondents concurred, suggesting a divide in opinion regarding the necessity of AI in academic pursuits. Almost all, sixteen participants, believed that AI would revolutionize academic writing and publishing, potentially transforming the way research is conducted and disseminated. This revolutionary change may include enhanced efficiency in data analysis and the ability to generate comprehensive literature reviews in a fraction of the time it currently takes. Consequently, sixteen of them advocated for universities to provide training and support for staff wishing to engage with AI, emphasizing the importance of equipping educators and researchers with the necessary skills to leverage these advanced tools effectively. This training could ensure that they remain competitive and relevant in their fields, ultimately benefiting the academic community as a whole.

In response to an open-ended question regarding how AI assists them in writing academic papers in English, one participant noted that they believed AI boosted their confidence in English writing, while another remarked that it even facilitated higher-level English proficiency. AI aids them in selecting the appropriate lexicon, structuring language correctly, and maintaining controlled academic language. For those less proficient in English, one individual asserted that AI is invaluable for translating Malay articles into English. Another participant noted that AI significantly expedited and reduced the cost of proofreading, as they previously employed proofreaders charging between BND $50 and BND $150 for a ten-page article. In contrast, AI offers unlimited use for only USD 20 per month. Lastly, one individual emphasized the importance of personal responsibility and integrity, advising against taking AI-generated content at face value.

CONCLUSION

This research holds profound significance in exploring the multifaceted perspectives of OML regarding the integration of artificial intelligence in the intricate domain of academic writing. Despite their relatively modest representation within the scholarly community, the contributions of OML unequivocally play a crucial role in shaping and influencing the university's Key Performance Indicators (KPI), which are essential for assessing institutional success and advancement. The academic landscape is often characterized by the relentless mantra of "publish or perish" (Moskovkin, 2024), a phrase that encapsulates the intense pressures faced by scholars to consistently generate research outputs. This pressure is particularly acute in the context of ongoing internationalization initiatives, which aim to enhance the university's global standing and competitive edge. Consequently, many OML may find themselves feeling disheartened or undervalued, fostering a perception that their contributions are trivial in the broader academic narrative. Nevertheless, with the advent of AI technologies, these OML are now positioned to articulate their thoughts and insights more effectively, nurturing a dialogue that resonates with the wider academic community at UBD. This technological innovation not only provides new avenues for collaboration and expression but also serves to amplify the voices of OML, enabling them to engage meaningfully in discussions that will shape the future of academic writing and research.

Surprisingly, the findings of this research contradict our initial hypothesis, which suggested that the OML, or the organization of modern learners, would maintain the status quo and passively accept the changing times without much resistance or adaptation. In reality, however, they are actively embracing AI technologies and using them diligently in their scholarly pursuits, showcasing a remarkable willingness to innovate and evolve. This proactive approach not only reflects their commitment to staying relevant in an increasingly digital landscape but also highlights their determination to enhance their research capabilities. This marks the beginning of their journey into the realm of artificial intelligence, with the hope that they will eventually compete on a global scale and produce top-tier papers that are accepted in esteemed journals. Such a transformation could significantly elevate their academic standing and influence in the field, paving the way for future breakthroughs and collaborations that integrate cutting-edge technology with traditional research methodologies.

The support and recognition from the university are crucial for the advancement of academic staff, serving as a foundational element in fostering a thriving educational environment. This phenomenon observed at UBD is not isolated but rather a global trend seen in universities worldwide that operate in languages other than English as their medium of instruction. For instance, institutions in regions like Europe and Asia have similarly recognized the importance of providing robust support systems for their faculty. Particularly, ambitious institutions aiming to enhance their international standing emphasize the cultivation and readiness of their academic personnel. They understand that a well-prepared faculty not only contributes to the institution's reputation but also significantly impacts student success. To achieve these goals, they invest substantial resources and time to prepare their staff, offering professional development programs, workshops, and mentorship opportunities. Alternatively, some universities adopt a more accelerated approach by encouraging faculty members to engage with artificial intelligence, integrating modern technology into their teaching and research methodologies. This proactive strategy not only equips faculty with essential skills for the digital age but also enhances their ability to contribute meaningfully to their fields, ultimately benefiting the academic community as a whole.

REFERENCES

Abramo, G., D'Angelo, C. A., & Solazzi, M. (2011). The relationship between scientists' research performance and the degree of internationalization of their research. *Scientometrics*, *86*, 629–643. DOI: 10.1007/s11192-010-0284-7

Bakay, M. (2023). Multicultural Classrooms in European Higher Education: Findings from Interviews with International Students and Teaching Staff. *International Journal on Lifelong Education and Leadership*, *9*(1), 1–17. DOI: 10.25233/ijlel.1341935

Bing, Z., & Leong, W. (2025). AI on Academic Integrity and Plagiarism Detection. *ASM Science Journal*, *20*(1), 1–9. DOI: 10.32802/asmscj.2025.1918

Cherian, M., Jeevitha, M, & Manoja, K., & Abhinav. (2025). Leveraging AI for Predicting Marketing and Customer Insights-An Overview. *Journal of Informatics Education and Research*, *5*(1), 705–711. DOI: 10.52783/jier.v5i1.2050

de Wit, H. (2020). Internationalization of Higher Education: The Need for a More Ethical and Qualitative Approach. *Journal of International Students*, *10*(1), i–iv. DOI: 10.32674/jis.v10i1.1893

Deng, L. (2018). Artificial Intelligence in the Rising Wave of Deep Learning: The Historical Path and Future Outlook [Perspectives]. *IEEE Signal Processing Magazine*, *35*(1), 180–177. DOI: 10.1109/MSP.2017.2762725

Dimari, A., Tyagi, N., Davanageri, M., Kukreti, R., Yadav, R., & Dimari, H. (2024). AI-Based Automated Grading Systems for open book examination system: Implications for Assessment in Higher Education. *2024 International Conference on Knowledge Engineering and Communication Systems (ICKECS)*, 1-7. https://doi .org/DOI: 10.1109/ICKECS61492.2024.10616490

Dwivedi, Y., Hughes, L., Ismagilova, E., Aarts, G., Coombs, C., Crick, T., Duan, Y., Dwivedi, R., Edwards, J., Eirug, A., Galanos, V., Ilavarasan, P., Janssen, M., Jones, P., Kar, A., Kizgin, H., Kronemann, B., Lal, B., Lucini, B., & Williams, M. (2019). Artificial Intelligence (AI): Multidisciplinary perspectives on emerging challenges, opportunities, and agenda for research, practice and policy. *International Journal of Information Management*, *57*, 1–47. DOI: 10.1016/J.IJINFOMGT.2019.08.002

Egbuhuzor, N., Ajayi, A., Akhigbe, E., Agbede, O., Ewim, C., & Ajiga, D. (2025). AI and data-driven insights: Transforming customer relationship management (CRM) in financial services. *Gulf Journal of Advance Business Research*, *3*(2), 483–511. DOI: 10.51594/gjabr.v3i2.93

Fadhil, M., Septiana, E., Sawitri, O., Abdillah, R., Syahfutra, W., & Nurafrezah, N. (2025). The Role of English Proficiency in Supporting Academic Achievement and Career Opportunities of University Students. *Widya Accarya, 16*(1), 36–41. DOI: 10.46650/wa.16.1.1645.36-41

Gacel-Ávila, J. (2005). The Internationalisation of Higher Education: A Paradigm for Global Citizenry. *Journal of Studies in International Education, 9*(2), 121–136. DOI: 10.1177/1028315304263795

Galloway, N., & Mckinley, J. (2021). *Englishization of Higher Education. Research Questions in Language Education and Applied Linguistics*. Springer., DOI: 10.1007/978-3-030-79143-8_123

Gligorea, I., Cioca, M., Oancea, R., Gorski, A., Gorski, H., & Tudorache, P. (2023). Adaptive Learning Using Artificial Intelligence in e-Learning: A Literature Review. *Education Sciences, 13*(12). Advance online publication. DOI: 10.3390/educsci13121216

Haenlein, M., & Kaplan, A. (2019). A Brief History of Artificial Intelligence: On the Past, Present, and Future of Artificial Intelligence. *California Management Review, 61*(4), 5–14. DOI: 10.1177/0008125619864925

Hamdan, M. (2017). Transformational Change in Brunei Higher Education: A Case Study of Universiti Brunei Darussalam. *Higher Education Research, 2*(3), 76–85. DOI: 10.11648/j.her.20170203.11

Heyard, R., & Hottenrott, H. (2021). The value of research funding for knowledge creation and dissemination: A study of SNSF Research Grants. *Humanities & Social Sciences Communications, 8*, 217. DOI: 10.1057/s41599-021-00891-x

Ishamina, A., & Deterding, D. (2017). English Medium Education in a University in Brunei Darussalam: Code-Switching and Intelligibility. In Fenton-Smith, B., Humphreys, P., & Walkinshaw, I. (Eds.), *English Medium Instruction in Higher Education in Asia-Pacific. Multilingual Education, 21, 281-298*. Springer., DOI: 10.1007/978-3-319-51976-0_15

Izharuddin, M., Hadi, F., & Lestyorini, M. (2023). Evaluation of World University Rankings of Southeast Asia HEIs Image. *Wiga: Jurnal Penelitian Ilmu Ekonomi, 13*(1), 38–45. DOI: 10.30741/wiga.v13i1.941

Jiang, F., Jiang, Y., Zhi, H., Dong, Y. S., Wang, Y., Dong, Q., Shen, H., & Wang, Y. (2017). Artificial intelligence in healthcare: Past, present and future. *Stroke and Vascular Neurology, 2*(4), 230–243. DOI: 10.1136/svn-2017-000101

Jones, G. M. (2015). Bilingual and Multilingual Education in Brunei and Malaysia: Policies and Practices. In Wright, W. E., Boun, S., & Garcia, O. (Eds.), *The Handbook of Bilingual and Multilingual Education* (pp. 531–541). Wiley-Blackwell.

Kawakibi, M. A., & Indrawan, F. (2024). The Importance of English in International Science Research Collaboration. *Teaching English as Foreign Language. Literatura y Lingüística*, *4*(1), 40–44.

Ko, G. (2025). Artificial Intelligence and Your Life: The Problems We Must Face. *Science Insights*, *46*(4), 1799–1803. DOI: 10.15354/si.25.pe211

Kurusu, K., Oda, C., Go, M., Wu, D., Saure, K., & Narang, S. (2024). English in the internationalization of higher education and international student mobility. *AILA Review*, *37*(1), 120–136. DOI: 10.1075/aila.24002

Kwiek, M. (2020). What large-scale publication and citation data tell us about international research collaboration in Europe: Changing national patterns in global contexts. *Studies in Higher Education*, *46*(12), 2629–2649. DOI: 10.1080/03075079.2020.1749254

Li, J., & Xue, E. (2022). Exploring High-Quality Institutional Internationalization for Higher Education Sustainability in China: Evidence from Stakeholders. *Sustainability*, *14*(13). Advance online publication. DOI: 10.3390/su14137572

Liang, S. (2024). Advancing Internationalization in the Formal Curriculum: Content Integration and Inclusive Design. *Journal of Studies in International Education*, *28*(5), 743–760. DOI: 10.1177/10283153241262461

Mahafdah, R., Bouallegue, S., & Bouallègue, R. (2024). Enhancing e-learning through AI: Advanced techniques for optimizing student performance. *PeerJ. Computer Science*, *10*. Advance online publication. DOI: 10.7717/peerj-cs.2576

Malik, A., Pratiwi, Y., Andajani, K., Numertayasa, I., Suharti, S., & Darwis, A., &, M. (2023). Exploring Artificial Intelligence in Academic Essay: Higher Education Student's Perspective. *International Journal of Educational Research Open*, *5*. Advance online publication. DOI: 10.1016/j.ijedro.2023.100296

Mittelmeier, J., Lomer, S., Al-Furqani, S., & Huang, D. (2024). Developing Meaningful Internationalisation that Impacts Students' Outcomes in Higher Education: A Scoping Review of the Literature 2011–2022. *Journal of Studies in International Education*, *28*(4), 503–525. DOI: 10.1177/10283153231222278

Mittelmeier, J., Rienties, B., Gunter, A., & Raghuram, P. (2020). Conceptualizing Internationalization at a Distance: A "Third Category" of University Internationalization. *Journal of Studies in International Education, 25*(3), 266–282. DOI: 10.1177/1028315320906176

Moskovkin, V. (2024, July 15th). Tracing the origins of 'publish or perish'. https://blogs.lse.ac.uk/impactofsocialsciences/2024/07/15/tracing-the-origins-of-publish-or-perish/

Muhammad, N., & Petra, M. (2021). *History and Development of the Brunei Education.*

Naseer, F., Khan, M., Tahir, M., Addas, A., & Aejaz, S. (2024). Integrating deep learning techniques for personalized learning pathways in higher education. *Heliyon, 10*(11). Advance online publication. DOI: 10.1016/j.heliyon.2024.e32628

Nast, C., Llopis, O., Yankova, D., & D'Este, P. (2025). Sourcing insights elsewhere: The positive influence of academic engagement on scientific impact. *Technovation, 139*. Advance online publication. DOI: 10.1016/j.technovation.2024.103112

Owan, V., Abang, K., Idika, D., Etta, E., & Bassey, B. (2023). Exploring the potential of artificial intelligence tools in educational measurement and assessment. *Eurasia Journal of Mathematics, Science and Technology Education, 19*(8). Advance online publication. DOI: 10.29333/ejmste/13428

Pratiwi, A. (2021). Communication and International Language. *KOMUNIKASIA: Journal of Islamic Communication and Broadcasting, 1*(1), 69–84. DOI: 10.32923/kpi.v1i1.1909

Queirós, A., Carvalho, T., Manatos, M., & Diogo, S. (2023). Internationalization of Portuguese Academia: The impact on academic engagement and collaboration with society. *Higher Education, 87*, 1769–1789. DOI: 10.1007/s10734-023-01090-0

Rashid, A., & Kausik, M. (2024). AI revolutionizing industries worldwide: A comprehensive overview of its diverse applications. *Hybrid Advances, 7*. Advance online publication. DOI: 10.1016/j.hybadv.2024.100277

Rathore, R., Nayeem, T., Agarwal, A., Kumar, S. P., & Bhardwaj, A. (2024). AI System for Autonomous Vehicles. *International Journal For Multidisciplinary Research, 6*(6), 1–9. DOI: 10.36948/ijfmr.2024.v06i06.28785

Rosyidah, N., Matin & Rosyidi, U. (2020). Internationalization in Higher Education: University's Effective Promotion Strategies in Building International Trust. *European Journal of Educational Research, 9*(1), 351–361. DOI: 10.12973/eu-jer.9.1.351

Shakirova, J., & Smolnikova, E. (2017). Influence of university internationalization on their positions in world rankings. *QUID Investigacion. Ciencia y Tecnologia, 1*, 859–864.

Shrivastava, A. (2024). Artificial Intelligence (AI): Evolution, Methodologies, and Applications. *International Journal for Research in Applied Science and Engineering Technology, 12*(IV), 5501–5505. DOI: 10.22214/ijraset.2024.61241

Sofyan, N. (2021). The role of English as global language. *Edukasi, 19*(1). Advance online publication. DOI: 10.33387/J.EDU.V19I1.3200

Stockemer, D., & Wigginton, M. (2019). Publishing in English or another language: An inclusive study of scholar's language publication preferences in the natural, social and interdisciplinary sciences. *Scientometrics, 118*, 645–652. DOI: 10.1007/s11192-018-2987-0

Taşkın, M. (2025). Artificial Intelligence in Personalized Education: Enhancing Learning Outcomes Through Adaptive Technologies and Data-Driven Insights. *Human-Computer Interaction, 8*(1). Advance online publication. DOI: 10.62802/ygye0506

Thenmozhi, V. (2025). A Study on the Effectiveness of AI-Powered Customer Service Enhancement at Syena Logistics PVT LTD. *International Journal of Scientific Research in Engineering and Management, 9*(4), 1–9. DOI: 10.55041/ijsrem45403

Tulasi, S., & Ahamed, I. (2025). AI-Powered E-Commerce Personalization and Recommendation Systems. *Journal of Informatics Education and Research, 5*(1), 2092–2099. DOI: 10.52783/jier.v5i1.2197

UBD. (2025). *UBD Moves Up in QS World University Rankings, Reinforces Global Standing.* Retrieved September 16, 2025, from https://ubd.edu.bn/2025/07/05/ubd-moves-up-in-qs-world-university-rankings-reinforces-global-standing/

Walters, W. (2017). Citation-Based Journal Rankings: Key Questions, Metrics, and Data Sources. *IEEE Access : Practical Innovations, Open Solutions, 5*, 22036–22053. DOI: 10.1109/ACCESS.2017.2761400

Wu, M.-J., Zhao, K., & Fils-Aime, F. (2022). Response rates of online surveys in published research: A meta-analysis. *Computers in Human Behavior Reports, 7*. Advance online publication. DOI: 10.1016/j.chbr.2022.100206

Wy, I., & Kurniawan, A. (2024). *English Language and Its Importance as Global Communication. Samā Jiva Jnānam.* International Journal of Social Studies., DOI: 10.25078/ijoss.v2i1.3920

Yildiz, T. (2021). Degree Mobility in the Higher Education Context: Could the English Language be a Key Driver of Internationalisation? *Journal of University Research*, *4*(3), 220–227. DOI: 10.32329/uad.876946

Yunos, R. (2013). *The Golden Warisan Brunei Darussalam (volume 1)*. Bandar Seri Begawan: Dewan Bahasa dan Pustaka.

Chapter 3
AI and Low-Resource Languages:
Bridging the Linguistic Divide

Nanlir Sallau Mullah
https://orcid.org/0000-0002-2048-5119
School of Computer Sciences, Universiti Sains Malaysia, Penang, Malaysia

Wan Mohd Nazmee Wan Zainon
Universiti Sains Malaysia, Penang, Malaysia

ABSTRACT

Artificial intelligence (AI) is rapidly transforming global communication, learning, and access to services, yet its benefits remain concentrated in high-resource languages, leaving most low-resource languages (LRLs) digitally marginalised. This chapter examines the linguistic digital divide, where large language models overwhelmingly prioritise dominant languages. Drawing on insights from computational linguistics, AI ethics, and community initiatives, we analyse socio-technical and cultural barriers to inclusion and present case studies from Africa, Latin America, and the Pacific that highlight both challenges and opportunities. We argue that bridging this gap requires multilingual pretraining, participatory data governance, sustainable funding, and culturally grounded design. By foregrounding marginalised voices and embedding linguistic justice in AI development, this chapter envisions a future where every language, regardless of resource level, secures a meaningful digital presence.

DOI: 10.4018/979-8-3373-4397-6.ch003

1. INTRODUCTION

In today's rapidly evolving digital age, artificial intelligence (AI) technologies are transforming key sectors such as communication, education, healthcare, and governance, to mention just a few. As artificial intelligence technologies continue to permeate everyday life, they have the potential to significantly enhance access to information, services, and opportunities across diverse populations (Sambasivan et al., 2021). However, language plays a dual role in this transformation, acting as a powerful bridge for speakers of high-resource languages, while simultaneously becoming a barrier for those whose native tongues are underrepresented or entirely absent in AI systems (P. Joshi et al., 2020; Nekoto et al., 2020). While AI systems, especially large language models (LLMs), have made significant progress in understanding and generating human language, their performance remains highly skewed toward high-resource languages such as English, Mandarin, and Spanish (Huang et al., 2024; Hasan et al., 2024). These systems are trained on massive datasets, most of which are dominated by content in a handful of global languages, thereby excluding thousands of others from full participation in the digital age (Joshi et al., 2020; Nekoto et al., 2020).

This asymmetry has led to what is now referred to as the linguistic digital divide, a growing gap in how different language communities benefit from AI technologies. Speakers of low-resource languages (LRLs), which are languages that lack sufficient digitised data, computational tools, or linguistic documentation, are systematically excluded from many of the advancements driven by artificial intelligence (Bird, 2022). These benefits include real-time translation, voice-activated assistants, education tools, healthcare chatbots, and more (Sambasivan et al., 2021). Without deliberate intervention, the linguistic digital divide risks reinforcing pre-existing global inequalities, limiting opportunities for billions of people to participate meaningfully in digital ecosystems.

Low-resource languages are not marginal by any measure. There are over 7,000 languages spoken worldwide, yet the vast majority of AI applications focus on fewer than 100 (Leben, 2018; Zhong et al., 2024). LRLs often serve as cultural and historical vessels, preserving unique epistemologies, oral traditions, and community identities (Bird, 2020). Their neglect of AI systems therefore not only hinders technological access but also poses existential threats to linguistic diversity and cultural preservation. The United Nations Educational, Scientific and Cultural Organisation (UNESCO) has warned that 40% of the world's languages are at risk of extinction, many of them low-resource and digitally invisible[1] (UNESCO, 2024).

Addressing this issue is both a technological challenge and an ethical imperative. The development of AI systems that support LRLs requires innovative approaches, including transfer learning, data augmentation, multilingual pretraining, community-driven data collection, and inclusive evaluation benchmarks (Conneau et al., 2019; Ruder et al., 2021). Moreover, equitable language AI must go beyond model accuracy to ensure community ownership, fairness, and cultural sensitivity in the design and deployment of AI tools (Hovy & Spruit, 2016).

This chapter explores the intersection of AI and low-resource languages with the goal of bridging the linguistic divide. It offers a critical examination of the technological barriers, socio-political dynamics, and innovative strategies shaping the development of inclusive language technologies. By doing so, it highlights the path forward toward a more linguistically just and technologically equitable world.

1.1 Contextualising the Linguistic Digital Divide

The digital divide traditionally refers to unequal access to information and communication technologies (ICT). This disparity spans physical, economic, and skills-based dimensions (Adam & Dzang Alhassan, 2020; Aydin, 2021). While significant efforts have been made to expand internet access worldwide, stark discrepancies persist: for instance, as of 2023, internet penetration reached 89% in Europe, yet only 37% in Africa—demonstrating deeply entrenched global inequalities. Emerging alongside this is the linguistic digital divide, a gap in which digital and AI technologies disproportionately serve high-resource languages (e.g., English, Spanish, Mandarin, French, Arabic), while low-resource languages (LRLs) remain underserved (Zhong et al., 2024). This extends the traditional digital divide into the domain of language representation and technological equity.

Modern AI methods, particularly large language models (LLMs), rely heavily on extensive digital and diverse training data. As a result, they often perform poorly with low-resource languages (LRLs), which suffer from insufficient or entirely absent digitised corpora (Tahir & Mehmood, 2021). Consequently, speakers of LRLs are systematically excluded from AI-driven benefits such as translation tools, healthcare diagnostics, economic opportunities, and educational aids, amplifying existing inequalities.

1.2 Importance of Low-Resource Languages in the Global Linguistic Ecosystem

LRLs are far from marginal; they are repositories of unique cultures, knowledge systems, and worldviews. Globally, there are over 7,000 languages, yet only a few hundred are significantly represented in digital data, leading to systematic linguistic bias in AI (Litre et al., 2022). Addressing this imbalance is not merely a technical challenge but a moral and cultural imperative. UNESCO and the United Nations underscore the issue, highlighting that nearly one-third of the world remains unconnected, often due to language exclusion (Adeyeye, 2024).

Furthermore, the digital and AI exclusion of LRLs undermines efforts toward cultural preservation, linguistic diversity, and equitable development. Research emphasises that inclusive AI systems must address not only representation in datasets but also underlying power dynamics and biases, ensuring that LRL communities are empowered as co-creators, rather than merely data providers (Choudhury et al., 2026). In short, LRLs form a critical part of the global linguistic tapestry. Ensuring their inclusion in AI not only preserves cultural heritage but also promotes fairness and democratizes access to technology for billions of speakers worldwide.

2. DEFINING LOW-RESOURCE LANGUAGES IN THE AI ERA

In the context of artificial intelligence (AI) and natural language processing (NLP), the term low-resource language (LRL) has gained significant traction. Yet, the designation is more complex than the label implies. Traditionally understood in linguistic and socio-political terms, "low-resource" now also entails the availability (or lack thereof) of machine-readable data, technical tools, and AI-ready infrastructure. This section provides a multidimensional definition of LRLs and explores illustrative cases from various global language families.

2.1 Linguistic, Cultural, and Technical Criteria

Low-resource languages cannot be defined by a single parameter. Rather, they exist along a spectrum of linguistic vitality, cultural visibility, and technical inclusion. The intersection of these dimensions shapes how languages are perceived and prioritised in AI development.

a. Linguistic Criteria

From a linguistic perspective, LRLs may be:

Orally dominant (with limited or no standardised writing system),

Characterised by high dialectal variation,

Under-documented in terms of grammatical structure, lexicons, and corpora.

These features make computational modelling challenging, especially when there is little consensus on orthography or a lack of linguistic expertise available for annotation and validation (Bird, 2020). For instance, tonal languages like Yoruba (Nigeria) or Hmong (China) require accurate tone marking to preserve meaning, yet digital texts often omit these marks, creating ambiguity in model training.

b. Cultural and Political Criteria

Culturally, many LRLs are associated with indigenous, minorities, or postcolonial communities, often marginalised in official policies and media representation. These languages may have millions of speakers (e.g., Hausa, Tarok, Quechua) yet still lack institutional support, funding, or global visibility. Some may be official languages in name only, with few resources allocated to their preservation or digital use.

The label "low-resource" thus reflects socio-political power structures more than speaker population. (Coupland, 2015) As Coupland (2015) argues, linguistic marginalisation is often a product of colonial histories and national language ideologies that prioritise hegemonic tongues. Hegemonic tongues mean languages that hold dominant power and influence over others, usually because they are tied to political control, economic strength, military power, or cultural prestige. For example, English is a hegemonic tongue today because it dominates international trade, science, technology, and diplomacy. Historically, Latin played a similar role in medieval Europe, and Arabic did in the Islamic Golden Age.

c. Technical and Infrastructural Criteria

Technically, a language is considered low-resource when it lacks the digital data and tools necessary to support AI tasks, such as:

- Pretrained language models,
- Tokenisers and morphological analysers,
- Machine translation corpora,
- Speech datasets,
- POS-tagged or syntactically annotated corpora.

Even if the language is widely spoken, it may be low-resource if it is absent from web content, underrepresented in public datasets, or excluded from multilingual AI benchmarks (Joshi et al., 2020). For instance, many Indian and African languages fall into this category despite having millions of daily speakers.

How technically low-resourced a language is can be measured in several ways (Ruder et al., 2021):

- Data quantity (amount of available parallel or monolingual text),
- Data quality (cleanliness, balance, genre diversity),
- Toolchain availability (tokenisers, speech synthesis, translation APIs),
- Model performance (accuracy, bias, hallucination rate in NLP tasks).

A language might perform poorly even in a multilingual model because its representation in training data is sparse or biased.

To better understand the disparities in AI readiness across global languages, it is essential to differentiate between high-resource, low-resource, and ultra-low-resource languages. These categories reflect varying levels of digital presence, availability of linguistic data, community involvement, and support within AI systems. While high-resource languages dominate digital infrastructures and benefit from extensive computational resources, low-resource and ultra-low-resource languages often remain excluded from mainstream AI development. Table 1 provides a comparative overview of these language categories, highlighting key dimensions such as data availability, model support, community engagement, and ethical considerations. This comparison sets the stage for designing targeted, inclusive, and context-aware AI interventions for underrepresented language communities.

Table 1. Comparative Overview of High-Resource, Low-Resource, and Ultra-Low-Resource Languages in the Context of AI Development

Criteria	High-Resource Languages	Low-Resource Languages	Ultra-Low-Resource Languages
Examples	English, Mandarin, Spanish, French	Yoruba, Igbo, Swahili, Quechua	Tarok, Ainu, San, Koyra Chiini
Digital Presence	Extensive and dominant in online platforms, media, and academic texts	Moderate presence; limited to certain domains	Extremely limited or nearly absent from digital platforms
Available NLP Resources	Abundant datasets, pretrained models, benchmarks (e.g., GLUE, SQuAD)	Small-scale corpora, partial embeddings, limited benchmarks	Minimal or no structured corpora; often no pretrained models available
Support in LLMs	Fully supported in models like GPT, BERT, and BLOOM	Often indirectly supported via multilingual models (e.g., XLM-R)	Typically unsupported or misrepresented; outputs prone to hallucinations
Annotated Linguistic Data	Rich in labelled datasets for translation, NER, POS, QA, etc.	Sparse and task-specific; may exist in academic silos	Rare or unavailable; often reliant on manual, community-led annotation
Community Involvement	Global collaborations, corporate and academic research networks	Increasingly, community-led projects (e.g., Masakhane, AI4Bharat)	Limited capacity and funding; heavily reliant on NGO or volunteer efforts

continued on following page

Table 1. Continued

Criteria	High-Resource Languages	Low-Resource Languages	Ultra-Low-Resource Languages
Script and Orthographic Standardisation	Standardised, globally recognised orthographies	Partial standardisation; dialectal variation is common	Often unstandardized or primarily oral; scripts may be underdeveloped
Ethical and Cultural Risks	Well-established content norms and consent frameworks	Risks of exclusion and misrepresentation	High risk of exploitation, data colonialism, and cultural erasure

*Low-resource status here refers to the availability of substantial digital corpora, NLP tools, and AI datasets. Some languages have large speaker bases but are still considered low-resource due to limited computational resources (e.g., Javanese, Malagasy).

2.2 Examples from Global Language Families

To illustrate the diversity of low-resource languages, we consider examples across major global language families, highlighting their AI challenges and potentials.

a. Niger-Congo Language Family (Africa)

The Niger–Congo language family is the largest in Africa (and one of the largest in the world) in terms of number of languages and geographic spread. Examples include (Creissels, 2019; Good, 2014):

Swahili (Kiswahili) – widely spoken in East Africa (Kenya, Tanzania, Uganda, parts of the DRC).

Yoruba – spoken mainly in Nigeria, Benin, and Togo.

Igbo – spoken in southeastern Nigeria.

Zulu – spoken in South Africa.

Shona – spoken in Zimbabwe and Mozambique.

Fula (Fulani, Pulaar, Fulfulde) – spoken across West and Central Africa.

Wolof – spoken mainly in Senegal, The Gambia, and Mauritania.

Akan (Twi, Fante) – spoken in Ghana and Côte d'Ivoire.

Bemba – spoken in Zambia and the Democratic Republic of Congo.

Lingala – spoken in the Democratic Republic of Congo and the Republic of Congo.

b. Austroasiatic and Austronesian Families (Asia-Pacific)

The Austroasiatic and Austronesian language families are two of the most significant in the Asia-Pacific region, both in terms of geographic spread and cultural diversity (Sallabank, 2011). Although they share the region, their historical origins, linguistic structures, and sociolinguistic contexts differ considerably. These two language families collectively illustrate the linguistic richness and complexity of the Asia-Pacific region. They also highlight the imbalance in technological representation, where a few dominant languages thrive while many others face the risk of further marginalisation without or with little targeted preservation and technological inclusion efforts.

Austroasiatic Language Family

The Austroasiatic language family comprises over 160 languages spoken by an estimated 120 million people across South and Southeast Asia. Its speakers are found in countries such as Cambodia, Vietnam, Laos, Thailand, Myanmar, and parts of India and Bangladesh. Prominent languages include Khmer, the official language of Cambodia, Vietnamese, widely spoken in Vietnam, and Mon, with historical significance in Myanmar and Thailand. Austroasiatic languages are among the earliest in the region, with archaeological and linguistic evidence pointing to deep historical roots and long-standing influence on surrounding language families (R. M. Blench, 2017).

Geographic distribution: Mainland Southeast Asia, parts of Northeast India, and scattered areas of southern China.

Key features: Many languages are spoken by minority groups; several are endangered; some have long historical records (e.g., Khmer, Vietnamese).

Examples:

- Vietnamese – the national language of Vietnam.
- Khmer (Cambodian) – the national language of Cambodia.
- Mon – spoken in Myanmar and Thailand.
- Santali – spoken in eastern India (Jharkhand, West Bengal, Odisha).
- Khasi – spoken in Meghalaya, India.
- Mundari – spoken in eastern India.

Austronesian Language Family

The Austronesian language family comprises over 1,200 languages spoken by more than 400 million people. It is the most widely dispersed language family in premodern history, stretching from Madagascar to Easter Island, and from Taiwan to New Zealand (Wellen, 2019; Greenhill et al., 2010).

Geographic distribution: Island Southeast Asia, the Pacific Islands, and parts of Madagascar.

Key features: One of the world's largest language families by geographic range, stretching from Madagascar in the west to Easter Island in the east.

Examples:

- Malay / Indonesian – official languages of Malaysia, Indonesia, and Brunei.
- Tagalog / Filipino – national language of the Philippines.
- Javanese – spoken on Java, Indonesia.
- Cebuano – spoken in the southern Philippines.
- Hawaiian – an indigenous language of Hawaii (USA).
- Malagasy – spoken in Madagascar.
- Māori – spoken in New Zealand.
- Samoan – spoken in Samoa and American Samoa.

The Austroasiatic and Austronesian language families represent two distinct yet culturally significant linguistic groups in Asia and the Pacific. While both encompass languages with millions of speakers, they also include many smaller, marginalised tongues that face challenges in digital representation and resource availability. The Austroasiatic family, found mainly in mainland Southeast Asia and parts of South Asia, includes both national languages like Vietnamese and Khmer and minority languages with limited technological support. The Austronesian family, spread across Island Southeast Asia, the Pacific, and Madagascar, contains globally prominent languages like Malay/Indonesian and Tagalog, as well as endangered and revitalising languages such as Hawaiian and Maori. Table 2 summarises key subgroups, representative languages, approximate speaker populations, and the extent to which these languages are considered low-resource in computational linguistics.

Table 2. Comparison Table: Austroasiatic vs. Austronesian Language Families

Language Family	Main Subgroups	Example Languages	Approx. Speaker Count (Millions)	Low-Resource Status
Austroasiatic	Vietic	Vietnamese	~86	No – major national language
	Khmer	Khmer (Cambodian)	~18	No – national language
	Monic	Mon	~1	Yes – minority, declining
	Munda	Santali, Mundari	Santali ~7, Mundari ~2	Yes – limited digital resources
	Khasic	Khasi	~1.5	Yes – minority, limited resources
	Other Minor Branches	Wa, Palaung, Nicobarese	<1 each	Yes – endangered
Austronesian	Malayo-Polynesian	Malay / Indonesian	~300 (combined)	No – strong digital presence
	Philippine	Tagalog / Filipino, Cebuano	Tagalog ~28, Cebuano ~20	No – widely used

continued on following page

Table 2. Continued

Language Family	Main Subgroups	Example Languages	Approx. Speaker Count (Millions)	Low-Resource Status
	Javanic	Javanese	~82	Partial – large speakers, underrepresented online
	Polynesian	Hawaiian, Maori, Samoan	Hawaiian ~0.024, Maori ~0.15, Samoan ~0.5	Yes – revitalisation ongoing
	Malagasy Branch	Malagasy	~25	Partial – moderate digital resources
	Micronesian	Chamorro, Kiribati	<0.1–0.12	Yes – low digital presence

c. Indo-Aryan and Dravidian Families (South Asia)

The Indo-Aryan and Dravidian language families dominate the linguistic landscape of South Asia. Indo-Aryan languages, including Hindi, Bengali, and Punjabi, are spoken by the majority, while Dravidian languages such as Tamil, Telugu, Kannada, and Malayalam are primarily concentrated in southern India and parts of Sri Lanka (Alam Monisha & Sultana, 2022). Marathi, Telugu, and Odia are spoken by tens of millions but have limited high-quality NLP datasets. The AI4Bharat project is working to address this gap by producing open-source models for Indian languages (Kakwani et al., 2020). Issues include inconsistent orthography, dialectal variation, and a lack of standardised evaluation benchmarks.

d. Quechuan and Mayan Families (Latin America)

The Quechuan and Mayan language families represent two of the most prominent Indigenous linguistic traditions in the Americas. Quechuan languages, such as Quechua, are spoken primarily in the Andean regions of South America, while Mayan languages, including K'iche' and Yucatec Maya, are native to parts of Mexico, Guatemala, Belize, and Honduras (AnderBois, 2019; Q. Emlen, 2025). Quechua: Spoken by over 8 million people across the Andes, Quechua is often low-resource due to a lack of standardised orthography and limited online presence (Q. Emlen, 2025).

K'iche': A Mayan language of Guatemala, K'iche' has minimal digitised resources. Some recent efforts focus on speech transcription and Bible translation, but there is almost no AI infrastructure in place. Recent research efforts, such as speech transcription models in AmericasNLP competitions and exploratory machine translation experiments, have made notable technical strides. However, the broader Quechua

AI ecosystem remains extremely underdeveloped, with virtually no foundational AI infrastructure in place (Romero et al., 2024).

e. Uralic and Turkic Families (Eurasia)

The Uralic and Turkic language families cover vast regions of Eurasia. Uralic languages, such as Finnish, Estonian, and Hungarian, are spoken from Northern Europe to Western Siberia, while Turkic languages, including Turkish, Uzbek, and Kazakh, are found across Central Asia, Eastern Europe, and parts of the Middle East (Johanson, 2021; Tambets et al., 2018).

Tatar and Chuvash: These Turkic languages are underrepresented in mainstream NLP tools, especially in comparison to Turkish. Challenges include limited corpora, Cyrillic orthography, and absence from multilingual LLMs like GPT and XLM-R.

Low-resource languages are not inherently underdeveloped; they are under-resourced by design, shaped by centuries of marginalisation and a lack of targeted investment in language technologies. Whether spoken by a few hundred or millions, these languages face common barriers to digital inclusion in AI systems. By recognising the linguistic, cultural, and technical nuances that define low-resource status, we can more accurately prioritise interventions and design ethical, community-centred AI solutions that bridge the linguistic divide.

3. TECHNICAL LANDSCAPE

As the field of natural language processing (NLP) advances, the challenges of supporting low-resource languages (LRLs) are increasingly addressed through multilingual models and cross-lingual techniques. Additionally, the creation of benchmarks and repositories is helping to better accommodate linguistic diversity. This section explores the major technological approaches and resources shaping the current landscape.

3.1 Overview of Multilingual and Zero-Shot Models

Multilingual language models aim to learn representations that are effective across multiple languages. These models have significantly improved performance for many underrepresented languages, particularly through zero-shot and few-shot learning techniques. One of the earliest and most influential multilingual models is mBERT (Multilingual BERT), introduced by Devlin et al. (2019). mBERT is trained on 104 languages using masked language modelling, enabling it to generalise linguistic patterns across languages despite no explicit cross-lingual objectives during training (Libovický et al., 2019). However, its performance on LRLs can be inconsistent due to uneven data representation in the training corpus.

XLM-R (Cross-lingual RoBERTa), developed by Conneau et al. (2019), improves upon mBERT by being trained on significantly larger and more diverse data from CommonCrawl. It incorporates 100 languages and achieves state-of-the-art results in several cross-lingual benchmarks like XNLI and MLQA. XLM-R is more robust for zero-shot and transfer learning scenarios, making it particularly useful for applications in LRL contexts.

Another major advancement is BLOOM (BigScience Large Open-science Open-access Multilingual Language Model), a multilingual autoregressive model trained on 46 languages and 13 programming languages using over 350 billion tokens (Scao et al., 2022). BLOOM was developed through an open science initiative to provide an accessible alternative to proprietary LLMs and is notable for its deliberate inclusion of African and Asian languages with low digital representation. These models enable zero-shot learning, where a model trained on high-resource languages can make predictions in a low-resource language without having seen any examples during training. Although promising, zero-shot capabilities are often limited by structural differences between languages, especially those outside the Indo-European family (Hu et al., 2020).

3.2 Transfer Learning, Synthetic Data, and Cross-Lingual Embeddings

To mitigate data scarcity in LRLs, researchers leverage several strategies:

a. Transfer Learning

Transfer learning enables models trained on high-resource languages or tasks to be adapted to new, low-resource settings. A common approach is to fine-tune pretrained multilingual models like mBERT or XLM-R on limited LRL data. This drastically reduces the amount of annotated data required to build effective NLP systems (Ruder et al., 2019). Cross-lingual transfer is more effective when there is typological similarity between the source and target languages.

b. Synthetic Data Generation

Synthetic data generation involves creating artificial examples to augment small datasets. Techniques include:

- Back-translation, where sentences are translated into another language and back to generate paraphrased variants.
- Machine translation bootstrapping, where monolingual corpora are translated using existing models to generate pseudo-parallel corpora (Artetxe et al., 2020). These methods are crucial for languages with no available labelled data, enabling the training of classifiers or translation models in a low-data regime.

c. Cross-Lingual Embeddings

Cross-lingual word embeddings project words from multiple languages into a shared vector space, allowing models to reason across languages. Early methods like MUSE (Lample et al., 2018) use bilingual dictionaries and adversarial training, while newer methods integrate these embeddings into transformer architectures. These embeddings enable tasks like bilingual lexicon induction, zero-shot classification, and multilingual information retrieval, especially when combined with transfer learning.

3.3 Data Repositories and Benchmarks

A key barrier to progress in LRL research is the lack of open, curated datasets. Several initiatives have emerged to fill this gap by providing diverse resources and evaluation tools:

- Masakhane: A grassroots research effort focused on African languages, Masakhane provides parallel corpora, translation models, and research on responsible AI practices in the Global South (Nekoto et al., 2020).
- FLORES-101 and FLORES-200: Developed by Facebook AI, these benchmarks offer parallel datasets across over 100 and 200 languages, respectively, enabling evaluation of multilingual translation models on truly low-resource languages (Goyal et al., 2022).
- XTREME and XTREME-R: These are comprehensive benchmarks for evaluating cross-lingual generalisation across a wide range of tasks, including question answering, NER, and sentence retrieval (Hu et al., 2020; Ruder et al., 2021). They provide performance metrics for over 40 languages and emphasise generalisation beyond English-centric datasets.
- Common Voice: Mozilla Common Voice, a volunteer-led, open-source project, actively gathers speech data from volunteers across the globe and extends its coverage to many underserved and Indigenous languages[2].

- AI4Bharat: A project aimed at advancing Indian language technologies, AI4Bharat provides benchmarks, pretrained models, and datasets for over 20 Indian languages, many of which are low-resource (Kakwani et al., 2020). Together, these resources are instrumental in democratizing NLP research and improving representation for marginalised language communities.

4. CASE STUDIES

Real-world applications of artificial intelligence in low-resource languages (LRLs) reveal both the promises and complexities of inclusive language technology. This chapter explores case studies from Africa and other regions, highlighting community-driven initiatives, technological advances, and the persistent barriers of linguistic standardisation and orthographic variation.

4.1 African Languages

Africa is home to over 2,000 languages, many of which are low-resource despite their high speaker populations (Muhammad et al., 2023). Languages such as Yoruba, Swahili, Tarok, Hausa and Igbo illustrate the dynamic efforts being made to include African linguistic heritage in AI development.

a. Yoruba

Yoruba, a tonal Niger-Congo language spoken by over 20 million people primarily in Nigeria, has seen increased attention from the AI community. The Masakhane project, a grassroots NLP initiative across Africa, has developed Yoruba–English machine translation models using openly contributed parallel corpora (Nekoto et al., 2020). Researchers have also developed Yoruba speech recognition models using the Mozilla Common Voice dataset, despite the challenges posed by tone and under-representation in textual data (Ogueji et al., 2021).

b. Swahili

Swahili (Kiswahili), one of Africa's most widely spoken and standardised languages, has benefited from relatively better resources than many others. It is often included in multilingual models like mBERT and XLM-R, which helps facilitate zero-shot learning in neighbouring languages. Notable projects include AI4D–Africa, which has developed Swahili datasets for part-of-speech tagging and named entity recognition, with active collaborations involving universities and NGOs (Adelani et al., 2021).

c. Igbo

Igbo, another major Nigerian language, is notably underrepresented in digital resources. Community initiatives have led to the creation of small-scale corpora and syntactic tools (Onyenwe, 2017). However, the diversity of dialects and the lack of standardised resources complicate model training and evaluation. Despite these limitations, neural machine translation efforts and morphological tagging systems are in development, often building on transfer learning from better-resourced African languages (Inuwa-Dutse, 2025; Onyenwe, 2017).

d. Hausa

Hausa is one of the most widely spoken languages in Africa, with over 50 million native speakers and an additional 20 million second-language speakers across Nigeria, Niger, Ghana, Cameroon, and Sudan (Adelani et al., 2021; Muhammad et al., 2023). Despite its wide usage and role as a lingua franca in West Africa, Hausa remains relatively underrepresented in high-quality NLP resources and benchmark datasets.

Although Hausa is better resourced than many African languages, its natural language processing ecosystem remains underdeveloped in key areas such as speech-to-text, Automated Speech Recognition (ASR), and morphological tagging (Muhammad et al., 2025). One compounding challenge is its use of the Ajami script, an Arabic-based orthography used alongside the Latin Boko script, which most existing NLP tools do not support, thereby excluding a significant portion of native written content. Initiatives like Masakhane and Lanfrica have begun compiling Hausa corpora for machine translation and named entity recognition (NER), but much of the development remains fragmented and lacks large-scale, community-owned infrastructure. Efforts to expand Hausa NLP are growing through partnerships with universities in Northern Nigeria and the African Institute for Mathematical Sciences (AIMS), but challenges remain around dataset standardisation, licensing, and cross-border language variation.

e. Tarok

Tarok language is spoken predominantly in the southern senatorial zone of Plateau State, Nigeria (R. Blench, 2020). Tarok language is classified as a minority language with an estimated speaker population of around 1,000,000. Langtang-North, Langtang-South and Wase local government areas are predominantly occupied by the

Tarok people. They are also found in large numbers in Mikang, Shendam, Kangke, and Quanpan LGAs. Unlike Yoruba, Hausa, or Igbo, Tarok suffers from extreme resource scarcity, lacking in digitised corpora, orthographic standardisation, and formal inclusion in national educational curricula. While the language is vibrant in oral use, it has very limited representation in digital spaces, making it nearly invisible in the context of modern NLP and AI systems.

The few existing linguistic resources for Tarok are mainly in the form of academic field studies and religious translations, such as the Tarok Bible and oral histories. There is currently no large-scale Tarok corpus, no publicly available tokeniser, and no pretrained models supporting the language. This places Tarok in the category of ultra-low-resource languages, for which zero-shot or meta-learning approaches are likely the only short-term paths toward inclusion. Developing NLP tools for Tarok would require not only technological interventions but also significant community engagement, documentation efforts, and government or NGO backing to create digital orthographies, standardise usage, and build speaker-led language revitalisation programs. Importantly, any effort must be inclusive and participatory, respecting the cultural context and linguistic ownership of the Tarok-speaking community.

4.2 Other Under-Represented Languages (e.g., Quechua, Māori)

Efforts to include Indigenous and minority languages beyond Africa also illustrate both the progress and the persistent challenges of bridging the linguistic digital divide.

a. Quechua

Quechua, an Indigenous language family of the Andes spoken by over 8 million people, suffers from extremely limited digital representation. Past initiatives such as the AILLA archive (Archive of the Indigenous Languages of Latin America) and the ALT project have provided text and audio corpora. Quechua NLP development is hindered by orthographic inconsistency and the general scarcity of digital and community-generated data (Court & Elsner, 2024). Some models have experimented with translating Quechua using unsupervised machine translation, though with limited accuracy.

b. Māori

Māori, an official language of New Zealand, has benefited from substantial government support in revitalisation efforts, including digital initiatives. Te Hiku Media, a Māori-led non-profit, has worked on building large Māori speech and text datasets through community partnerships, emphasising data sovereignty and cultural integrity[3]. Their "Papa Reo" project aims to create tools such as automatic speech recognition (ASR) and text-to-speech (TTS) models that reflect the values and needs of Māori speakers.

These cases emphasise the importance of community engagement, data sovereignty, and policy support in making AI beneficial for LRLs.

4.3 Challenges of Standardisation and Orthography

One of the most persistent technical and socio-linguistic barriers in building AI systems for LRLs is the lack of standardisation in grammar, spelling, and orthography.

Dialectal Variation: Many LRLs are not monolithic. For instance, Igbo has more than 20 dialects, and efforts to create unified resources often privilege certain dialects over others, leading to community resistance or reduced adoption (Onyenwe, 2017). In Quechua, dialects differ so widely that mutual intelligibility is not always guaranteed.

Orthographic Ambiguity: Some languages have multiple writing systems or inconsistent spelling conventions. For example, Yoruba utilises tone marks (diacritics), which are often omitted in informal writing, yet these tonal differences change meaning drastically. This makes it difficult for AI models to disambiguate meaning without robust, standardised datasets (Ogueji et al., 2021).

Historical Colonisation of Scripts: In some cases, colonial legacies introduced orthographies that were not reflective of the phonological systems of Indigenous languages. This mismatch can hinder accurate tokenisation, morphological analysis, and pronunciation modelling.

Lack of Official Recognition or Policy Support: Many languages are not officially recognised in their home countries, leading to minimal investment in linguistic development or digitisation efforts. For example, some Nigerian languages are not part of the official educational curriculum, reducing institutional motivation to develop AI tools.

Efforts to address these challenges include participatory corpus development, community-driven orthography standardisation, and the integration of human-in-the-loop methodologies that allow native speakers to correct AI outputs (Bird, 2020). In addition, language policy reform and decolonial approaches to NLP are essential to ensure that LRLs are not merely included for technological completeness but empowered with autonomy and dignity.

5. ETHICAL CONSIDERATIONS AND RISKS

As efforts to integrate low-resource languages (LRLs) into artificial intelligence (AI) systems expand, ethical challenges emerge concerning who collects the data, who benefits from it, and whose values shape the resulting technologies. While the technological inclusion of LRLs promises benefits like improved communication, language preservation, and access to services, it also raises concerns related to consent, ownership, exclusion, and bias. Without deliberate ethical safeguards, attempts to close the linguistic digital divide may instead reinforce colonial power dynamics and marginalisation.

5.1 Consent, Ownership, and Data Sovereignty

One of the foremost ethical concerns in LRL AI development is data sovereignty—the right of a language community to control how their linguistic data is collected, used, and distributed. For many Indigenous or marginalised groups, language is not just a means of communication but a sacred and communal cultural asset (Bird, 2020). Extracting texts or recordings from these communities without informed consent or a clear framework for benefit-sharing constitutes a form of data colonialism.

Most large language models are trained on publicly scraped internet data, often without transparency about what languages or content were included. In the case of LRLs, this raises the risk of misrepresentation, especially if datasets are derived from religious texts, colonial literature, or poorly translated documents (Sambasivan et al., 2021). Furthermore, once incorporated into proprietary AI systems, these datasets may be commercialised without any feedback or compensation mechanisms for the communities involved.

Projects like Te Hiku Media in New Zealand, which pioneered a Māori data licensing agreement known as the Kaitiakitanga License, provide a compelling counterexample. Their approach ensures that Indigenous communities retain ownership and control over how their data is used in AI applications, reflecting culturally grounded principles[4]. This model of participatory consent and data governance must be expanded globally, especially in African and Latin American contexts where many LRLs originate.

5.2 Technological Imperialism and Community Exclusion

Another ethical pitfall in AI for LRLs is technological imperialism, the imposition of external technologies and values onto marginalised communities without their active participation or alignment with local contexts (Hovy & Spruit, 2016). This occurs when AI systems are built for LRLs by external researchers or corporations without meaningful engagement with native speakers, linguists, or cultural stewards. When communities are excluded from decisions about model design, priorities, or use cases, the result is often misaligned tools that fail to reflect the social realities or needs of their users. For example, deploying machine translation for a language like Igbo or Hausa, or Yoruba, without consulting speakers may overlook dialectal differences, informal expressions, or cultural nuances—leading to miscommunication or erasure of identity.

Moreover, this kind of top-down development reinforces a colonial hierarchy of knowledge production, where Western institutions and English-speaking researchers are positioned as the default gatekeepers of AI innovation. It also risks linguistic flattening, in which the complex diversity of a language is reduced to standardised or "acceptable" forms based on external norms. To counteract technological imperialism, ethical AI development for LRLs must adopt a decolonial approach, centring the voices of affected communities and supporting co-creation models. Initiatives like Masakhane demonstrate this well by prioritising African researchers, enabling distributed collaboration, and focusing on context-aware language tasks (Nekoto et al., 2020).

5.3 Bias and Hallucination in LLMs

Large Language Models (LLMs) like GPT, Claude, DeepSeeks, etc, present additional ethical risks in their application to LRLs, particularly through bias propagation and hallucination.

a. Bias Propagation

LLMs are trained on vast datasets that often reflect the biases, stereotypes, and social hierarchies embedded in those corpora. For LRLs, where data is scarce, models tend to rely on imperfect or unrepresentative samples. For instance, if Hausa or Quechua training data comes primarily from religious texts or academic translations, the resulting model may overfit to formal styles and omit colloquial or culturally important expressions (Abid et al., 2021).

This also opens the door to ethnolinguistic bias, where the representation of one dialect or script is privileged over others. In the case of Hausa, models may prioritise the Latin alphabet while ignoring the Ajami script, leading to digital exclusion of certain demographics (Muhammad et al., 2025). Gender and religious bias can also be exacerbated when LLMs echo patriarchal or colonial content from historical sources.

b. Hallucination

Hallucination refers to the phenomenon where an LLM generates fluent but factually incorrect or misleading outputs. This is particularly dangerous in LRLs where:

Ground-truth data is lacking,

Evaluation benchmarks are immature,

And native speakers may rely on these outputs in educational or public-service contexts.

For example, a hallucinated translation of a health advisory into an Indigenous language could result in misinformation with real-world consequences. Moreover, since LLMs are probabilistic and not inherently grounded in truth, they may fabricate grammar rules, vocabulary, or idioms in LRLs simply to complete a sentence structure plausibly.

To address these issues, AI systems must integrate human-in-the-loop validation, community evaluation benchmarks, and bias detection audits tailored to the cultural and linguistic characteristics of the target LRL (Raji et al., 2021). Transparency in model architecture and training data is also critical to empowering external audits and corrections.

6. TOWARD AN INCLUSIVE AI-LANGUAGE FRAMEWORK

The current state of AI development remains significantly skewed toward high-resource languages, leaving thousands of languages at the margins of technological relevance (Zhong et al., 2024). To bridge this gap, an inclusive AI-language framework must prioritise equity, empowerment, and shared benefit. This requires not only technical innovation but also a shift in ownership, policy design, and collaborative practice. Inclusivity in language AI must be rooted in the values of accessibility, cultural sensitivity, community sovereignty, and global linguistic justice.

6.1 Community-Driven NLP Development

A cornerstone of inclusive AI-language development is the meaningful involvement of language communities in every stage of the AI pipeline, from data collection and annotation to evaluation and deployment. Community-driven NLP ensures that technologies are built with and for the people who speak the languages in question, rather than being externally imposed by corporations or academic institutions (Shi et al., 2024).

Projects such as Masakhane exemplify this approach by enabling African researchers, 1linguists, and technologists to co-create NLP tools for their own languages.

Masakhane operates on decentralised collaboration and peer mentoring, allowing contributors to shape research questions, gather culturally relevant corpora, and publish results that directly serve their communities (Nekoto et al., 2020). Similarly, Te Hiku Media in Aotearoa (New Zealand) empowers Māori speakers to control how their speech and text data is used, while promoting language revitalisation through AI[5].

Key principles of community-led NLP include:

1. Participatory Data Governance and Informed Consent

In community-led NLP, the people whose language data is being used are not treated as passive sources but as active decision-makers. This means communities have a say in how their data is collected, stored, and shared, ensuring it aligns with their values and privacy norms. Informed consent goes beyond signing a form; it involves explaining clearly how the data will be used, what risks may exist, and how benefits (such as tools or datasets) will be returned to the community. This principle helps counter exploitative "data scraping" approaches that often ignore the rights of speakers of marginalised languages.

2. Cultural Contextualization in Task Selection and Model Interpretation

Many NLP tasks, like sentiment analysis or part-of-speech tagging, were originally designed for high-resource, Western languages. Community-led NLP adapts or redefines tasks to reflect local communication practices and cultural priorities. For example, politeness markers, oral storytelling, or proverbs may carry more weight in African or Indigenous languages than in English. Similarly, interpreting a model's outputs requires cultural insight; what seems like a "neutral" sentence to an outsider may carry strong connotations locally. By embedding cultural knowledge, models become not only more accurate but also more authentic to the lived experience of speakers.

3. Capacity Building for Local Researchers

Community-led NLP emphasises equitable participation by ensuring that local scholars and practitioners are not just data annotators but also co-authors and leaders. This involves providing access to training (e.g., workshops on NLP techniques), resources (computational tools, GPUs, datasets), and pathways to visibility (conference participation, joint publications). Such investment empowers communities to sustain their own NLP projects instead of relying indefinitely on outside experts.

In the long term, this fosters research sovereignty, where communities take charge of the future of their language technologies.

4. Feedback Loops with Native Speakers

Models trained on low-resource or endangered languages often make mistakes due to limited data. Community-led approaches implement iterative feedback systems, where native speakers validate translations, classifications, or generated text, and then refine the outputs. This creates a cycle of continuous improvement while also instilling trust; users see that their expertise is valued and their corrections matter. Beyond technical gains, these loops promote a sense of linguistic pride and ownership.

When these principles come together, NLP becomes more than just a technical exercise; it becomes a means of cultural preservation and empowerment. Bottom-up participation ensures that AI systems are accurate, trusted, and sustainable. Moreover, it challenges the imbalance of power where only high-resource languages dominate the digital space, thereby contributing to linguistic justice and diversity in AI.

6.2 Policy and Funding Mechanisms

While grassroots efforts are vital, large-scale impact requires structural support in the form of policy frameworks and targeted funding mechanisms. Governments, intergovernmental organisations, and philanthropic entities must recognise linguistic inclusion as a digital rights issue and a cornerstone of sustainable development.

National and regional governments play a critical role in:

Recognising low-resource languages officially in education and media

Funding digitisation and corpus development efforts

Supporting bilingual and multilingual NLP research hubs

For example, India's Ministry of Electronics and IT has funded projects under AI4Bharat, resulting in NLP tools for over 20 Indian languages (Kakwani et al., 2020). Similarly, South Africa's Centre for Artificial Intelligence Research (CAIR) supports the development of language technologies for South African languages (Butcher et al., 2021; Mlambo & Matfunjwa, 2024).

To be effective, funding mechanisms must:

1. Prioritise community-led proposals

Funding should empower the very communities whose languages are underrepresented. This means placing local researchers, linguists, and cultural stakeholders at the centre of project leadership, rather than positioning them as peripheral data annotators. Prioritising such proposals ensures that research agendas emerge from community needs and aspirations, not only from external academic or commercial interests.

2. Reduce bureaucratic barriers for grassroots applicants

Grassroots organisations and community researchers often face administrative hurdles when applying for grants, such as complex documentation requirements or high entry thresholds. Simplifying application processes, providing multilingual guidance, and offering support structures (like mentorship in proposal writing) can level the playing field, allowing smaller but impactful initiatives to secure funding.

3. Encourage long-term infrastructure building rather than short-term prototypes

Many grants focus on quick deliverables or one-off models that cannot be sustained. To achieve lasting impact, funding should support infrastructure development, including open datasets, digital archives, training pipelines, and community-managed repositories. Long-term investment shifts the focus from "demo systems" to sustainable ecosystems that nurture ongoing innovation, adaptability, and ownership.

4. Include evaluation criteria for inclusion, ethics, and sustainability

Traditional grant evaluations often prioritise technical novelty or citation impact. For LRL projects, funders must also consider whether proposals promote inclusivity, respect ethical norms, and build sustainable outcomes. This means asking: Does the project avoid data exploitation? Does it create opportunities for community participation? Does it leave behind resources that others can build on? Embedding these criteria ensures that funding not only advances technical progress but also aligns with social justice and linguistic equity.

Without such equitable policies and investments, low-resource languages will continue to be underserved, even when the technical feasibility of building AI tools for them already exists. A restructured funding model that values community leadership, accessibility, long-term vision, and ethical responsibility is essential to closing the digital divide and ensuring that AI serves all languages, not just the dominant ones.

6.3 Open-Source Initiatives and Participatory Design

Open-source software and participatory design methodologies provide powerful levers for democratizing AI development and bridging the resource gap for LRLs.

a. Open-Source Ecosystems

Platforms like Hugging Face, AllenNLP, spaCy, and Fairseq have enabled public access to model architectures, training scripts, and multilingual pretrained weights. The release of BLOOM (Scao et al., 2022), a multilingual large language model trained in over 46 natural languages, was driven by an open science consortium and explicitly included many low-resource languages, particularly from Africa and Asia.

Similarly, Common Voice by Mozilla offers crowdsourced speech data from over 100 languages, including Slovenian, Irish, Chuvash, Kabyle, Catalan, Welsh, Breton, Tatar and many others (Ardila et al., 2020). This democratisation of resources enables small teams, researchers, and community members to build speech recognition and synthesis systems even in the absence of corporate backing.

b. Participatory Design

Participatory design emphasises inclusive co-creation and iterative development based on user feedback. In language technology, this involves:

- Co-designing interfaces and use cases with native speakers

Testing tools in real-world, culturally relevant scenarios (e.g., community radio, oral history apps, or local health campaigns)

- Allowing users to report errors and influence updates

One key example is MakondoLab's work with Indigenous communities in Latin America, where NLP tools are designed for language revitalisation and storytelling, not just translation or classification (Mager et al., 2018). Importantly, open-source and participatory efforts must guard against extractive research, where data is collected without benefit-sharing. Licensing frameworks like the Kaitiakitanga License or Creative Commons with community restrictions can help ensure that open data remains community-governed. Creating a truly inclusive AI-language ecosystem demands more than technical solutions. It calls for a transformative shift, where LRLs are not treated as peripheral challenges but as central to the future of equitable AI. Through community-led development, inclusive policy, and open, participatory infrastructure, we can ensure that every language, regardless of digital footprint, has the opportunity to thrive in the AI era.

7. LINGUISTIC DIVERSITY AND THE DIGITAL DIVIDE

The digital era has transformed how languages are used, preserved, and developed. However, the benefits of this transformation have not been equitably distributed. Global linguistic diversity is vast, yet access to digital tools and resources is highly uneven. This disparity, often referred to as the linguistic digital divide, shapes which languages flourish in AI systems and which remain marginalised. Understanding its roots, metrics, and consequences is crucial to designing equitable AI solutions.

7.1 Global Linguistic Diversity: A Rich Yet Uneven Landscape

Humanity speaks an estimated 7,000 languages, yet their presence in digital environments is far from balanced (Calude & Pagel, 2011; Pratap et al., 2024). A small number of high-resource languages, such as English, French, Italian, German, Finnish and Arabic, dominate the internet, digital media, and AI training datasets (Bonab et al., 2019). By contrast, thousands of other languages exist primarily in oral form, lack standardised orthographies, or have minimal online content. While multilingualism is a global norm, the digital space tends to reflect a linguistic monoculture driven by economic power, technological capacity, and historical dominance. This unevenness has profound implications for cultural preservation, education, and equitable access to AI technologies.

7.2 Metrics for Resource Classification or How Languages Are Classified by Resource Level in NLP

In AI contexts, languages are often considered high-resource, low-resource, or ultra-low-resource based on several measurable criteria. Studies have shown that these criteria meaningfully affect model performance, inclusion in benchmarks, and tool availability.

1. Size and Quality of Digital Corpora

One of the key dimensions is how much data (both labelled and unlabelled) exists for a language, and how clean, representative, and well-formatted that data is. Joshi et al. (2020) show that even languages with many speakers (e.g. Igbo, Burmese) can still be low-resource if their digital presence and representation in research are small, leading to low mean reciprocal rank (MRR) scores (P. Joshi et al., 2020).

2. Availability of NLP Tools

This includes whether tokenisers, embeddings, pretrained models, part-of-speech taggers, etc., are available for a language. Truly low-resource or ultra-low-resource languages often lack these tools, which limits what NLP tasks can be performed or how well they can be done. For example, weak supervision methods (e.g. POS tagging using dictionaries or transfer from other languages) are shown to perform poorly when core tools (high-coverage dictionaries, language resources) are missing (Kann et al., 2020). For example, Hausa has tens of millions of speakers but remains underrepresented in multilingual NLP models due to a lack of digitised corpora and uneven orthographic standards (Blachman et al., 2022).

3. Presence in Multilingual Benchmarks (e.g. XTREME, XTREME-R, FLORES)

Benchmarks are a strong indicator of how "resourced" a language is in practice. Ruder et al. (2021) in XTREME-R extended earlier multilingual benchmarks to include more typologically diverse languages and showed that many languages still lag behind in benchmark inclusion. Languages not included in (or poorly represented in) major benchmarks are often in the low- or ultra-low category.

4. Amount of Web Content / Indexed Digital Text

The volume of text for a language on the internet (web pages, Wikipedia articles, news sites, etc.) correlates strongly with how many NLP resources exist for it. If web content is minimal or poorly indexed, this restricts both unsupervised data for pre-training and supervised data for specific tasks. Joshi et al. (2020) report that despite millions of speakers, some languages have very limited web presence, contributing to their classification as low-resource (P. Joshi et al., 2020).

5. Speech Data Availability for ASR (Automatic Speech Recognition) and TTS (Text-to-Speech)

Many high-resource languages benefit from large, annotated speech corpora, which is often missing for low-resource languages. Benchmarks like XTREME-UP explicitly include tasks for ASR, Optical Character Recognition (OCR), and OCR plus speech to test model performance in "scarce-data" settings for underrepresented languages (Ruder et al., 2023).

Why These Criteria Matter in NLP

Performance gaps: Studies (e.g. weakly supervised POS taggers) demonstrate that for truly low-resource languages, even the best methods perform poorly without basic resources (Kann et al., 2020).

Benchmark bias: Benchmarks often select languages based on available data and infrastructure, reinforcing disparity. Languages with fewer tools or less pretraining data get excluded. (Joshi et al., Ruder et al., others).

Model generalisation: When pretraining or evaluation ignores speech data or lacks in-language fine-tuning, generalisation to real-world, diverse language settings suffers. Some benchmarks (XTREME, XTREME-UP) attempt to address this by including multiple modalities and data-scarce tasks (Ruder et al., 2023).

7.3 Colonial Histories and Language Hierarchies

The roots of the linguistic digital divide are deeply entwined with colonial histories. Colonial administrations often imposed a dominant language, such as English, French, or Spanish, at the expense of Indigenous languages (Pennycook, 2010). This created entrenched language hierarchies that persist today, influencing education systems, media policies, and resource allocation. Colonial languages typically became the default mediums for technology development, while Indigenous languages were relegated to informal or domestic spheres. This historical imbalance is mirrored in AI: dominant languages enjoy early adoption in large language models (LLMs), while many others remain absent from AI datasets and benchmarks.

7.4 Language Politics and Policy Gaps

National and regional language policies can either reinforce or mitigate the digital divide. In some countries, official language recognition does not guarantee digital inclusion, as seen with Māori in New Zealand and many African languages (Ogueji et al., 2021). Policies may focus on preservation without allocating resources for technological integration. In other contexts, language politics exacerbate inequalities when governments prioritise a small set of languages for e-governance, education, and public services, sidelining minority or Indigenous tongues. The absence of clear digital language strategies leads to ad hoc and fragmented initiatives that fail to address structural disparities.

7.5 Socioeconomic Consequences of the Linguistic Digital Divide

1. Economic marginalisation

Communities whose languages are underrepresented or excluded from digital technologies often miss out on opportunities in e-commerce, remote work, and digital entrepreneurship (Alabdali et al., 2023). For example, the NYU Pathfinders report "Digital Equity as an Enabling Platform for Equality and Economic Opportunity" stresses that low-income workers in underserved regions are frequently excluded from digital economies, limiting their employment and business growth options (Bailey & Nyabola, 2021). Similarly, "The AI Gap: How Socioeconomic Status Affects Language Technology Interactions" (Bassignana et al., 2025) documents how users from lower socioeconomic backgrounds have less frequent access to, and derive fewer benefits from, language technologies, which restricts their ability to engage in digital business or remote work effectively.

2. Educational inequality

When AI-driven educational tools (such as adaptive learning platforms, online tutoring, or reading assistants) are unavailable or under-trained for native and minority languages, literacy, and skill development suffer. Students from disadvantaged socioeconomic backgrounds are less likely to have access to effective digital tools in their home or school language, leading to persistent attainment gaps (Raihan et al., 2025). Another study, "Influence of Socioeconomic Factors on Access to Digital Resources for Education", shows that lack of internet access and digital literacy skills correlates strongly with lower academic performance in digital learning environments among students in lower-income households (B. M. Joshi et al., 2024).

3. Healthcare disparities

Language matters for health information access. Digital health resources, online guidance, telemedicine platforms, and chatbots are often offered only in more dominant languages. The "Digital Health Technologies and Inequalities: A Scoping Review" (Badr et al., 2024) shows that communities speaking less-resourced languages get less benefit from digital health tools, which can reduce timely access to critical medical information and services. (Badr et al., 2024). Similarly, "A Framework for Digital Health Equity" (Richardson et al., 2022) emphasises how having health content translated into local languages, or the absence thereof, affects uptake of services and patient understanding of treatment, thereby worsening health outcomes in linguistically marginalised populations.

4. Civic exclusion

Language barriers make it difficult or impossible for speakers of some languages to engage with digital government or political platforms (e-government, public consultations, legal documents). Lack of access or understanding excludes them from participating fully in civic life. For instance, the NYU Pathfinders report notes that for marginalised groups, digital platforms are often less accessible or less usable, not just due to a lack of translation but because design does not account for linguistic diversity (Bailey & Nyabola, 2021). Also, broader studies of digital inequality (such as "Bridging the Digital Divides: a Literature Review" by Vassilakopoulou et al., 2021) find that disparities in access, usage, and control of ICTs often align with socioeconomic stratification, which includes civic participation metrics.

These impacts, economic, educational, health, and civic, form reinforcing cycles of disadvantage. Without targeted policy, investment, and inclusive design, low- and ultra-low-resource language communities remain underserved even when technical capability exists. Thus, linguistic digital equity is not just a technical challenge; it is a core issue of socioeconomic justice.

7.6 Inclusion as a Digital Right

Inclusion in the digital sphere is increasingly recognised as a human right. The United Nations has emphasised the principle of linguistic diversity in cyberspace, arguing that digital participation should not be constrained by language barriers (UN, 2022). This aligns with the concept of digital language rights, which frames language inclusion in AI systems as essential to cultural survival, self-determination, and equitable access to opportunities. From this perspective, the exclusion of low-resource languages from AI development is not merely a technical issue; it is a violation of fundamental rights to participation in the global digital commons.

7.7 Toward Measuring and Closing the Gap

Closing the Linguistic Digital Divide Requires a Multi-Pronged Strategy. To effectively reduce disparities in how languages are represented and served in AI/NLP systems, several coordinated actions are needed:

1. Comprehensive Metrics that Combine Technical Resource Levels with Sociolinguistic Vitality Indices.

It's not enough to measure only how much data a language has; we also need to track its sociolinguistic health, how many speakers, whether children are learning it, how used it is in daily life, its presence in education, etc. Studies such as those assessing language vitality in indigenous or minoritised languages (e.g. Nahuatl, Bisaya) combine measures like absolute speaker population, inter-generational transmission, and domains of use to determine risk or vitality (Dani & Seruji, 2024; Gomashie & Terborg, 2021). Instruments like UNESCO's Language Vitality and Endangerment scale, EGIDS, EuLaViBar, etc., are often used to this end (Drude et al., 2018; Karjalainen et al., 2013).

2. Community-Driven Data Creation to Ensure Cultural Authenticity and Consent.

Data for AI should be collected with the active participation of the language community, ensuring both that data reflects the linguistic and cultural norms of speakers and that they consent to how their data will be used. For example, "Toward Community-Based NLP" describes participatory research where community members help define tasks, collect data, and evaluate outputs (Pillai et al., 2023). This helps avoid misrepresentations, culturally insensitive mistakes, and exploitation.

3. Policy Integration, where Governments Embed Linguistic Inclusion into National AI Strategies

Governments need formal policies that require or encourage inclusion of languages beyond just the major or official ones ("Leveraging Artificial Intelligence to Foster Tolerance and Coexistence in Multicultural Societies: A Framework for Inclusive Policy and Social Integration," 2025). This can include mandates or incentives for datasets, models, or digital services to support multiple languages, especially under-represented ones. While explicit examples are fewer, some countries are beginning to include language preservation and digital inclusion in their national strategies.

4. Targeted Funding for Ultra-Low-Resource Languages, Especially Those at Risk of Extinction.

Languages with very few speakers, or that are losing speakers rapidly, often lack even basic digital resources. These need special attention: grants, infrastructure, tools, and projects designed specifically to help them not only catch up but preserve linguistic heritage. The work on the vitality of Nahuatl (Gomashie & Terborg, 2021) and Bisaya (Dani & Seruji, 2024) highlight how vulnerable languages show markers of endangerment and thus illustrate urgency.

5. Open-Source Infrastructure to Reduce Barriers for Underrepresented Language Developers.

Infrastructure includes not only data (corpora), but tools (tokenisers, pre-trained models, speech corpora, annotation tools) that are openly available. Open source lowers cost, increases transparency, and allows local communities to adapt, improve, and extend models. A number of community-based NLP initiatives advocate for this. Pillai et al. (2023) describe how open methods, tools, and community transparency contribute to more equitable NLP development.

To measure progress in closing the linguistic digital divide, researchers, funders, and policymakers should regularly review which languages are represented in AI models and datasets. They also need to assess how well these languages are supported, including the availability of tools for text, speech, and translation. Equally important is ensuring that dataset composition is made transparent in research publications and product claims. This means clearly stating which languages are included, how much data is used, and the quality of that data so that progress can be tracked and accountability maintained.

8. NATURAL LANGUAGE PROCESSING FOR LOW-RESOURCE LANGUAGES

8.1 Introduction

Natural Language Processing (NLP) sits at the heart of human-AI interaction. It encompasses a broad set of computational techniques that allow machines to analyse, understand, and generate human language (Chowdhury, 2003). The field has witnessed dramatic advances in recent years, largely propelled by neural networks, large-scale pretraining, and massive multilingual datasets. Despite this progress, the benefits have not been evenly distributed. Languages with abundant digital resources, referred to as high-resource languages (HRLs), have been the primary beneficiaries of NLP advancements (Maurya & Desarkar, 2023). In contrast, low-resource languages (LRLs) remain underrepresented, and often invisible, in both research and application domains.

This disparity is not a result of linguistic inferiority but rather reflects historical, socio-political, and infrastructural inequalities. LRLs often lack digitised corpora, annotated datasets, standardised orthographies, and consistent linguistic tools, elements critical for building robust NLP models (Pakray et al., 2025). Furthermore, their exclusion has downstream effects on access to AI-powered services such as machine translation, chatbots, education platforms, and e-governance. As NLP becomes an indispensable tool in modern life, addressing these gaps is not just a technical challenge; it is a moral imperative. This section explores the current state of NLP for LRLs, reviewing key tasks, existing barriers, and innovative techniques aimed at bridging this divide. Through a combination of technical detail and real-world examples, we aim to provide a roadmap for researchers, developers, and policymakers committed to inclusive language technology.

8.2 Overview of NLP Tasks Relevant to LRLs

To understand how Low-Resource Languages (LRLs) can benefit from NLP, it is essential to examine the typical tasks within the NLP pipeline. These foundational tasks enable higher-level applications, but each depends heavily on linguistic data (often human-annotated); deficiencies in any of these tasks significantly hamper development for LRLs. Below are several critical tasks and how they pose challenges for LRLs, with recent work illustrating these difficulties and potential strategies.

8.2.1 Tokenisation and Morphological Analysis

Tokenisation is the process of breaking text into smaller units (such as words, subwords, or morphemes). Languages without clear word boundaries (e.g., Thai, Lao) or with rich morphological structure (e.g., Turkish, Amharic) expose weaknesses in standard tokenisers. For instance, recent work on Turkish has developed evaluation metrics tailored for morphologically rich, low-resource languages, showing that vocabulary-specific token percentages correlate strongly with downstream performance more than pure token-purity metrics (Bayram et al., 2025). Similarly, in the context of Indic languages, Pattnayak et al. (2025) compare tokenisation strategies (Byte-Pair Encoding, SentencePiece, character-level) for NER tasks in low-resource Indic languages (e.g., Santali, Manipuri, Sindhi) and demonstrate that SentencePiece better preserves morphological integrity and entity consistency than BPE in zero-shot settings.

8.2.2 Part-of-Speech (POS) Tagging

POS tagging assigns grammatical categories (noun, verb, adjective, etc.) to tokens. For LRLs, this task suffers from a scarcity of annotated corpora and tagsets that capture the diversity of grammar across languages. Eskander, Muresan, and

Collins (2020) show that in truly low-resource scenarios, even weakly supervised POS taggers achieve well under 50% accuracy for many languages, highlighting how insufficient cross-lingual transfer or limited dictionary coverage sharply reduces performance (Eskander et al., 2020). These findings indicate that robust POS tags require both annotated data and typology-aware design.

8.2.3 Named Entity Recognition (NER)

NER identifies names of persons, places, organisations, etc. Cultural and orthographic differences complicate NER in LRLs. In low-resource Indic languages, entity boundaries are ambiguous, and naming conventions may diverge from those in high-resource languages. The study by Pattnayak, Patel, and Agarwal (2025) also demonstrates that in languages like Santali, Manipuri, and Sindhi, the tokenisation strategy (character-level vs. subword) strongly influences whether pre-trained models correctly identify entity spans in zero-shot transfer.

8.2.4 Dependency Parsing

Dependency parsing involves mapping the syntactic structure of sentences by identifying dependencies among words (e.g., subject-verb, verb-object relationships). For many LRLs, treebanks (annotated syntactic datasets) are extremely limited. For example, recent work on Amharic using an expanded and cleaned UD (Universal Dependencies) treebank (Seyoum et al., 2019) has shown that while performance improves when datasets are cleaned and expanded, many morpho-syntactic features (captured by MLAS, BLEX metrics) remain weak, especially when data is scarce or noisy.

8.2.5 Machine Translation (MT)

MT, especially neural machine translation (NMT), is perhaps the most visible high-impact NLP task for LRLs. However, NMT's dependence on large bilingual or parallel corpora means that many LRLs are underserved or produce low-quality translations. Where corpora exist, alignments may be poor or domain-mismatched. For instance, in the cross-lingual transfer work by Rahman et al. (2023), models with segmentation-based representations outperform image-based or character-level models for translation tasks when the source and target share script or morphology, but overall quality remains much lower for truly low-resource pairs than for high-resource ones.

8.3 Techniques for Addressing Data Scarcity

The lack of annotated and unannotated data is the defining challenge for NLP in Low-Resource Languages (LRLs). Overcoming this bottleneck requires a diverse toolkit of strategies aimed at maximising the utility of existing data and minimising the need for large corpora. Below are key techniques that have shown promise, along with recent empirical evidence.

8.3.1 Transfer Learning and Multilingual Pretraining

Transfer learning enables models pretrained on resource-rich languages to be adapted to LRLs. Models like Multilingual BERT (mBERT) and XLM-RoBERTa (XLM-R) are foundational in this domain. For example, From Zero to Hero: On the Limitations of Zero-Shot Cross-Lingual Transfer with Multilingual Transformers (Lauscher et al., 2020) shows that mBERT and XLM-R perform well when transferring to languages that are either typologically close to high-resource source languages or have some pretraining data, but their performance drops significantly for typologically distant LRLs or when target-language pretraining data is minimal. Transfer quality is also affected by tokenisation granularity and the degree of linguistic similarity between source and target (Bankula & Bankula, 2025; Xu et al., 2022).

8.3.2 Cross-Lingual Annotation Projection

Cross-lingual annotation projection is a technique that uses parallel or aligned corpora to map annotations (such as POS tags, Named Entity Recognition (NER) labels, or dependency relations) from high-resource languages (s) (HRLs) to low-resource languages (LRLs) (Huck et al., 2019). The process involves annotating text in the HRL where reliable taggers exist, aligning sentences (at word or character level) with their translations in the LRL, and transferring the annotations via alignment tools. Despite introducing noise (due to alignment errors, translation mismatches, or structural divergences), projection enables the creation of weakly-supervised training data at low cost, bootstrapping supervised models for many fundamental NLP tasks.

Recent work confirms the continued effectiveness of annotation projection for LRLs. For example, Revisiting Projection-based Data Transfer for Cross-Lingual Named Entity Recognition in Low-Resource Languages (Politov et al., 2025) shows that refining word alignments via back-translation and using formalised matching strategies for source entity projection significantly improves NER accuracy over standard projection baselines across 57 languages. Similarly, Low-Resource Named Entity Recognition with Cross-Lingual Annotation Projection (Cotterell & Duh, 2024) demonstrates that character-level information combined with projection from supported HRLs yields substantial gains in low-resource NER tasks. Another example, Zero Resource Cross-Lingual Part of Speech Tagging (Chopra, 2024), highlights that even hidden Markov models trained on projections with minimal supervision can surpass naive baselines in POS tagging for truly zero-resource settings.

8.3.3 Data Augmentation

Data augmentation creates additional training examples from limited datasets, helping models generalise better. One influential method is back-translation, translating from target to source and then back again to create paraphrases. For example, Fadaee, Bisazza, & Monz (2017) use data augmentation targeting rare words in low-resource neural machine translation; they show improvements of up to \~2.9 BLEU points over baseline and \~3.2 BLEU over naive back-translation settings (Fadaee et al., 2017). Also, recent work on African languages (MaFAND dataset) shows that combining back-translation with techniques like "switchout" and sentence concatenation improves translation robustness and generalisation for several low-resource language pairs (Adelani et al., 2021).

8.3.4 Unsupervised and Self-Supervised Learning

Unsupervised / self-supervised learning enables models to leverage unlabelled corpora via masked language modelling, next sentence prediction, or similar tasks. These methods reduce dependence on human annotations. For example, models like mBERT, XLM-R, and mT5 have shown that when unlabelled data for an LRL is available, even in smaller amounts, self-supervised pretraining can significantly improve downstream tasks. Cross-Linguistic Transfer in Multilingual NLP: The Role of Language Families and Morphology (Bankula & Bankula, 2025) investigates how morphology and language family distance influence performance in zero-shot transfer, indicating that these methods are more effective when some unlabelled text exists for the target language.

8.4 Challenges in NLP for LRLs

Technical innovation alone is not sufficient. Several structural and linguistic challenges continue to hinder NLP performance and accessibility for LRLs:

Script and Orthographic Diversity: Some languages lack a standardised script, use multiple writing systems, or have inconsistent spelling conventions. This complicates data collection and model training.

Morphological Richness: In highly inflected languages like Inuktitut or Amharic, the same root can generate thousands of word forms. Word-based models often struggle due to vocabulary sparsity.

Code-Switching and Multilingualism: LRL speakers often mix languages within a single conversation or text. Traditional monolingual models fail to handle this linguistic reality.

Bias and Representation: Pretrained models can amplify existing biases, misrepresenting or marginalising underrepresented communities. For instance, cultural terms may be ignored, misclassified, or replaced with irrelevant high-resource equivalents (Bender et al., 2021).

Limited Community Involvement: Many NLP projects are designed without consultation or participation from native speakers, resulting in tools that are linguistically inaccurate or culturally inappropriate.

9. FUTURE RESEARCH DIRECTIONS

Future work on LRLs in AI should expand across several dimensions:

1. Inclusive Benchmarks and Evaluation

Progress in multilingual NLP demands benchmarks that go beyond mere accuracy to also capture fairness, bias, and cultural appropriateness. For example, LinguaSafe (Ning et al., 2025) is a multilingual safety benchmark that spans 12 languages, including low-resource ones, and measures indirect safety, oversensitivity, and severity of unsafe content, revealing that models perform worse in underrepresented languages. Another recent study, The Bitter Lesson Learned from 2,000+ Multilingual Benchmarks (Wu et al., 2025), shows English is disproportionately represented in benchmarks and that translated benchmarks often misalign with human judgment in local contexts. Such work underscores the need for benchmarks designed with LRLs in mind from the start.

2. Hybrid Methodologies

Combining transfer learning, cross-lingual embeddings, and community-driven data creation is showing stronger results than single-approach methods. In Low-Resource Neural Machine Translation: A Benchmark for Five African Languages (Lakew et al., 2020), multilingual NMT coupled with transfer learning produced gains of up to +5 BLEU points in several African LRLs. Similarly, Language Fairness in Multilingual Information Retrieval (Yang et al., 2024) introduces a fairness metric (PEER) and shows that synthetic code-switching plus fine-tuning improves performance for LRLs without harming high-resource ones. These hybrid methods, especially when paired with community collaboration, offer a path forward for more robust, inclusive LRL systems.

3. Speech and Multimodal AI

Many LRLs are oral or have strong oral traditions with limited written text. Thus, advancing ASR, TTS, and multimodal (audio+visual+text) capabilities is essential. Papakyriakopoulos and Xiang (2023) discuss ethical considerations for speech datasets, including diversity in accents and metadata that supports inclusion (Papakyriakopoulos & Xiang, 2023). Such work emphasises that speech dataset design must account for demographic diversity and cultural norms to avoid bias. Also, nascent multimodal benchmarks are beginning to evaluate LLMs on safety and question-answering tasks with images or audio inputs in LRLs, revealing substantial gaps in performance.

4. Ethical Data Governance

Data sovereignty and ethical control become ever more important, especially when communities are the sources of linguistic data. Calac & Mackey (2025) highlight that in digital health and Indigenous community research, ethical frameworks around data ownership, consent, and governance significantly affect community trust and outcomes (Calac & Mackey, 2025). Urban American Indian and Alaska Native communities have also reported harms when data governance is not clearly addressed (Haozous et al., 2021). Institutionalising these frameworks (e.g., data use agreements, community advisory boards) ensures that LRLs are not simply sources of data but collaborators and decision-makers.

5. Policy and Funding Integration

Without structural support, many LRLs will continue to lag despite technical methods being available. Effective policies embed linguistic inclusion into national AI strategies, language education, and digital infrastructure. Recent evaluative work for Arabic cross-lingual NER (Qu et al., 2024) demonstrates that language-adaptive pretraining (LAPT) and funding toward language-specific preprocessing significantly boost performance when policy and institutional support exist (Qu et al., 2024). Furthermore, fairness benchmarking (e.g., FairLex) shows that differences in jurisdictional policies, legal mandates, or national attention result in major disparities across languages in model performance. Thus, targeted funding for ultra-low-resource and endangered languages, backed by national and international policy, is critical.

10. CONCLUSION

Key Insights and Call to Action

The transformative power of artificial intelligence (AI) in language technologies is undeniable. However, as this book has consistently shown, this transformation has not been equitably distributed. Thousands of low-resource languages (LRLs), spoken by billions globally, remain largely absent from AI systems, threatening to deepen digital inequalities and linguistic marginalisation in the age of intelligent machines.

One of the key insights from this work is that the linguistic digital divide is not a purely technical issue, but a deeply socio-political and ethical concern. The underrepresentation of LRLs in AI systems is a reflection of broader structural inequities, colonial histories, lack of institutional investment, and ongoing exclusion from the global digital economy. Addressing this divide requires a paradigm shift in how AI is developed, deployed, and governed.

Technically, there has been notable progress in the development of multilingual and zero-shot models such as mBERT, XLM-R, and BLOOM. These innovations offer pathways for LRL inclusion through transfer learning, cross-lingual embeddings, and synthetic data generation. However, these advances are insufficient if they are built on unrepresentative data, developed without community participation, or deployed without regard for local cultural and linguistic nuances.

This book has emphasised the importance of community-driven NLP, where native speakers and local researchers co-create tools that serve their linguistic and cultural realities. Through case studies on African languages like Yoruba, Igbo, Tarok, and Hausa, as well as Indigenous languages like Māori and Quechua, we have seen the challenges and promises of bringing underrepresented languages into the AI fold. These case studies highlight that language revitalisation, digital inclusion, and ethical AI development must go hand in hand.

Ethically, we must reckon with issues of data sovereignty, technological imperialism, and the risks of bias and hallucination in large language models (LLMs). Without consent, transparency, and fair benefit-sharing, AI interventions risk reproducing the very hierarchies they claim to disrupt.

Moreover, achieving true inclusion demands policy-level commitment and funding mechanisms. Governments, global institutions, and philanthropic organisations must recognise language inclusion as a form of digital justice. They must fund infrastructure for LRLs, support grassroots innovation, and create environments where linguistic diversity is not an afterthought, but a design principle.

Finally, open-source ecosystems and participatory design practices must be scaled and protected. Projects like Common Voice, Masakhane, Te Hiku Media, and AI4Bharat demonstrate that collaborative, decentralised, and community-first approaches are not only possible—they are essential.

Call to Action

To bridge the linguistic divide and ensure that no language is left behind in the AI era, we issue the following calls to action:

1. To Technologists:

Design with cultural context in mind. Collaborate directly with language communities, and prioritise transparency, modularity, and open access in your tools and datasets.

2. To Researchers and Institutions:

Invest in long-term partnerships with LRL communities. Shift from extractive to participatory research paradigms. Publish in local venues, train local scholars, and return value to the communities you study.

3. To Policymakers and Funders:

Elevate language inclusion as a national and global priority. Develop inclusive AI strategies, fund digital linguistic infrastructure, and embed ethical standards in AI procurement and deployment.

4. To Communities and Language Advocates:

Assert your linguistic rights in the digital space. Partner with technologists and scholars, build your own datasets, and demand voice, visibility, and control over your linguistic future.

5. To Global Organisations:

Support multilingualism as a public good. Incorporate language rights into broader digital inclusion agendas and facilitate cross-border cooperation for shared linguistic resources.

Final Reflection

AI presents both a threat and an opportunity for the future of the world's languages. If deployed without care, it will widen the gap in linguistic diversity. But if developed inclusively, ethically, and collaboratively, it can become a powerful ally in preserving, revitalising, and amplifying the voices of even the most marginalised language communities. The choice is ours. Let us ensure that the next era of AI is not only intelligent but also inclusive, equitable, and multilingual.

ACKNOWLEDGEMENT

AI tools were used to improve grammar, clarity, and style.

REFERENCES

Abid, A., Farooqi, M., & Zou, J. (2021). Persistent Anti-Muslim Bias in Large Language Models. *AIES 2021 - Proceedings of the 2021 AAAI/ACM Conference on AI, Ethics, and Society*, 298–306. DOI: 10.1145/3461702.3462624

Adam, I. O., & Dzang Alhassan, M. (2020). Bridging the global digital divide through digital inclusion: the role of ICT access and ICT use. *Transforming Government: People. Process and Policy*, *15*(4), 580–596. DOI: 10.1108/TG-06-2020-0114

Adelani, D. I., Abbott, J., Neubig, G., Daniel, D., Kreutzer, J., Lignos, C., Palen-michel, C., Buzaaba, H., Rijhwani, S., Ruder, S., Mayhew, S., & Azime, I. A. (2021). MasakhaNER : Named Entity Recognition for African Languages. *Transactions of the Association for Computational Linguistics*, *9*, 1116–1131.

Adeyeye, G. M. (2024). Social Isolation: Overcoming Linguistic Obstacles and Mitigating Social Isolation in Diverse Communities. *E-Journal of Humanities. Arts and Social Sciences*, *5*(11), 1970–1982. DOI: 10.38159/ehass.202451127

Alabdali, S. A., Pileggi, S. F., & Cetindamar, D. (2023). Influential Factors, Enablers, and Barriers to Adopting Smart Technology in Rural Regions: A Literature Review. *Sustainability (Switzerland)*, *15*(10), 1–38. DOI: 10.3390/su15107908

Alam Monisha, S. T., & Sultana, S. (2022). A Review of the Advancement in Speech Emotion Recognition for Indo-Aryan and Dravidian Languages. *Advances in Human-Computer Interaction, 2022*. Advance online publication. DOI: 10.1155/2022/9602429

AnderBois, S. (2019). Reportatives and Quotatives in Mayan Languages. *Proceedings of Form and Analysis in Mayan Linguistics, 5*(FAMLi 5), 1–16.

Ardila, R., Branson, M., Davis, K., Henretty, M., Kohler, M., Meyer, J., Morais, R., Saunders, L., Tyers, F. M., & Weber, G. (2020). Common voice: A massively-multilingual speech corpus. *LREC 2020 - 12th International Conference on Language Resources and Evaluation, Conference Proceedings*, 4218–4222.

Artetxe, M., Labaka, G., & Agirre, E. (2020). Translation artifacts in cross-lingual transfer learning. *EMNLP 2020 - 2020 Conference on Empirical Methods in Natural Language Processing, Proceedings of the Conference*, 7674–7684. DOI: 10.18653/v1/2020.emnlp-main.618

Aydin, M. (2021). Does the digital divide matter? Factors and conditions that promote ICT literacy. *Telematics and Informatics, 58*(November 2020), 101536. DOI: 10.1016/j.tele.2020.101536

Badr, J., Motulsky, A., & Denis, J. L. (2024). Digital health technologies and inequalities: A scoping review of potential impacts and policy recommendations. *Health Policy, 146*(October 2023), 105122. DOI: 10.1016/j.healthpol.2024.105122

Bailey, L. E., & Nyabola, N. (2021). Digital Equity as an Enabling Platform for Equality and Inclusion. In *Pathfinders for Peaceful, Just and Inclusive Societies* (Issue June). https://cic.nyu.edu/wp-content/uploads/1662/65/cic_pathfinders_digital_equity_as_an_enabling_platform_june_2021.pdf

Bankula, A., & Bankula, P. (2025). Cross-Linguistic Transfer in Multilingual NLP: The Role of Language Families and Morphology. *ArXiv Preprint ArXiv:2505.13908*. https://arxiv.org/pdf/2505.13908

Bassignana, E., Curry, A. C., & Hovy, D. (2025). The AI Gap: How Socioeconomic Status Affects Language Technology Interactions. *ArXiv Preprint ArXiv:2505.12158, February*, 18647–18664. DOI: 10.18653/v1/2025.acl-long.914

Bayram, M. A., Fincan, A. A., Gümüş, A. S., Karakaş, S., Diri, B., & Yıldırım, S. (2025). Tokenization Standards for Linguistic Integrity: Turkish as a Benchmark. *ArXiv Preprint ArXiv:2502.07057*, 1–10. https://arxiv.org/pdf/2502.07057

Bird, S. (2020). Decolonising Speech and Language Technology. *COLING 2020 - 28th International Conference on Computational Linguistics, Proceedings of the Conference*, 3504–3519. DOI: 10.18653/v1/2020.coling-main.313

Bird, S. (2022). Local Languages, Third Spaces, and other High-Resource Scenarios. *Proceedings of the Annual Meeting of the Association for Computational Linguistics, 1*, 7817–7829. DOI: 10.18653/v1/2022.acl-long.539

Blench, R. (2020). Research on the Plateau languages of Central Nigeria. *Hamburg University Press, 93*, 1–5. DOI: 10.15460/auue.2021.93.1.209

Blench, R. M. (2017). Origins of ethnolinguistic identity in southeast asia. In *Handbook of East and Southeast Asian Archaeology* (Issue 0). DOI: 10.1007/978-1-4939-6521-2_42

Bonab, H., Allan, J., & Sitaraman, R. (2019). Simulating CLIR translation resource scarcity using high-resource languages. *ICTIR 2019 - Proceedings of the 2019 ACM SIGIR International Conference on Theory of Information Retrieval*, 129–136. DOI: 10.1145/3341981.3344236

Butcher, N., Wilson-Strydom, M., & Baijnath, M. (2021). Artificial intelligence capacity in Sub-Saharan Africa: Compendium report. In *AI4D Africa*. https://idl-bnc-idrc.dspacedirect.org/items/27ea1089-760f-4136-b637-16367161edcc

Calac, A. J., & Mackey, T. K. (2025). A systematic review of responsible stewardship of research and health data from Indigenous communities. *NPJ Digital Medicine*, *8*(1). Advance online publication. DOI: 10.1038/s41746-025-01902-w PMID: 40764752

Calude, A. S., & Pagel, M. (2011). How do we use language? shared patterns in the frequency of word use across 17 world languages. *Philosophical Transactions of the Royal Society of London. Series B, Biological Sciences*, *366*(1567), 1101–1107. DOI: 10.1098/rstb.2010.0315 PMID: 21357232

Chopra, S. (2024). Zero Resource Cross-Lingual Part Of Speech Tagging. *ArXiv Preprint ArXiv:2401.05727*. https://arxiv.org/pdf/2401.05727v1

Choudhury, P., Kim, D. Y., & Kang, S. (2026). Beyond English-Centric AI: Strategic Frameworks for Developing Low-Resource Language Models. SSRN *Electronic Journal*, 1–45.

Chowdhury, G. (2003). Natural language processing. *Annual Review of Information Science & Technology*, *37*, 51–89.

Conneau, A., Khandelwal, K., Goyal, N., Chaudhary, V., Wenzek, G., Guzmán, F., Grave, E., Ott, M., Zettlemoyer, L., & Stoyanov, V. (2019). Unsupervised cross-lingual representation learning. *ACL 2019 - 57th Annual Meeting of the Association for Computational Linguistics, Tutorial Abstracts*, 31–38. DOI: 10.18653/v1/p19-4007

Cotterell, R., & Duh, K. (2024). Low-Resource Named Entity Recognition with Cross-Lingual, Character-Level Neural Conditional Random Fields. *ArXiv Preprint ArXiv:2404.09383*. https://arxiv.org/abs/2404.09383

Coupland, N. (2015). The Handbook of Language and Globalization. In *Etika Jurnalisme Pada Koran Kuning : Sebuah Studi Mengenai Koran Lampu Hijau* (Vol. 16, Issue 2).

Court, S., & Elsner, M. (2024). Shortcomings of LLMs for Low-Resource Translation: Retrieval and Understanding are Both the Problem. *Conference on Machine Translation - Proceedings, 2024-Novem*, 1332–1354. DOI: 10.18653/v1/2024.wmt-1.125

Creissels, D. (2019). Morphology in Niger-Congo Languages. In *Oxford Research Encyclopedia of Linguistics*. DOI: 10.1093/acrefore/9780199384655.013.535

Dani, N. A., & Seruji, Z. (2024). Language Vitality Among the Young Generation of a Minority Ethnic Group in Sabah. *International Journal of Humanities, Philosophy and Language, 7*(28), 01–18. DOI: 10.35631/ijhpl.728001

Devlin, J., Chang, M.-W., Lee, K., & Toutanova, K. (2019). BERT: Pre-training of Deep Bidirectional Transformers for Language Understanding. *Conference of the North American Chapter of the Association for Computational Linguistics: Human Language Technologies, Mlm*. https://arxiv.org/abs/1810.04805

Drude, S., Ingason, A. K., Kristinsson, A. P., Arnbjörnsdóttir, B., Sigurðsson, E. F., Rögnvaldsson, E., Nowenstein, I., & Sigurjónsdóttir, S. (2018). Digital resources and language use: Expanding the EGIDS scale for language development into the digital domains. *Conference: Communities in Control - Learning Tools and Strategie for Multilingual Endangered Language Communities, August 2019*. https://www.researchgate.net/publication/335244251_Digital_resources_and_language_use_Expanding_the_EGIDS_scale_for_language_development_into_the_digital_domains

Emlen, Q. N. (2025). The Quechuan-Aymaran relationship. In *The oxford guide to the languages of the central andes* (Issue December). DOI: 10.1093/oso/9780198849926.003.0026

Eskander, R., Muresan, S., & Collins, M. (2020). Unsupervised cross-lingual part-of-speech tagging for truly low-resource scenarios. *EMNLP 2020 - 2020 Conference on Empirical Methods in Natural Language Processing, Proceedings of the Conference*, 4820–4831. DOI: 10.18653/v1/2020.emnlp-main.391

Fadaee, M., Bisazza, A., & Monz, C. (2017). Data augmentation for low-Resource neural machine translation. *ACL 2017 - 55th Annual Meeting of the Association for Computational Linguistics, Proceedings of the Conference (Long Papers), 2*, 567–573. DOI: 10.18653/v1/P17-2090

Gomashie, G. A., & Terborg, R. (2021). Nahuatl, selected vitality indicators and scales of vitality in an Indigenous language community in Mexico. *Open Linguistics, 7*(1), 166–180. DOI: 10.1515/opli-2021-0010

Good, J. (2014). *Niger-Congo languages Niger-Congo*. The Cambridge handbook of areal linguistics.

Goyal, N., Gao, C., Chaudhary, V., Chen, P. J., Wenzek, G., Ju, D., Krishnan, S., Ranzato, M., Guzmán, F., & Fan, A. (2022). The FLORES-101 Evaluation Benchmark for Low-Resource and Multilingual Machine Translation. *Transactions of the Association for Computational Linguistics, 10*, 522–538. DOI: 10.1162/tacl_a_00474

Greenhill, S. J., Drummond, A. J., & Gray, R. D. (2010). How accurate and robust are the phylogenetic estimates of austronesian language relationships? *PLoS One, 5*(3), 1–6. DOI: 10.1371/journal.pone.0009573 PMID: 20224774

Haozous, E. A., Faan, J. L., & Soto, C. (2021). Urban American Indian and Alaska Native Data Sovereignty: Ethical Issues. *American Indian and Alaska Native Mental Health Research, 28*(2), 77–99. DOI: 10.5820/AIAN.2802.2021.77 PMID: 34586627

Hasan, M. A., Tarannum, P., Dey, K., Razzak, I., & Naseem, U. (2024). Do Large Language Models Speak All Languages Equally? A Comparative Study in Low-Resource Settings. *ArXiv Preprint ArXiv:2408.02237.* https://arxiv.org/abs/2408.02237

Hovy, D., & Spruit, S. L. (2016). The social impact of natural language processing. *54th Annual Meeting of the Association for Computational Linguistics, ACL 2016 - Short Papers*, 591–598. DOI: 10.18653/v1/p16-2096

Hu, J., Ruder, S., Siddhant, A., Neubig, G., Firat, O., & Johnson, M. (2020). XTREME: A massively multilingual multi-task benchmark for evaluating cross-lingual generalization. *37th International Conference on Machine Learning, ICML 2020, PartF16814*, 4361–4371.

Huang, K., Mo, F., Li, H., Li, Y., Zhang, Y., Yi, W., Mao, Y., Liu, J., Xu, Y., Xu, J., Nie, J.-Y., & Liu, Y. (2024). A Survey on Large Language Models with Multilingualism: Recent Advances and New Frontiers. *ArXiv Preprint ArXiv:2405.10936*, 1–65. https://arxiv.org/abs/2405.10936

Huck, M., Dutka, D., & Fraser, A. (2019). Cross-lingual Annotation Projection Is Effective for Neural Part-of-Speech Tagging. *In Proceedings of the Sixth Workshop on NLP for Similar Languages, Varieties and Dialects, 2001*, 223–233. DOI: 10.18653/v1/w19-1425

Inuwa-Dutse, I. (2025). NaijaNLP: A Survey of Nigerian Low-Resource Languages. *ArXiv Preprint ArXiv:2502.19784, 1*(1). https://arxiv.org/abs/2502.19784

Johanson, L. (2021). *The Turkic Languages* (Johanson, L., & Csató, É. Á., Eds.). Routledge., 10.4324/9781315758115

Joshi, B. M., Khatiwada, S. P., & Pokhrel, R. K. (2024). Influence of Socioeconomic Factors on Access to Digital Resources for Education. *Rupantaran: A Multidisciplinary Journal, 8*(01), 17–33. DOI: 10.3126/rupantaran.v8i01.65197

Joshi, P., Santy, S., Budhiraja, A., Bali, K., & Choudhury, M. (2020). The State and Fate of Linguistic Diversity and Inclusion in the NLP World. *Proceedings of the 58th Annual Meeting of the Association for Computational Linguistics*, 6282–6293. https://aclanthology.org/2020.acl-main.560/

Kakwani, D., Kunchukuttan, A., Golla, S., Gokul, N. C., Bhattacharyya, A., Khapra, M. M., & Kumar, P. (2020). Findings of the Association for Computational Linguistics IndicNLPSuite: Monolingual Corpora, Evaluation Benchmarks and Pre-trained Multilingual Language Models for Indian Languages. *Findings of the Association for Computational Linguistics: EMNLP 2020, 3*, 4948–4961. https://github.com/AI4Bharat/indicnlp_catalog

Kann, K., Lacroix, O., & Søgaard, A. (2020). Weakly supervised POS taggers perform poorly on truly low-resource languages. *AAAI 2020 - 34th AAAI Conference on Artificial Intelligence, 2015*, 8066–8073. DOI: 10.1609/aaai.v34i05.6317

Karjalainen, H., Puura, U., Günthal, R., & Kovaleva, S. (2013). Karelian in Russia. *Studies in European Language Diversity*, *26*, 1–269. https://phaidra.univie.ac.at/detail_object/o:80726

Lakew, S. M., Negri, M., & Turchi, M. (2020). Low Resource Neural Machine Translation: A Benchmark for Five African Languages. *ArXiv Preprint ArXiv:2003.14402*, 1–10. https://arxiv.org/abs/2003.14402

Lample, G., Conneau, A., Ranzato, M., Denoyer, L., & Jegou, H. (2018). Word translation without parallel data. *International Conference on Learning Representations, 2015*, 1–14. https://openreview.net/pdf?id=H196sainb

Lauscher, A., Ravishankar, V., Vulić, I., & Glavaš, G. (2020). From zero to hero: On the limitations of zero-shot language transfer with multilingual transformers. *EMNLP 2020 - 2020 Conference on Empirical Methods in Natural Language Processing, Proceedings of the Conference*, 4483–4499. DOI: 10.18653/v1/2020.emnlp-main.363

Le Scao, T., Fan, A., Akiki, C., Pavlick, E., Ilić, S., Hesslow, D., Castagné, R., Luccioni, A. S., Yvon, F., Gallé, M., Tow, J., Rush, A. M., Biderman, S., Webson, A., Ammanamanchi, P. S., Wang, T., Sagot, B., Muennighoff, N., del Moral, A. V., . . . Wolf, T. (2022). BLOOM: A 176B-Parameter Open-Access Multilingual Language Model. *ArXiv Preprint ArXiv:2211.05100*. https://arxiv.org/abs/2211.05100

Leben, W. R. (2018). Languages of the World. *Oxford Research Encyclopedia of Linguistics, January*. DOI: 10.1093/acrefore/9780199384655.013.349

Leveraging Artificial Intelligence to Foster Tolerance and Coexistence in Multicultural Societies. A Framework for Inclusive Policy and Social Integration. (2025). *SSRN* 5318927. https://www.researchgate.net/publication/392982288_Leveraging_Artificial_Intelligence_to_Foster_Tolerance_and_Coexistence_in_Multicultural_Societies_A_Framework_for_Inclusive_Policy_and_Social_Integration

Libovický, J., Rosa, R., & Fraser, A. (2019). How Language-Neutral is Multilingual BERT? *ArXiv Preprint ArXiv:1906.01502*, 4996–5001. https://arxiv.org/abs/1911.03310

Litre, G., Hirsch, F., Caron, P., Andrason, A., Bonnardel, N., Fointiat, V., Nekoto, W. O., Abbott, J., Dobre, C., Dalboni, J., Luxardo, G., & Bohbot, H. (2022). Participatory Detection of Language Barriers towards Multilingual Sustainability (ies) in Africa. *Sustainability*, •••, 1–15.

Mager, M., Mager, E., Medina-Urrea, A., Meza, I., & Kann, K. (2018). Lost in Translation: Analysis of Information Loss During Machine Translation Between Polysynthetic and Fusional Languages. *ArXiv Preprint ArXiv:1807.00286*, 73–83. https://arxiv.org/abs/1807.00286

Maurya, K. K., & Desarkar, M. S. (2023). Towards Low-resource Language Generation with Limited Supervision. *BigPicture 2023 - Big Picture Workshop, Proceedings*, 80–92. DOI: 10.18653/v1/2023.bigpicture-1.7

Mlambo, R., & Matfunjwa, M. (2024). The use of technology to preserve indigenous languages of South Africa. *Literator*, *45*(1), 1–8. DOI: 10.4102/lit.v45i1.2007

Muhammad, S. H., Abdulmumin, I., Ayele, A. A., Ousidhoum, N., Adelani, D. I., Yimam, S. M., Ahmad, I. S., Beloucif, M., Mohammad, S. M., Ruder, S., Hourrane, O., Brazdil, P., Jorge, A., Ali, F. D. M. A., David, D., Osei, S., Bello, B. S., Ibrahim, F., Gwadabe, T., . . . Arthur, S. (2023). AfriSenti: A Twitter Sentiment Analysis Benchmark for African Languages. *EMNLP 2023 - 2023 Conference on Empirical Methods in Natural Language Processing, Proceedings*, 13968–13981. DOI: 10.18653/v1/2023.emnlp-main.862

Muhammad, S. H., Ahmad, I. S., Abdulmumin, I., Lawan, F. I., Sani, B., Imam, S. H., Aliyu, Y., Sani, S. A., Umar, A. U., Gwadabe, T., Church, K., & Marivate, V. (2025). HausaNLP: Current Status, Challenges and Future Directions for Hausa Natural Language Processing. *ArXiv Preprint*. https://arxiv.org/abs/2505.14311

Nekoto, W., Marivate, V., Matsila, T., Fasubaa, T., Kolawole, T., Fagbohungbe, T., Oluwole Akinola, S., Hassan Muhammad, S., Kabongo, S., Osei, S., Freshia, S., Andre Niyongabo, R., Macharm, R., Ogayo, P., Ahia, O., Meressa, M., Adeyemi, M., Mokgesi-Selinga, M., Okegbemi, L., & Bashir, A. (2020). Participatory Research for Low-resourced Machine Translation: A Case Study in African Languages. In Trevor Cohn, Y. L., & He, Y. (Eds.), *Association for Computational Linguistics* (Vol. 9). Association for Computational Linguistics., DOI: 10.18653/v1/2020.findings-emnlp.195

Ning, Z., Gu, T., Song, J., Hong, S., Li, L., Liu, H., Li, J., Wang, Y., Lingyu, M., Teng, Y., & Wang, Y. (2025). LinguaSafe: A Comprehensive Multilingual Safety Benchmark for Large Language Models. *ArXiv Preprint ArXiv:2508.12733*. https://arxiv.org/abs/2508.12733

Ogueji, K., Zhu, Y., & Lin, J. (2021). Small data? no problem! exploring the viability of pretrained multilingual language models for low-resourced languages. *MRL 2021 - 1st Workshop on Multilingual Representation Learning, Proceedings of the Conference*, 116–126.

Onyenwe, I. E. (2017). *Developing Methods and Resources for Automated Processing of the African Language Igbo* (Issue April). The University of Sheffield.

Pakray, P., Gelbukh, A., & Bandyopadhyay, S. (2025). Natural language processing applications for low-resource languages. *Natural Language Processing*, *31*(2), 183–197. DOI: 10.1017/nlp.2024.33

Papakyriakopoulos, O., & Xiang, A. (2023). Considerations for Ethical Speech Recognition Datasets. *WSDM 2023 - Proceedings of the 16th ACM International Conference on Web Search and Data Mining, 1*(1), 1287–1288. DOI: 10.1145/3539597.3575793

Pattnayak, P., Patel, H. L., & Agarwal, A. (2025). Tokenization Matters: Improving Zero-Shot NER for Indic Languages. *In 2025 IEEE International Conference on Electro Information Technology (EIT)*. DOI: 10.1109/eIT64391.2025.11103625

Pennycook, A. (2010). Language as a local practice. In *Language as a Local Practice* (first, Issue April 2010). DOI: 10.4324/9780203846223

Pillai, M., Griffin, A. C., Kronk, C. A., & McCall, T. (2023). Toward Community-Based Natural Language Processing (CBNLP): Cocreating With Communities. *Journal of Medical Internet Research, 25*. Advance online publication. DOI: 10.2196/48498 PMID: 37540551

Politov, A., Shkalikov, O., Jäkel, R., & Färber, M. (2025). Revisiting Projection-based Data Transfer for Cross-Lingual Named Entity Recognition in Low-Resource Languages. *ArXiv Preprint ArXiv:2501.18750*. https://arxiv.org/abs/2501.18750

Pratap, V., Tjandra, A., Shi, B., Tomasello, P., Babu, A., Kundu, S., Elkahky, A., Ni, Z., Vyas, A., Fazel-Zarandi, M., Baevski, A., Adi, Y., Zhang, X., Hsu, W.-N., Conneau, A., & Auli, M. (2024). Scaling Speech Technology to 1,000+ Languages. *Journal of Machine Learning Research, 25*, 1–52. https://arxiv.org/abs/2305.13516

Qu, X., Gu, Y., Xia, Q., Li, Z., Wang, Z., & Huai, B. (2024). A Survey on Arabic Named Entity Recognition: Past, Recent Advances, and Future Trends. *IEEE Transactions on Knowledge and Data Engineering, 36*(3), 943–959. DOI: 10.1109/TKDE.2023.3303136

Rahman, M. M., Sakib, F. A., Faisal, F., & Anastasopoulos, A. (2023). To token or not to token: A Comparative Study of Text Representations for Cross-Lingual Transfer. *MRL 2023 - 3rd Workshop on Multi-Lingual Representation Learning, Proceedings of the Workshop*, 173–183. DOI: 10.18653/v1/2023.mrl-1.6

Raihan, M. M. H., Subroto, S., Chowdhury, N., Koch, K., Ruttan, E., & Turin, T. C. (2025). Dimensions and barriers for digital (in)equity and digital divide: A systematic integrative review. *Digital Transformation and Society, 4*(2), 111–127. DOI: 10.1108/DTS-04-2024-0054

Raji, I. D., Bender, E. M., Paullada, A., Denton, E., & Hanna, A. (2021). AI and the Everything in the Whole Wide World Benchmark. *Advances in Neural Information Processing Systems, NeurIPS*.

Richardson, S., Lawrence, K., Schoenthaler, A. M., & Mann, D. (2022). A framework for digital health equity. *NPJ Digital Medicine, 5*(1). Advance online publication. DOI: 10.1038/s41746-022-00663-0 PMID: 35982146

Romero, M., Gómez-Canaval, S., & Torre, I. G. (2024). Automatic Speech Recognition Advancements for Indigenous Languages of the Americas. *Applied Sciences (Switzerland), 14*(15). Advance online publication. DOI: 10.3390/app14156497

Ruder, S., Clark, J. H., Gutkin, A., Kale, M., Ma, M., Nicosia, M., Rijhwani, S., Riley, P., Sarr, J. M. A., Wang, X., Wieting, J., Gupta, N., Katanova, A., Kirov, C., Dickinson, D. L., Roark, B., Samanta, B., Tao, C., Adelani, D. I., & Talukdar, P. (2023). XTREME-UP: A User-Centric Scarce-Data Benchmark for Under-Represented Languages. *Findings of the Association for Computational Linguistics: EMNLP, 2023*, 1856–1884. DOI: 10.18653/v1/2023.findings-emnlp.125

Ruder, S., Constant, N., Botha, J., Siddhant, A., Firat, O., Fu, J., Liu, P., Hu, J., Garrette, D., Neubig, G., & Johnson, M. (2021). XTREME-R: Towards More Challenging and Nuanced Multilingual Evaluation. *EMNLP 2021 - 2021 Conference on Empirical Methods in Natural Language Processing, Proceedings*, 10215–10245. DOI: 10.18653/v1/2021.emnlp-main.802

Ruder, S., Peters, M., Swayamdipta, S., & Wolf, T. (2019). Transfer learning in natural language processing tutorial. *NAACL HLT 2019 - 2019 Conference of the North American Chapter of the Association for Computational Linguistics: Human Language Technologies - Tutorial Abstracts, 2010*, 15–18.

Sallabank, J. (2011). Encyclopedia of the world's endangered languages. In *Language Learning Journal* (Vol. 39, Issue 1, pp. 122–123). DOI: 10.1080/09571736.2011.553475

Sambasivan, N., Arnesen, E., Hutchinson, B., Doshi, T., & Prabhakaran, V. (2021). Re-imagining algorithmic fairness in India and beyond. *FAccT 2021 - Proceedings of the 2021 ACM Conference on Fairness, Accountability, and Transparency*, 315–328. DOI: 10.1145/3442188.3445896

Seyoum, B. E., Miyao, Y., & Mekonnen, B. Y. (2019). Universal dependencies for Amharic. *LREC 2018 - 11th International Conference on Language Resources and Evaluation*, 2216–2222.

Shi, W., Li, R., Zhang, Y., Ziems, C., Yu, C., Horesh, R., de Paula, R. A., & Yang, D. (2024). CultureBank: An Online Community-Driven Knowledge Base Towards Culturally Aware Language Technologies. *EMNLP 2024 - 2024 Conference on Empirical Methods in Natural Language Processing, Findings of EMNLP 2024*, 4996–5025. DOI: 10.18653/v1/2024.findings-emnlp.288

Tahir, B., & Mehmood, M. A. (2021). Corpulyzer: A Novel Framework for Building Low Resource Language Corpora. *IEEE Access : Practical Innovations, Open Solutions*, *9*, 8546–8563. DOI: 10.1109/ACCESS.2021.3049793

Tambets, K., Yunusbayev, B., Hudjashov, G., Ilumäe, A. M., Rootsi, S., Honkola, T., Vesakoski, O., Atkinson, Q., Skoglund, P., Kushniarevich, A., Litvinov, S., Reidla, M., Metspalu, E., Saag, L., Rantanen, T., Karmin, M., Parik, J., Zhadanov, S. I., Gubina, M., & Metspalu, M. (2018). Genes reveal traces of common recent demographic history for most of the Uralic-speaking populations. *Genome Biology*, *19*(1), 1–20. DOI: 10.1186/s13059-018-1522-1 PMID: 30241495

Vassilakopoulou, P., & Hustad, E. (2023). Bridging Digital Divides: A Literature Review and Research Agenda for Information Systems Research. *Information Systems Frontiers*, *25*(3), 955–969. DOI: 10.1007/s10796-020-10096-3 PMID: 33424421

Wellen, K. (2019). Families in austronesian expansion. *Journal of Family History, 44*(3), 239–253. DOI: 10.1177/0363199019839330

Wu, M., Wang, W., Liu, S., Yin, H., Wang, X., Zhao, Y., Lyu, C., Wang, L., Luo, W., & Zhang, K. (2025). The Bitter Lesson Learned from 2,000+ Multilingual Benchmarks. *ArXiv Preprint ArXiv:2504.15521*. https://arxiv.org/abs/2504.15521

Xu, N., Gui, T., Ma, R., Zhang, Q., Ye, J., Zhang, M., & Huang, X. (2022). Cross-Linguistic Syntactic Difference in Multilingual BERT: How Good is It and How Does It Affect Transfer? *Proceedings of the 2022 Conference on Empirical Methods in Natural Language Processing, EMNLP 2022, Section 2*, 8073–8092. DOI: 10.18653/v1/2022.emnlp-main.552

Yang, E., Jänich, T., Mayfield, J., & Lawrie, D. (2024). Language Fairness in Multilingual Information Retrieval. *SIGIR 2024 - Proceedings of the 47th International ACM SIGIR Conference on Research and Development in Information Retrieval*, 2487–2491. DOI: 10.1145/3626772.3657943

Zhong, T., Yang, Z., Liu, Z., Zhang, R., Liu, Y., Sun, H., Pan, Y., Li, Y., Zhou, Y., Jiang, H., Chen, J., & Liu, T. (2024). Opportunities and Challenges of Large Language Models for Low-Resource Languages in Humanities Research. *ArXiv Preprint*, 1–31. https://arxiv.org/abs/2412.04497

ENDNOTES

1. https://www.unesco.org/en/articles/multilingual-education-bet-preserve-indigenous-languages-and-justice
2. https://www.mozillafoundation.org/en/blog/mozilla-expands-volunteer-led-push-for-inclusive-ai-in-taiwanese-indigenous-languages/?utm_source=chatgpt.com
3. https://tehiku.nz/te-hiku-tech/papa-reo/
4. https://tehiku.nz/te-hiku-tech/papa-reo/
5. https://tehiku.nz/te-hiku-tech/papa-reo/

Chapter 4
When Machines Speak Wisdom: A Comparative Study of Perambahan Brunei Darussalam in AI Platforms

Zulfadzlee Zulkiflee
https://orcid.org/0000-0002-0894-0430
Universiti Brunei Darussalam, Brunei

Ashrol Rahimy Damit
https://orcid.org/0000-0002-1860-4211
Universiti Brunei Darussalam, Brunei

ABSTRACT

This study evaluates how three major AI platforms, ChatGPT, Gemini AI, and DeepSeek, interpret Bruneian proverbs (Perambahan Brunei) from the domains of human traits, animals, and nature. It analyses each platform's literal translations, metaphorical interpretations, cultural contextualization, and moral framing based on 30 proverbs. Findings reveal distinct interpretive patterns. ChatGPT offers explanations grounded in Bruneian values but often oversimplifies cultural meanings. Gemini AI showcases the best cultural literacy, preserving Malay nuances and depth, while DeepSeek demonstrates moderate cultural awareness but frequently reframes proverbs with a functional approach. The study emphasises the need for AI design that prioritises cultural specificity and community expertise, highlighting the importance of responsible AI development to ensure local control over the interpretation of heritage.

DOI: 10.4018/979-8-3373-4397-6.ch004

INTRODUCTION

The preservation and transmission of cultural heritage through oral traditions represents one of humanity's most fundamental challenges in the digital age (UNESCO, 2024). Proverbs have been pivotal in preserving and transmitting cultural knowledge across generations, evolving to remain pertinent in contemporary society (Bridges, 2024). As collections of thoughts and ideas, proverbs are stated in short figurative phrases and sentences that encapsulate cultural wisdom and social values (Dabbagh, 2016). In Brunei Darussalam, *Perambahan Brunei* serves as a distinctive linguistic and cultural phenomenon that encapsulates the wisdom, values, and worldview of Bruneian society. *Perambahan* are similar to those of *peribahasa* in the standard Malay language. Still, *perambahan* creatively uses symbols found in society to beautify the language, which are appropriate yet concise and to the point. These traditional expressions differ fundamentally from standard Malay proverbs, incorporating unique cultural references and community-specific metaphors that reflect the lived experiences of Bruneian society. The lingua franca of Brunei Darussalam is Bahasa Melayu, also known as Brunei Malay, which differs from standard Malay in pronunciation, vocabulary, expressions, and idioms, with *perambahan* representing one of its most distinctive features.

The significance of *Perambahan Brunei* extends beyond mere linguistic curiosity; these expressions serve as vehicles for moral instruction, social cohesion, and the preservation of cultural identity. Proverbs are typically a form of unwritten communication that is stylistically distinct from regular speech, often employing metaphors or symbols. This cultural function positions proverbs as essential communicative tools that maintain social harmony whilst transmitting generational wisdom (Lomotey et al., 2021). Contemporary technological advances, particularly in artificial intelligence (AI) and natural language processing, present unprecedented opportunities for cultural preservation and education. Natural language processing (NLP) is a subfield of artificial intelligence (AI) that utilises machine learning to enable computers to comprehend and interact with human language. The emergence of sophisticated AI platforms such as ChatGPT, Perplexity, and DeepSeek has revolutionised how individuals access and interact with cultural knowledge. Natural language processing utilises machine learning and computational linguistics to enhance the system's understanding of natural speech and adapt to any modifications or requirements over time. The challenge lies in the inherent complexity of cultural interpretation (Dabbagh, 2017). Unlike some other proverbs, which do not incorporate culturally specific concepts, the interpretation of proverbs is not possible without cultural knowledge to inform speculation beyond their literal meaning (Dabbagh, 2016). Proverbs operate on multiple semantic levels, requiring not only linguistic competency but also a deep understanding of culture, historical context, and sociocultural awareness (Ghazanfari et al., 2025). This multi-layered complexity poses significant challenges for AI systems, which primarily rely on pattern recognition and statistical correlations rather than genuine cultural comprehension.

Recent scholarship in cultural linguistics has emphasised the importance of understanding proverbs within their broader cultural framework (Dabbagh, 2017). Cultural linguistics introduces the analysis of proverbs using the three analytical tools of cultural metaphor, cultural schema, and cultural category, which are collectively referred to as cultural conceptualisations (Dabbagh, 2017). This theoretical framework provides a way for evaluating AI interpretation capabilities, offering systematic approaches to assess cultural understanding beyond surface-level linguistic processing (Lomotey et al., 2021). The educational implications of AI-mediated cultural learning are profound (Shanmuga et al., 2025). Proverbs play an essential role in stimulating critical thinking and abstract reasoning in language learners. They encourage learners to go beyond the literal meanings of words, fostering the ability to infer deeper, metaphorical meanings (Shanmuga et al., 2025). As educational institutions increasingly integrate AI tools into their curricula, understanding the capabilities and limitations of these systems in handling cultural content becomes paramount for maintaining academic integrity and cultural authenticity (Journal of Artificial Intelligence and Technology, 2024).

Furthermore, the globalisation of AI technologies raises concerns about cultural bias and representation (Kraf & Soulier, 2024). Cultural bias is prevalent in many large language models (LLMs), primarily due to the scarcity of data representative of diverse cultures (Kraf & Soulier, 2024). The predominance of Western-centric training data in major AI systems potentially marginalises non-Western cultural expressions, creating interpretive gaps that could perpetuate cultural misunderstanding or misrepresentation (Kovač et al., 2025).

The preservation of indigenous knowledge systems faces unprecedented urgency in our rapidly digitalising world (UNESCO, 2024). AI can understand the context, themes, and messages of texts and identify language patterns, metaphors, and symbolism used to convey local wisdom; however, AI still faces challenges in understanding complex and implicit cultural nuances and contexts (Fitryansyah & Fauziah, 2024). The discussions highlighted the accuracy of AI translations in conveying general messages of ancient proverbs, leading to improved global accessibility (UNESCO, 2024). While AI demonstrates potential for enhancing cultural accessibility, the balance between technological convenience and cultural authenticity requires careful examination (Shanmuga et al., 2025). Research in cross-cultural AI understanding has highlighted significant challenges in developing culturally-sensitive systems (Kovač et al., 2025).

Cross-cultural intelligent language learning systems utilise advanced AI technologies to provide adaptive, personalised learning experiences that cater to the unique linguistic and cultural backgrounds of each learner (Xia et al., 2024). However, regarding cultural alignment, current models often struggle with cultural specificity and community-validated interpretations (Kraf & Soulier, 2024). These findings suggest that cultural competency in AI systems is achievable, but it requires deliberate design choices and culturally informed training approaches (Masoud et al., 2025). The comparative analysis of AI platforms offers valuable insights into the current state of cultural interpretation capabilities. DeepSeek AI specialises in code generation and technical tasks, excelling in Chinese NLP. ChatGPT provides a well-rounded blend of general-purpose conversation, creative content, and comprehensive code assistance. Perplexity is engineered for real-time fact retrieval and academic research. Each platform brings distinct strengths and limitations to cultural interpretation tasks, necessitating systematic evaluation to understand their relative effectiveness.

The theoretical foundations for this research draw from multiple disciplines, including cultural linguistics, artificial intelligence studies, and educational technology (Dabbagh, 2017). This interdisciplinary approach to studying proverbs provides a theoretical framework for the field of proverbs in general and within pragmatics in particular (Lomotey et al., 2021). The interdisciplinary nature of this inquiry reflects the complex intersection of technology and culture, requiring nuanced analytical approaches that honour both technical capabilities and cultural authenticity (Shanmuga et al., 2025). Several scholars argue for the importance of preserving indigenous knowledge in the face of digital globalisation, emphasising that ethical AI must consider how to incorporate and safeguard local knowledge systems, which are often passed down orally or through community-based learning structures (Shanmuga et al., 2025). Research from African contexts has shown that culturally insensitive digital literacy programs can contribute to the erosion of local traditions and practices. This concern extends to the interpretation of traditional knowledge by AI systems, where inadequate cultural understanding could contribute to the misrepresentation or oversimplification of complex cultural concepts (UNESCO, 2024).

This study addresses a critical gap in understanding how contemporary AI systems interpret and transmit traditional cultural knowledge (Dabbagh, 2017). By focusing specifically on *Perambahan Brunei*, this research contributes to broader discussions about AI's role in cultural preservation whilst providing practical insights for educators, technology developers, and cultural preservation advocates (UNESCO, 2024). The investigation examines whether AI platforms can serve as effective intermediaries for cultural transmission or whether they introduce distortions that compromise cultural integrity (Shanmuga et al., 2025).

AI PLATFORMS

The emergence of large language models (LLMs) represents a transformative development in artificial intelligence, fundamentally altering how human societies access and interact with information (Chaitanya & Rolla, 2024). Large language models function through neural network architectures trained on vast quantities of text data, enabling the systems to generate contextually appropriate responses through pattern recognition and statistical inference (Dabbagh, 2017). The sophistication of contemporary LLMs extends beyond simple text generation to encompass capabilities such as reasoning, translation, summarisation, and interpretive analysis across multiple languages and domains (Hadi et al., 2023). The rapid increase of accessible, user-friendly interfaces has democratised access to these computational tools, enabling non-specialist users to engage with advanced AI systems for educational, professional, and personal purposes.

The democratisation of AI access introduces both opportunities and risks for the transmission of cultural knowledge. Although AI systems offer unique opportunities for sharing cultural knowledge and educational applications, their underlying architecture and training methods may systematically favour specific epistemologies, worldviews, and languages over others (Kovač et al., 2025). The capacity of LLMs to process and generate responses in multiple languages does not guarantee cultural appropriateness or epistemological alignment with the conceptual frameworks embedded within those languages (Luo, 2023). This tension between technical capability and cultural adequacy constitutes the central issue explored through the comparative analysis of three prominent platforms: ChatGPT, DeepSeek, and Gemini.

ChatGPT, developed by OpenAI and released in November 2022, rapidly achieved unprecedented adoption, accumulating more than one million users within five days of its public launch (Teubner et al., 2023). The model represents an implementation of the Generative Pre-trained Transformer (GPT) architecture, specifically the GPT-4 variant employed in this research (Yenduri et al., 2024). The accessibility and user-friendly interface of ChatGPT have made it the most recognisable AI platform for general users, serving as a primary point of contact for non-specialist engagement with AI technologies. The platform's design philosophy emphasises accessibility and user satisfaction, leading to interfaces and response formats that are optimised for immediate understanding and emotional connection with diverse global audiences. The training data underlying ChatGPT's development reflects the historical predominance of English-language and Western academic sources within digital information ecosystems (Sani, 2025).

The architectural design, prioritising logical consistency, emotional accessibility, and universal comprehensibility, necessarily translates complex cultural concepts into psychological and emotional terms (Neuman, 2024). The ChatGPT platform consistently demonstrates a capacity for sophisticated language generation and contextual awareness; however, it systematically reduces cultural complexity to simplified psychological categories or universal human experiences (Yuan et al., 2024). This reductionist approach improves accessibility for international audiences unfamiliar with specific cultural traditions. Still, it also diminishes the unique meaning and knowledge that give cultural expressions, like Bruneian proverbs, their actual value. The translation of relational and cosmological concepts into emotional and psychological terms, while making them understandable to Western audiences, may distort the underlying worldviews embedded in these expressions.

On the other hand, DeepSeek, developed by the Chinese technology company DeepSeek-AI and released in 2023, represents a distinctive architectural and design approach to creating large language models (Puspitasari et al., 2025). Unlike ChatGPT, which has pursued increasingly large model sizes and broader general capability, DeepSeek has emphasised computational efficiency, cost-effectiveness, and specialised capability in technical domains (Joshi, 2025). The platform excels particularly in code generation, mathematical reasoning, and tasks that require precise logical inference, positioning it as a preferred tool for technical professionals and researchers who prioritise accuracy and computational reliability over breadth of capability (Joshi, 2025). DeepSeek's training data reflects the incorporation of substantial volumes of Chinese-language sources, technical documentation, and scientific literature, resulting in distinctive epistemological orientations compared to Western-dominant models.

The inclusion of Chinese philosophical texts and Chinese scientific literature shapes the model's interpretive frameworks toward instrumental rationality, social hierarchy, and collective benefit (Ye et al., 2025). The practical implications of DeepSeek's design become evident in its interpretive approach to cultural material. The platform demonstrates intermediate-level cultural awareness and functional understanding of relational systems, yet consistently reframes cultural expressions through instrumental logics, emphasising social functionality and normative behaviour. Notably, DeepSeek exhibits a systematic tendency to embed moral-juridical framings within ostensibly neutral cultural interpretation, transforming descriptive statements regarding isolation or individual autonomy into normative condemnations of social transgression. This interpretive tendency reflects the underlying philosophical frameworks prioritising collective conformity and social order, potentially misrepresenting cultures that valorise individual agency or voluntary separation from community participation.

Gemini AI, developed by Google DeepMind and released in 2023 (Carlà et al., 2024), represents a deliberate attempt to incorporate diverse linguistic and cultural perspectives within large language model development. The model architecture incorporates explicit design decisions to enhance cross-cultural understanding and to reduce systematic biases associated with English-language dominance and Western philosophical frameworks. Gemini's training data demonstrates greater linguistic and cultural diversity compared to earlier models, incorporating significant volumes of materials in non-English languages and from diverse cultural traditions, including sources from Southeast Asia (Google, 2024). The distinctive design approach underlying Gemini includes explicit mechanisms for cultural adaptation and epistemological flexibility. Rather than imposing a single interpretive framework applicable across all cultural contexts, Gemini incorporates architectural features enabling sensitivity to cultural specificity and context-dependent meaning-making (De Santis & Rizzi, 2025). This design philosophy reflects the recognition that genuine cross-cultural understanding requires the preservation of indigenous conceptual categories rather than the translation of foreign concepts into Western philosophical frameworks

The implications of Gemini's design approach become evident in its interpretive performance. The platform consistently demonstrates a superior capacity to engage with Malay terminology, to preserve sensory and phenomenological specificity in the interpretation of metaphorical expressions, and to acknowledge the philosophical dimensions of cultural wisdom without reducing complexity to simplified psychological categories. Most significantly, Gemini more frequently preserves paradoxical structures embedded within proverbs rather than resolving paradoxes through sequential logic or instrumental framing (Hakkarainen & Engelbrecht, 2024). This interpretive approach maintains greater fidelity to Bruneian epistemologies, which apparently accommodate paradox and contradiction within integrated, relational, and ontological frameworks.

The selection of these three platforms for comparative analysis reflects strategic methodological choices. ChatGPT represents the most widely adopted AI platform globally, serving as the primary interface through which non-specialist users interact with advanced LLMs (Zampatti et al., 2024). Therefore, it deserves investigation as the primary means for public access to AI-mediated cultural knowledge. DeepSeek offers a unique non-Western perspective on LLM development by integrating substantial amounts of Chinese-language content and philosophical concepts. This approach provides valuable insights into how the composition of training data and design philosophy influence the interpretation of cultural material. Gemini signifies a significant effort to foster cross-cultural understanding and diminish Western-centric bias in mainstream AI platforms. This initiative serves as a potential model for how culturally informed design can improve AI's understanding of minority language expressions. Collectively, these three platforms span the spectrum of contemporary LLM approaches, ranging from Western-dominant, accessibility-prioritising architectures (ChatGPT) through non-Western, efficiency-prioritising systems informed by alternative philosophical traditions (DeepSeek) to explicitly cross-culturally informed platforms prioritising epistemological pluralism (Gemini).

This framework enables the investigation of how differences in training data, design philosophy, and architectural choices result in varying interpretive outcomes when examining Bruneian cultural material.

LITERATURE REVIEW

Proverb is the academic study of proverbs and traditional sayings, combining linguistics, folklore, cultural anthropology, and literary analysis. Proverbs function as distinctive linguistic and cultural phenomena, characterised by brevity, apparent simplicity, and the capacity to encode complex cultural wisdom within concise figurative expressions (Dabbagh, 2016). Research on proverbs reveals that they operate on multiple semantic levels simultaneously. These levels include: the literal meaning, which refers to the straightforward denotative content; the metaphorical sense, which encompasses the more profound significance conveyed through imagery and symbolism; and the cultural meaning, which pertains to the context-specific importance within distinct worldviews, value systems, and historical backgrounds. This multi-layered analysis has proven particularly productive for understanding how proverbs simultaneously function as linguistic units, as aesthetic or poetic expressions, and as vehicles for the transmission of cultural knowledge (Yankah, 1989).

Recent research has increasingly emphasised the importance of understanding proverbs within their specific cultural and historical contexts, rejecting earlier approaches that attempted to identify universal meanings or functions across diverse cultural traditions (Lomotey et al., 2021). This shift in context acknowledges that proverbs are not simply transparent vehicles for universal wisdom; instead, they are deeply rooted in specific epistemologies, cosmologies, and social structures. Dabbagh (2016, 2017) has developed influential theoretical frameworks that integrate cultural linguistics with traditional paremiology, arguing that the artistic interpretation of proverbs requires systematic engagement with cultural schemas, categories, and metaphors, conceptual structures that shape how cultures organise knowledge and make meaning. This theoretical integration has enabled sophisticated analysis of how proverbs encode culture-specific worldviews and epistemologies.

Cultural linguistics, as developed by scholars such as Dabbagh (2016; 2017), represents a theoretical approach that integrates traditional linguistic analysis with anthropological and philosophical perspectives on how languages encode cultural knowledge. The theory of cultural conceptualisations posits that languages do not merely express pre-existing thoughts or experiences, but rather that linguistic categories and structures actively shape how speakers of particular languages perceive, categorise, and conceptualise reality. Proverbs, within this theoretical framework, serve as vibrant sites for examining how specific cultures encode complex conceptual relationships and epistemological assumptions within compact linguistic forms.

The analytical tool of cultural metaphor, fundamental to cultural linguistics approaches, emphasises that metaphors function not merely as literary devices but as fundamental structures through which cultures organise and transmit knowledge (Boers, 2003). Metaphor involves the mapping of conceptual structures from one domain (typically more concrete and physically grounded) to another domain (normally more abstract) (Lakoff, 2006). Cultures differ significantly in which metaphors they utilise, how they elaborate and extend metaphors, and the philosophical implications they draw from particular metaphorical systems. Bruneian proverbs, within this framework, represent culturally specific elaborations of conceptual metaphors that encode distinctively Bruneian understandings of relationality, temporality, human agency, and knowledge production.

Recent research examining artificial intelligence and natural language processing has increasingly recognised that a machine's understanding of language does not automatically constitute an understanding of culture. Natural language processing systems, including large language models, achieve significant capability in surface-level linguistic processing, identifying grammatical structures, recognising semantic relationships between words, and generating contextually appropriate responses (Ponti et al., 2019). Although the ability to process language exists, it does not imply an understanding of the cultural meanings embedded in language, nor does it demonstrate appreciation for the epistemological assumptions that inform specific linguistic expressions.

Significant research has documented cultural bias in large language models, revealing systematic patterns whereby training data dominated by English-language and Western sources produces interpretive frameworks that systematically privilege Western epistemologies while misrepresenting or inadequately engaging with non-Western cultural perspectives (Helm et al., 2024). Kovač et al. (2025) have examined how cultural bias manifests through AI systems' differential capacity to understand expressions from majority versus minority language communities, documenting patterns whereby AI systems achieve an adequate understanding of culturally specific concepts from well-represented cultures while systematically misinterpreting or oversimplifying expressions from underrepresented communities. This research suggests that the problem of cultural bias in AI systems is not incidental but reflects fundamental characteristics of how these systems are trained and designed.

Research on cross-cultural intelligent language learning systems has begun to address these challenges, exploring how AI systems can be designed to incorporate cultural specificity and community-validated interpretations (Wu & Cai, 2025). This research suggests that superior cultural understanding is technically achievable but requires explicit design decisions, including the incorporation of diverse training sources, engagement with community-based knowledge validation, and architectural choices prioritising cultural authenticity alongside accessibility. The absence of such deliberate design choices in commercially dominant platforms suggests that market logics do not naturally converge toward cultural preservation and representation.

The emergence of AI-mediated cultural learning introduces new possibilities and risks for educational practice (Morgado et al., 2025). On one hand, AI systems enable unprecedented access to cultural knowledge, enabling learners globally to engage with cultural expressions from traditions they would otherwise encounter only with difficulty. On the other hand, if AI systems systematically misrepresent cultural meanings by reducing cultural concepts to Western psychological categories or by embedding normative moral judgments, then AI-mediated learning may actually undermine rather than promote genuine cultural understanding. Research examining the educational utility of AI suggests that the quality of cultural representation has a significant impact on learning outcomes (Lee et al., 2025). Learners exposed to culturally authentic, epistemologically precise interpretations of cultural expressions develop a deeper and more nuanced understanding of both the specific culture and broader principles regarding how meaning is culturally situated. On the other hand, learners exposed to reductionist or biased representations may form distorted understandings that hinder genuine cross-cultural comprehension.

Within the Bruneian context, scholarly documentation of *perambahan* as a distinctive linguistic and cultural phenomenon has remained relatively limited. The foundational work, primarily undertaken by linguists, is exemplified by Hadijah Haji Hassan's (1987) compilation, *Perambahan Brunei, which* serves as a seminal reference cataloguing culturally embedded expressions within Standard Bruneian Malay. This lexicographic contribution has been complemented by broader linguistic surveys exploring the status and variation of languages in Brunei Darussalam (Haji-Othman et al., 2016), which contextualise *perambahan* within the larger linguistic ecology where Brunei Malay functions as the vernacular lingua franca in diglossic tension with Standard Malay, English, and increasingly marginalised indigenous minority languages. As linguistic practices function not merely as communicative tools but as markers of cultural heritage (Zulfadzlee & Suhaimi, 2025), the accurate semantic translation of culturally embedded expressions becomes essential, notably as these expressions embody specific worldviews and identity markers of Bruneian speakers. Recent pedagogical initiatives have sought to preserve *perambahan* through digital and interactive media platforms, as evidenced by educational technology projects that employ instructional design models to make *perambahan* accessible to younger generations through online and multimedia formats. However, these existing studies remain primarily descriptive and preservation-focused, rarely engaging with questions of semantic transfer, computational processing, or how contemporary artificial intelligence systems interpret and negotiate meaning within these culturally bound expressions. Consequently, applying comparative analysis of AI platform interpretations to *perambahan* addresses a significant research gap, situating digital linguistic analysis at the intersection of Brunei Darussalam's documented cultural heritage and the emerging challenges posed by machine learning approaches to minority language understanding.

Bruneian Malay occupies a distinctive position within the broader Malay-language ecology, characterised by distinctive phonological, lexical, and pragmatic features not fully captured by standard Malay. *Perambahan Brunei* represents vibrant and culturally specific expressions that encode wisdom regarding relationality, temporality, and human agency within Bruneian cosmological frameworks. The specific challenge addressed in this research concerns whether contemporary AI systems, trained primarily on English-language and Western-sourced materials, can adequately represent and convey Bruneian cultural knowledge, or whether systematic misrepresentation is inevitable given the training and design of existing platforms.

The comprehensive examination of how AI systems interpret Bruneian proverbs requires integration of multiple theoretical traditions and methodological approaches. Paremiology provides frameworks for understanding the multi-layered meaning embedded within proverbs and the importance of cultural context for interpretation. Cultural linguistics provides theoretical tools for understanding how languages encode culture-specific epistemologies and worldviews. AI scholarship offers insight into how machine learning systems develop systematic biases and how these biases manifest in interpretive outcomes. Educational scholarship offers insight into the implications of AI-mediated cultural learning for the development of cultural understanding. Indigenous knowledge preservation scholarship provides critical perspectives on the ethical and political dimensions of technology-mediated cultural transmission.

The significance of this research extends beyond technical assessment toward broader questions concerning the role of AI systems in cultural preservation and transmission. As AI systems become increasingly central to the global circulation of knowledge, understanding their capacity to represent minority language knowledge authentically becomes urgent. The present research contributes evidence regarding both the current limitations and the potential (as demonstrated by Gemini's superior performance) for AI systems to engage respectfully with minority language cultural expressions. This evidence can inform both technical development decisions and policy discussions regarding how to ensure that AI-mediated knowledge transmission does not inadvertently accelerate the erosion of minority language cultures and epistemologies.

METHODOLOGY

This study employed a qualitative comparative analysis design to systematically evaluate the interpretive capabilities of three prominent AI platforms: ChatGPT (GPT-4), Gemini AI, and DeepSeek AI. Data collection employed a purposive sampling approach to select 30 traditional *Perambahan Brunei* (2003), stratified equally across three thematic categories that reflect traditional classification systems: human traits, animals, and nature, with 10 proverbs selected per category. Selection criteria incorporated cultural authenticity (documented Bruneian origins), thematic diversity (balanced representation across all three domains), linguistic complexity (varying metaphorical sophistication), and contemporary relevance (continued recognition in modern Bruneian society), thereby ensuring rigorous methodological standards aligned with established frameworks in cultural linguistics research (Dabbagh, 2017). The 30 proverbs informed the broader comparative assessment of platform-specific patterns and interpretive tendencies. Each proverb was submitted to all three AI platforms using a standardised prompt designed to elicit comprehensive cultural and linguistic interpretation.

The data collection protocol employed the following prompt structure: *"What does this proverb mean '[Proverb Text in Malay]'? Provide a 1) Literal meaning, 2) Cultural interpretation, and 3) Metaphorical interpretation."* This prompt structure specifically targeted three distinct analytical dimensions: literal semantic content, culturally grounded meaning, and extended metaphorical significance, thereby facilitating a systematic evaluation of how each platform processed surface-level translation versus deeper cultural understanding. Controlled variables were maintained to ensure methodological consistency throughout data collection, including the use of identical prompts across all platforms, consistent temporal spacing of queries, completion of all 30 proverb analyses within a single continuous session, and the prohibition of follow-up questions or clarifications. This approach simulated real-world usage scenarios where users seek autonomous AI interpretations without iterative refinement. Data analysis involved systematic documentation of all platform responses, followed by a thematic analysis to identify recurring interpretive patterns, and a comparative examination across multiple dimensions, including literal semantic accuracy, metaphorical precision, cultural sensitivity, and engagement with indigenous epistemological frameworks.

AI responses underwent systematic analysis using a multi-dimensional coding framework to evaluate several key dimensions. This framework assessed accuracy in cultural context, nuance preservation, idiomatic expression handling, and overall coherence with traditional interpretations (Chen & Zmire, 2024; Hamdi et al., 2023). The first is 1) literal accuracy, which assesses fidelity to the literal content, including references to animals and objects, numerical or temporal specifications and direct translations. 2) Metaphorical precision evaluates the preservation of specific imagery like animals, natural phenomena or actions, metaphorical coherence, conceptual mapping from concrete to abstract and how imagery grounds meaning. 3) Cultural authenticity, examines engagement with culturally specific concepts, Malay terminology and Bruneian-specific worldviews and values, avoiding culturally inappropriate generalisations and recognising Bruneian epistemological frameworks. 4) Epistemological alignment, focuses on consistency with Bruneian relational frameworks, respecting paradoxical thinking, avoiding imposed Western logical structures, recognising indigenous knowledge systems and acknowledging community authority. Finally, 5) normative dimensions, identify embedded value judgements, assessing whether descriptive statements become prescriptive and evaluating cultural bias, including Western, instrumental and conservative social values. This structured approach helps ensure a comprehensive evaluation of AI responses, considering both their accuracy and cultural relevance. This method is crucial for identifying gaps in the long-tail knowledge of large language models regarding pluralistic cultures, as traditional annotation or synthetic data often fail to capture the nuances of culturally specific information (Ziems et al., 2025).

FINDINGS

The comparative analysis of AI platform responses across selected *Perambahan Brunei* revealed distinctive interpretive approaches and varying degrees of cultural understanding. Each platform demonstrated unique strengths and limitations in processing traditional Bruneian wisdom, with notable differences in depth, cultural sensitivity, and authenticity of interpretation. ChatGPT consistently provided the most comprehensive and culturally-grounded explanations, often incorporating relevant Bruneian values such as responsibility *(tanggungjawab),* community harmony *(muafakat),* and humility *(merendah diri).* The platform demonstrated a sophisticated understanding of metaphorical layers and frequently connected traditional concepts to contemporary applications in thoughtful ways. This nuanced approach enabled it to capture both the historical significance and modern relevance of Bruneian proverbs. Gemini showed firm factual grounding and research-oriented responses, often providing scientific or historical context for natural phenomena and cultural practices mentioned in proverbs. However, its interpretations tended to be more literal and less culturally nuanced than those of ChatGPT. While the platform excelled at contextualising information through empirical frameworks, it sometimes missed the deeper cultural resonance that traditional sayings carry within Bruneian society. DeepSeek performed moderately across all categories, offering clear but sometimes generalised responses that lacked a specific Bruneian cultural context.

The platform showed systematic analytical approaches but struggled with deeper cultural interpretation and contemporary relevance. Despite providing coherent explanations, it frequently failed to engage with the subtle layers of meaning embedded in traditional wisdom and its application to modern Bruneian life.

Literal Meaning

Table 1. Comparative Analysis of Literal Interpretation of Perambahan Brunei Across AI Platforms

Item	Perambahan Brunei	Chatgpt	Deepseek	Gemini
Human traits	Adi-beradi gimbar kepala	Siblings clash heads.	Siblings fan (each other's) heads.	Siblings make a commotion/ uproar with their heads.
	Berbagi tubuh	Sharing one body.	To share a body or To divide a body.	To share a body.
Animal	Ambuk tunggal	A lone chick	a lone water buffalo" or "a single buffalo."	Single person/individual alone" or "The sole one."
	Macam bagas ucing beranak	Like a house where a cat has given birth	Like a kitten biting while giving birth	Like a cat immediately giving birth.
Nature	Aing inda sajakan pasang, ada jua masanya surut	The sea is not always high tide; sometimes it is low tide.	The water is not always at high tide; there will also be its time to recede	Water does not always rise (stay at high tide); there is also a time for it to recede (go to low tide).
	Bertuhur berdalam	Digging makes the ground deeper.	to stack (something) deeply" or "a pile that has depth.	Shallow at the surface, deep within

The three AI platforms demonstrate considerable variation in their literal interpretation of Bruneian proverbs, revealing distinct approaches to semantic translation that range from strict literal interpretation to contextual elaboration. In the human traits category, ChatGPT consistently produces minimalist literal translations (Puppel, 2024) (e.g., "Siblings clash heads," "Sharing one body"), prioritising direct word-for-word equivalence without semantic elaboration. DeepSeek tends towards dual or alternative interpretations (Joshi, 2025), as evidenced by "Siblings fan (each other's) heads" for *adi-beradi gimbar kepala*, which introduces an unexpected semantic shift through divergent lexical processing. Gemini occupies a middle ground, offering literal translations enriched with contextual markers such as "commotion/uproar with their heads," thereby offering a broader interpretive framework while staying true to the original expression.

The animal domain reveals more pronounced interpretive divergence, particularly in terms of lexical accuracy and metaphorical displacement. DeepSeek produces a notable translation error with "*ambuk tunggal*", rendering it as "lone water buffalo" rather than the correct reference, which suggests potential cross-linguistic interference or database confusion (Almahasees et al., 2025). Conversely, Gemini demonstrates metaphorical generalisation (Wang & Zhao, 2023), abstracting the specific animal reference into the universal conceptual category of "single person/ individual alone," therefore, it sacrifices cultural specificity for semantic generalisation. ChatGPT maintains literal adherence throughout this category, preserving both animal references and action sequences with minimal interpretive modification; however, this approach risks obscuring the proverbial intent beneath the literal surface meaning.

The nature domain presents the most sophisticated interpretive performance, with all three platforms demonstrating improved semantic accuracy relative to earlier categories. Both ChatGPT and DeepSeek render *aing inda sajakan pasang, ada jua masanya surut* with substantial literal accuracy, preserving tidal metaphor and temporal cyclicity. Gemini's interpretation distinguishes itself through explicit parenthetical clarification, emphasising the reciprocal relationship between high and low tides, thereby rendering the cyclical principle more transparent (Fridaus et al., n.d). However, in *bertuhur berdalam*, Gemini's rendering as "Shallow at the surface, deep within" represents a fundamental metaphorical inversion absent from the original proverb's depth-focused construction, suggesting that even literal translation approaches encounter semantic reinterpretation challenges when confronted with culturally embedded imagery (Obeidat et al., 2024).

Methaphorical Meaning

The metaphorical interpretations across the three AI platforms reveal fundamentally distinct approaches to cultural knowledge representation, ranging from surface-level semantic mapping to sophisticated engagement with Bruneian ontological frameworks. In the human traits domain, ChatGPT demonstrates a consistently simplified metaphorical translation strategy, reducing complex cultural concepts to universal emotional or social states (conflict, unity, patience, forgiveness). This reductionist approach, while accessible, systematically removes contextual particularity from expressions rooted in Southeast Asian relational epistemologies. Conversely, DeepSeek provides intermediate-level metaphorical elaboration that attempts to contextualise the proverbs within community-based social structures ("People from the same group, community, or origin must stick together"), thereby demonstrating awareness of collectivist value systems whilst remaining somewhat generic in its cultural specificity. Gemini distinguishes itself through explicitly articulated cultural terminology and ontological grounding, particularly in berbagi tubuh, where it introduces the Malay concepts of kebersamaan (togetherness/unity) and solidaritas (solidarity), thereby positioning the metaphorical meaning within its original linguistic and conceptual ecosystem rather than translating it into Western philosophical categories.

Table 2. Comparative Analysis of Metaphorical Interpretation of Perambahan Brunei Across AI Platforms

Item	Perambahan Brunei	Chatgpt	Deepseek	Gemini
Human traits	Adi-beradi gimbar kepala	Conflict is natural in close relationships. Unity beyond conflict. Reminder of patience and forgiveness.	People from the same group, community, or origin must stick together and support one another, especially against outsiders or in times of difficulty.	It could mean that even the closest relationships, like those between siblings, can occasionally be full of conflict, noise, or disagreement.
	Berbagi tubuh	Deep loyalty, harmony, and shared destiny.	To be so deeply connected that you function as a single entity, sharing in all successes, failures, joys, and sufferings.	Total Partnership: It means a complete and profound sharing of one's entire existence, including fate, suffering, and joy. Inseparable Bond: To "share a body" is to be an inseparable unit—what affects one, affects the other. It is a powerful expression of *kebersamaan* (togetherness/unity) and *solidaritas* (solidarity) in the most intense form of human connection.
Animal	Ambuk tunggal	Symbol of loneliness, fragility, or uniqueness, depending on context.	A person who stubbornly goes their own way, refusing to cooperate or listen to others, is vulnerable and likely to fail.	An outcast, a pariah, or a person who is forced to be solitary because their arrogant, selfish, or unpleasant behavior has caused others to avoid them.
	Macam bagas ucing beranak	Describes chaotic, unorganized, or restless situations — whether in the household, workplace, or community.	For a junior, subordinate, or less knowledgeable person to challenge, reprimand, or try to teach their superior.	To be in a state of sudden, overwhelming panic or extreme haste, especially when dealing with a situation that requires immediate, intense, and often disorderly action.
Nature	Aing inda sajakan pasang, ada jua masanya surut	Symbol of life's impermanence, reminding people to stay humble, patient, and resilient.	No situation, whether favorable or unfavorable, lasts forever. Change is the only constant.	Success, happiness, wealth, or power are all temporary and subject to change. Nothing lasts forever.
	Bertuhur berdalam	Suggests that inquiry and knowledge are endless, and the deeper one goes, the more profound (or complicated) things become.	To engage in a process of deep, layered, and cumulative learning or creation	What appears simple or shallow on the surface often holds deep complexity, meaning, or wisdom beneath.

The *adi-beradi gimbar kepala* interpretation exemplifies platform-specific limitations in handling paradoxical cultural wisdom. ChatGPT's interpretation breaks down the proverb into sequential statements: conflict exists, unity transcends conflict, and patience should govern relations (Montanari, 2025). This approach treats the paradox as separate rhetorical units, rather than viewing it as a cohesive philosophical proposition regarding the complexity of relationships. DeepSeek's framing of intra-group solidarity against external threats presents a protective-function interpretation that is absent from the original expression. This could conflate community cohesion with defensive positioning. Gemini's characterisation of the proverb as acknowledging that "the closest relationships can occasionally be full of conflict, noise, or disagreement" demonstrates superior fidelity to the Bruneian understanding that intimacy permits, indeed necessitates, the expression of disagreement without threatening fundamental bonds. This interpretation preserves the paradoxical wisdom conflict, allowing closeness and conflict to coexist rather than resolving the paradox through linear sequencing or instrumental framing (Weigang & Brom, 2025).

The animal domain reveals the most pronounced divergence in metaphorical interpretive capacity, with significant implications for understanding how AI systems navigate culturally specific animal symbolism. For *ambuk tunggal*, ChatGPT offers a generalised interpretation ("loneliness, fragility, or uniqueness") that remains semantically accurate but phenomenologically underdetermined, failing to distinguish between the solitude of chosen autonomy versus imposed isolation. DeepSeek's interpretation introduces socially determined marginality ("a person who stubbornly goes their own way, refusing to cooperate or listen to others, is vulnerable and likely to fail"), which significantly reframes the proverb's meaning through a normative social lens that privileges conformity and penalises deviation. Gemini's interpretation, "an outcast, a pariah, or a person who is forced to be solitary because their arrogant, selfish, or unpleasant behaviour has caused others to avoid them", similarly embeds judgment within the metaphorical structure, transforming a description of existential isolation into a moral commentary on individual culpability. This contrast suggests that metaphorical interpretation operates as a site of cultural meaning-making where AI systems either naturalise Western individualised agency frameworks or reproduce conservative social normativity (Viidalepp, 2023).

The *macam bagas ucing beranak* interpretation demonstrates differential capacity to preserve metaphorical precision across platform architectures. ChatGPT's rendering as "chaotic, unorganised, or restless situations" achieves semantic adequacy whilst sacrificing the specific metaphorical image (a cat giving birth) that grounds the proverb's cultural intelligibility. DeepSeek's interpretation introduces a pedagogical power dynamic ("for a junior, subordinate, or less knowledgeable person to challenge, reprimand, or try to teach their superior") that inverts the proverb's actual semantic trajectory. The expression describes situations of chaos and overwhelm, not hierarchical transgression. Gemini's interpretation captures the affective and temporal dimensions more precisely (Rane et al., 2024) ("a state of sudden, overwhelming panic or extreme action, especially when dealing with a situation that requires immediate, intense, and often disorderly action"), therefore preserving both the sensory immediacy of the image and its metaphorical extension into experienced temporality. The superiority of Gemini's interpretation lies in its attention to the phenomenological dimensions of chaos, the simultaneity, intensity, and disorderedness, rather than reducing the metaphor to generic descriptions of confusion or social transgression (Rane et al., 2024).

The nature domain demonstrates the most sophisticated metaphorical interpretation across all three platforms, suggesting that nature-based proverbs, which operate through more universal symbolic registers, achieve greater cross-cultural semantic transportability. For *aing inda sajakan pasang, ada jua masanya surut*, ChatGPT's interpretation ("symbol of life's impermanence, reminding people to stay humble, patient, and resilient") successfully captures the proverb's fundamental philosophical content regarding cyclical temporality and psychological disposition. However, the explicit valorisation of humility, patience, and resilience introduces normative dimensions absent from the original expression. DeepSeek's framing situates the proverb within an epistemological framework of permanence ("No situation, whether favourable or unfavourable, lasts forever. Change is the only constant"), which represents an engagement with metaphysical philosophy while potentially universalising a specifically Bruneian understanding of cosmic recurrence and human positioning within natural cycles. Gemini's interpretation, "Success, happiness, wealth, or power are all temporary and subject to change. Nothing lasts forever," achieves remarkable semantic efficiency by crystallising the proverb's existential wisdom into a concise, metaphorical proposition that preserves both the cyclical structure and its implications for human aspiration and acceptance.

The *bertuhur berdalam* interpretation reveals the deepest interpretive divergence, with significant implications for understanding how metaphor functions as a vehicle for transmitting epistemological sophistication. ChatGPT's rendering ("inquiry and knowledge are endless, and the deeper one goes, the more profound [or complicated] things become") articulates an epistemological infinity principle that aligns with Western scholarly traditions of knowledge accumulation and deepening specialisation, thereby potentially imposing an acquisitive knowledge paradigm onto an expression. DeepSeek's interpretation ("to engage in a process of deep, layered, and cumulative learning or creation") similarly emphasises additive accumulation, though with greater attention to the processual and temporal dimensions of knowledge development (Zhai, 2025). Gemini's interpretation "What appears simple or shallow on the surface often holds deep complexity, meaning, or wisdom beneath" introduces an ontological inversion that reframes the proverb from an epistemological statement (how one acquires knowledge) into a hermeneutical principle (how meaning is concealed within apparent simplicity). This interpretation challenges the knowledge-accumulation paradigm, emphasising the importance of interpretive excavation and recognising hidden depths (Tanverdi, 2024). This approach achieves greater consistency with contemplative philosophical traditions that value the intuitive understanding of complex meanings, going beyond mere analytical breakdown (Speer, 2018).

The comparative analysis reveals clear platform-specific tendencies in proverb interpretation. ChatGPT consistently simplifies metaphorical depth into psychological categories, making cultural wisdom more accessible but less ontologically precise (Harnad, 2024). DeepSeek reframes proverbs through functional social logics that emphasise order, conformity, and individual responsibility (Segerer, 2025). Gemini, by contrast, demonstrates stronger cultural indexing (Team et al., 2023), retaining Malay conceptual nuance and preserving philosophical and experiential dimensions. These differing epistemic orientations reflect each model's training data, design priorities, and interpretive frameworks. Collectively, the findings highlight broader implications for cultural knowledge preservation, equitable representation in digital systems, and the capacity of AI to transmit minority language wisdom without distorting its underlying worldview.

Cultural Meaning

Table 3. Comparative Analysis of Cultural Interpretation of Perambahan Brunei Across AI Platforms

Item	Perambahan Brunei	Chatgpt	Deepseek	Gemini
Human traits	Adi-beradi gimbar kepala	In Malay/Brunei culture, family and kinship are highly valued. However, conflicts, arguments, or rivalries within the family are natural — even among siblings.	Siblings must always protect, comfort, and defend each other's honor and well-being.	This interpretation would suggest that when siblings argue, it is a very noisy and public affair (a "commotion"), or it could be a sarcastic statement that siblings, despite their bond, often cause trouble or conflict for each other (or their parents).
	Berbagi tubuh	Unity in kinship or relationships, inseparable bond.	To have an inseparable bond, where the well-being of one is intrinsically tied to the well-being of the other	Unity in Matrimony/Love Deep Intimacy/Sacrifice Parent-Child Bond
Animal	Ambuk tunggal	Describes an only child or someone isolated in kinship, highlighting vulnerability in a society that values family support.	An individual who operates alone, willfully separating themselves from the community and its collective wisdom.	Social Isolation: The cultural interpretation is used to describe a person who is not liked by others, has poor social habits, and is consequently shunned or ostracized by the community. They are left alone because of their own actions. Bad Reputation: It often implies that the person is an *ambuk tunggal* (a sole outcast) because of negative traits like arrogance (*suka membesar-besarkan diri* - liking to boast about oneself) or other socially unacceptable behavior.
	Macam bagas ucing beranak	Symbol of domestic mess and noisy disorder, reflecting lived village experiences.	An action or situation that is completely the wrong way round; an absurd reversal of the natural order.	Sudden Urgency: It is culturally used to describe a state of extreme, sudden panic or haste where someone rushes or acts without any prior planning or warning. Overwhelming Speed: It emphasizes an action that is done extremely quickly and in a state of commotion or emotional distress, resembling an uncontrolled animal instinct.

continued on following page

Table 3. Continued

Item	Perambahan Brunei	Chatgpt	Deepseek	Gemini
Nature	Aing inda sajakan pasang, ada jua masanya surut	A reflection of how traditional communities drew wisdom from natural cycles, linking tides to human fate.	Life is cyclical. Good fortune and bad fortune are both temporary states that will inevitably change.	Acceptance of Cycles: It reflects the cultural wisdom that all things in life are cyclical. Prosperity (*pasang*) is never permanent, and loss or hardship (*surut*) is an inevitable part of existence.
	Bertuhur berdalam	Reflects the cultural value of persistence in searching for knowledge or truth, rooted in agrarian life.	To diligently and progressively build upon one's knowledge, skills, or social standing over time.	Patience in Observation: It encourages patience when evaluating a situation, a relationship, or a body of work. A person's character or a project's quality may seem simple or unremarkable at first (*tuhur*), but true value and complexity are revealed upon closer inspection or continued experience (*dalam*). Gradual Understanding: In the context of learning or knowledge, it suggests that genuine understanding and wisdom are not achieved quickly but require continuous effort to delve beneath the surface.

The cultural domain reveals the most significant interpretive disparities between platforms, with fundamental implications for how AI systems represent and transmit Southeast Asian relational epistemologies. In the analysis of *adi-beradi gimbar kepala*, ChatGPT demonstrates surface-level cultural acknowledgement, recognising that "in Malay/Brunei culture, family and kinship are highly valued," yet systematically subordinates this cultural insight to Western individualistic frameworks by emphasising conflict as a natural social phenomenon rather than as a culturally-specific relational paradox. The platform's interpretation positions conflict naturalisation within universal psychological registers rather than within Bruneian cosmologies of familial obligation and relational complexity. DeepSeek advances beyond this reductionism by explicitly articulating the protective function of sibling bonds ("Siblings must always protect, comfort, and defend each other's honour and well-being"), thereby demonstrating awareness of honour-based relational systems prevalent in Southeast Asian kinship structures. However, this interpretation privileges defensive solidarity whilst potentially obscuring the proverb's deeper philosophical proposition regarding the coexistence of intimacy and contestation.

Gemini's interpretation demonstrates superior cultural literacy by acknowledging both the "noisy and public affair" dimension of sibling conflict and its irrelevance to fundamental bonds, thereby preserving the paradoxical wisdom that Bruneian proverbs encode. Most significantly, Gemini contextualises the interpretation within lived experience ("sibling relationships cause trouble or conflict for each other or their parents"), grounding cultural meaning in phenomenological reality rather than abstract relationality. For *berbagi tubuh*, Gemini explicitly engages with the Malay terminology of matrimonial and filial bonds, situating the proverb within specific kinship relationships rather than treating it as a generalised statement about unity. The invocation of "Deep Intimacy" and "Sacrifice" as distinct cultural dimensions illustrates a nuanced understanding that Bruneian conceptualisations of shared existence encompass both emotional depth and material responsibilities. DeepSeek's formulation "an inseparable bond, where the well-being of one is intrinsically tied to the well-being of the other" articulates the ontological interdependence whilst remaining somewhat generic in its cultural specificity, failing to distinguish between romantic union, parental connection, and siblinghood as culturally differentiated forms of berbagi tubuh.

The animal domain reveals the most troubling patterns of AI cultural interpretation (Hagendorff et al., 2023), where all three platforms demonstrate a significant capacity to impose normative social judgments on culturally specific animal symbolism. For *ambuk tunggal*, ChatGPT's rendering as "an only child or someone isolated in kinship, highlighting vulnerability in a society that values family support" remains descriptively accurate, while naturalising a deficit-based interpretation of solitude that may not inherently reside in the original Bruneian expression. The platform treats isolation as inherently pathological within collectivist frameworks rather than as a morally neutral existential condition (D'Amato, 2025). DeepSeek's interpretation significantly distorts the cultural meaning through moral-juridical framing, characterising the *ambuk tunggal* as "an individual who operates alone, wilfully separating themselves from the community and its collective wisdom," thereby transforming a description of isolation into a normative condemnation of individualism. This interpretation reflects a dominant collectivist ideology that punishes deviation and values conformity, potentially misrepresenting Bruneian cultural attitudes toward individual autonomy and community participation.

Gemini's interpretation "Social Isolation: The cultural interpretation is used to describe a person who is not liked by others, has poor social habits, and is consequently shunned or ostracised by the community" explicitly names the social mechanism of exclusion whilst simultaneously invoking "Bad Reputation" and character deficiency (suka membesar-besarkan diri) as causal explanations. This interpretation demonstrates explicit cultural literacy (Nanduri, 2024) through the use of Malay terminology whilst simultaneously embedding judgmental frameworks that may distort the proverb's phenomenological content. Critically, all three platforms transform what may be a descriptive statement about isolation (the state of being alone) into a normative condemnation (the person deserves to be alone because of a character deficiency). This systematic transformation reveals how AI systems encode conservative social normativity into ostensibly neutral cultural interpretation.

The *macam bagas ucing beranak* analysis demonstrates parallel normative distortions across all three platforms. ChatGPT describes "domestic mess and noisy disorder, reflecting lived village experiences," which provides a culturally grounded description whilst potentially romanticising rural chaos. DeepSeek's interpretation inverts the proverb's semantic content entirely, framing the expression as describing "an action or situation that is completely the wrong way round; an absurd reversal of the natural order," thereby introducing an ontological transgression framework that is absent from expressions regarding sudden, overwhelming situations. Most significantly, Gemini's multi-faceted interpretation demonstrates superior semantic richness (Ye et al., 2025) through articulating "Sudden Urgency," "Overwhelming Speed," and the affective dimension of "uncontrolled animal instinct," thereby preserving the sensory and temporal immediacy of the image while grounding it in culturally recognisable experiences of panic and haste.

The nature domain demonstrates the most promising cross-platform alignment, though significant interpretive disparities persist regarding the philosophical and epistemological dimensions of nature-based proverbs. For *aing inda sajakan pasang, ada jua masanya surut,* ChatGPT explicitly situates the proverb within "how traditional communities drew wisdom from natural cycles, linking tides to human fate," thereby showing a deep understanding of the connection between celestial and tidal phenomena and human time, which is a key aspect of much Southeast Asian cosmology (Loeffen, 2025). DeepSeek's interpretation advances this recognition through explicit philosophical framing ("Life is cyclical. Good fortune and bad fortune are both temporary states that will inevitably change"), thereby extracting the metaphysical principle from its natural grounding. Gemini's interpretation "Acceptance of Cycles: It reflects the cultural wisdom that all things in life are cyclical. Prosperity (*pasang*) is never permanent, and loss or hardship *(surut)* is an inevitable part of existence" demonstrates explicit cultural terminology, thereby grounding the philosophical principle in Malay conceptual categories (*pasang dan surut)* that extend beyond literal tidal reference into metaphorical domains of fortune and adversity. Critically, the *aing inda sajakan pasang* interpretation across all three platforms emphasises acceptance, resignation, or humility as appropriate psychological responses to cyclical reality. While this interpretation may align with Buddhist or Islamic philosophical traditions prevalent in Brunei Darussalam, it potentially imposes normative psychological framings onto what may function primarily as a descriptive ontological statement regarding the nature of temporality rather than as prescriptive guidance regarding emotional disposition. The proverb describes cyclical reality; the platforms systematically interpret this description as implying prescribed psychological attitudes (such as humility, patience, and acceptance), thereby transforming cosmological observation into an ethical injunction.

The *bertuhur berdalam* interpretation reveals the deepest philosophical divergence. ChatGPT's characterisation "Reflects the cultural value of persistence in searching for knowledge or truth, rooted in agrarian life" appropriately contextualises the proverb within epistemological and occupational domains, whilst potentially oversimplifying the depth-related metaphor. DeepSeek's interpretation frames the proverb as prescribing progressive knowledge accumulation ("To diligently and progressively build upon one's knowledge, skills, or social standing over time"), thereby imposing a linear developmental epistemology onto an expression potentially more aligned with hermeneutical or contemplative philosophical traditions. Gemini's interpretation demonstrates remarkable sophistication through explicitly introducing the Malay temporal-epistemological framework of tahur (surface appearance) and dalam (depth/interior), distinguishing between initial superficial assessment and achieved deeper understanding through sustained engagement (*dalam* as both depth and continued dwelling). The invocation of "Patience in Observation" and "Gradual Understanding", grounded in continued experiential engagement *(dalam),* positions the proverb within contemplative epistemologies that valorise incremental revelation through sustained attention rather than systematic knowledge accumulation.

Several overarching patterns crystallise from the cultural domain analysis. ChatGPT consistently localises cultural meaning through explicit cultural acknowledgement (Spennemann, 2023) ("In Malay/Brunei culture") whilst simultaneously translating cultural significance into Western psychological and emotional registers (conflict naturalisation, vulnerability, individual identity). DeepSeek demonstrates intermediate-level cultural awareness through functional and instrumental framings that emphasise social order, protective mechanisms, and progressive development. The platform's interpretations, whilst acknowledging cultural specificity, consistently embed conservative social normativity and progressive developmentalism that may distort rather than illuminate Bruneian cultural wisdom. Most troublingly, DeepSeek exhibits a systematic tendency toward moral-juridical framing, transforming descriptive statements about social isolation or reversed natural order into normative condemnations of individual transgression.

Gemini demonstrates superior capacity for cultural literacy through explicit engagement (Sellami, 2025) with Malay terminology (*kebersamaan, solidaritas, suka membesar-besarkan diri, pasang, surut, tahur, dalam),* thereby grounding interpretation within indigenous conceptual categories rather than imposing external frameworks. Gemini's interpretations preserve phenomenological specificity (affective intensity, temporal urgency, sensory immediacy) whilst engaging with philosophical dimensions of Bruneian proverbs. However, even Gemini's sophisticated interpretations occasionally embed normative dimensions (characterising emotional intensity as "uncontrolled animal instinct") and demonstrate a systematic tendency to emphasise acceptance, patience, and gradual understanding as appropriate psychological responses to cosmological realities, potentially misrepresenting Bruneian cultural attitudes toward agency, temporality, and knowledge production.

These platform-specific epistemologies have profound implications for preserving cultural knowledge in digital contexts. If minority language wisdom is systematically filtered through AI interpretive frameworks that naturalise Western psychological individualism (ChatGPT), impose instrumental social normativity (DeepSeek), or occasionally embed conservative moral judgments (all platforms), the transmission of cultural knowledge occurs not through transparent representation but through systematic distortion that reproduces hegemonic epistemologies whilst appearing to preserve cultural specificity. The superior performance of Gemini in this domain does not resolve the fundamental challenge: the extent to which AI systems can function as vehicles for transmitting minority language wisdom without fundamentally altering it through the interpretive frameworks embedded within their training data, architectural priorities, and design philosophies. The proverbs *themselves adi-beradi gimbar kepala, berbagi tubuh, ambuk tunggal, macam bagas ucing beranak, aing inda sajakan pasang, bertuhur berdalam* represent centuries of accumulated Bruneian cultural wisdom regarding relationality, community participation, temporality, and epistemology. That all three platforms consistently impose external interpretive frameworks, whether psychological, instrumental, or normative, suggests that AI-mediated cultural knowledge transmission requires explicit attention to the question of representational justice and epistemological sovereignty for minority language communities.

DISCUSSION

The comparative analysis of literal, metaphorical, and cultural interpretations across the three AI platforms reveals fundamentally distinct epistemological frameworks that systematically shape how each system engages with Bruneian proverbs. Rather than functioning as neutral translation mechanisms, these platforms constitute active sites of meaning-making where underlying training data distributions, architectural priorities, and design philosophies determine interpretive outcomes. ChatGPT demonstrates a consistent reductionist strategy, systematically abbreviating complex cultural expressions into psychological dispositions, emotional states, or universal human experiences (Yuan et al., 2024). This approach privileges accessibility and cognitive simplification, rendering proverbs comprehensible to international audiences whilst simultaneously stripping them of their ontological specificity and cultural particularity. The platform's interpretation of *berbagi tubuh* as "unity in kinship or inseparable bond," for instance, subsumes the intricate relational ontologies of Malay matrimonial and filial systems into a generalised Western romantic-dyadic model of paired emotional intimacy.

DeepSeek exhibits a distinctive epistemological orientation toward instrumental social functionality, interpreting proverbs through frameworks of protective solidarity, hierarchical respect, and progressive individual development (Deng et al., 2025). The platform consistently reframes proverbs, prescribing specific social functions: sibling bonds protect against external threats, individual autonomy must subordinate itself to community welfare, and knowledge accumulates progressively through determined effort. This interpretive strategy implicitly privileges social order, conformity, and normative behaviour, systematically penalising deviation and valorising alignment with collective norms (Kiaos, 2023). The characterisation of *ambuk tunggal* as describing "an individual who operates alone, wilfully separating themselves from the community and its collective wisdom" exemplifies this tendency, transforming what may be a morally neutral description of isolation into a normative condemnation of individualistic transgression. Gemini, by contrast, demonstrates superior capacity for cultural indexing and phenomenological precision (Zhao et al., 2024), explicitly engaging with Malay conceptual categories and preserving the sensory immediacy, temporal complexity, and philosophical sophistication embedded within proverbs. The platform's interpretation of *macam bagas ucing beranak* through the framework of "Sudden Urgency" and "Overwhelming Speed" maintains fidelity to both the experiential dimensions and the cultural specificity of the expression.

These platform-specific epistemologies do not emerge randomly but reflect systematic differences in training data composition, architectural design choices, and explicit or implicit design philosophies. The prevalence of Western academic and literary sources in training datasets, the dominance of English-language content, the architectural emphasis on logical consistency and universal applicability, and the design priority of generating outputs acceptable to global audiences collectively shape how each platform navigates the relationship between literal surface meaning and cultural depth (Egdom, 2024). The implications extend beyond technical assessment toward fundamental questions of how minority language knowledge circulates within digital contexts dominated by platforms designed and trained within hegemonic linguistic and epistemological frameworks.

The Semantic Gap: Navigating Literal, Metaphorical, and Cultural Registers

The comprehensive analysis reveals significant interpretive discontinuities as the analytical framework moves from literal meaning toward metaphorical and cultural dimensions. Literal interpretation, while varying in precision across platforms, generally achieves reasonable semantic adequacy by privileging direct word-for-word equivalence. DeepSeek's rendering of *ambuk tunggal* as "lone water buffalo" rather than the correct avian reference represents a notable exception, suggesting potential cross-linguistic interference or database confusion. However, at the metaphorical level, interpretive divergence becomes pronounced, with platforms demonstrating differential capacity to preserve the specific animal imagery that grounds metaphorical meaning whilst extending that imagery into abstract conceptual domains (Goatly, 2022). ChatGPT's reduction of *macam bagas ucing beranak* to "chaotic, unorganised, or restless situations" sacrifices the specific phenomenological image of a cat giving birth, thereby severing the metaphor from its sensory grounding. DeepSeek's inversion of the proverb into a hierarchical transgression narrative fundamentally distorts its semantic content.

The cultural domain demonstrates even more dramatic interpretive divergence, with platforms employing fundamentally distinct frameworks for understanding the social, relational, and philosophical significance of expressions. This progression from adequate literal translation toward increasingly distorted cultural interpretation suggests that the challenge of AI-mediated cultural knowledge transmission is not merely one of linguistic accuracy but of epistemological framework alignment. Literal meaning operates primarily within systems of referential correspondence (Allott & Textor, 2022) (does the translation correctly identify the denotative content?). Metaphorical meaning requires preservation of imagistic specificity and the logical connections between image and abstraction (does the translation preserve the sensory and conceptual structure through which meaning is generated?). Cultural meaning, however, demands engagement with epistemologies, ontologies, and cosmologies that are fundamentally embedded within Bruneian worldviews and frameworks, which may differ significantly from the underlying philosophical assumptions in AI training data.

Normative Dimensions and the Problem of Interpretive Authority

A critical insight emerging from the cultural domain analysis concerns the systematic embedding of normative judgments within ostensibly descriptive cultural interpretation. All three platforms, when interpreting proverbs regarding isolation (*ambuk tunggal*), overwhelming chaos (*macam bagas ucing beranak*), or individual autonomy, systematically transform descriptive statements into normative prescriptions. The *ambuk tunggal* does not merely describe solitude; it is interpreted as describing merited social exclusion resulting from character deficiency. The *macam bagas ucing beranak* does not merely represent chaotic simultaneity; it is interpreted as condemning disorderly conduct or natural transgression. This systematic translation of description into prescription reveals how AI systems, despite operating within frameworks of ostensible neutrality, actively constitute moral and social norms. These embedded normative dimensions do not represent neutral cultural translation but rather constitute active participation in the reproduction of conservative social values, emphasising conformity, hierarchical respect, social inclusion, and emotional restraint as culturally appropriate responses.

The authority of AI platforms in making these interpretive judgments warrants particular attention within minority language contexts. When international users consult these platforms regarding the cultural meaning of Bruneian proverbs, they encounter not neutral translation but active meaning-making shaped by platforms' underlying epistemologies and embedded normativities. The platforms' apparent authority, derived from computational sophistication, international accessibility, and ostensible neutrality, lends credibility to interpretations that may significantly distort indigenous meanings (Ofosu-Asare, 2025). The fact that Gemini, through superior engagement with Malay terminology and cultural specificity, produces interpretations more aligned with Bruneian cultural understanding suggests that interpretive accuracy is achievable but requires explicit design priorities favouring cultural specificity over universal accessibility.

Implications for Language Knowledge Transmission in Digital Contexts

The findings reveal fundamental tensions between the globalising logic of large language models and the particularising logic of language knowledge transmission. AI platforms designed to serve international audiences through accessible, simplified, universally comprehensible outputs systematically subordinate the epistemological specificity and cultural particularity that constitute the distinctive value of language expressions (Dervin, 2025). Bruneian proverbs encode centuries of accumulated cultural wisdom regarding relationality, temporality, knowledge production, and human agency within specific cosmological frameworks. When these expressions are filtered through AI interpretive frameworks designed for global accessibility, the transmission of knowledge occurs not through transparent representation but through systematic transformation that simultaneously preserves surface meanings whilst fundamentally altering underlying epistemologies.

The implications extend beyond technical questions of translation accuracy toward fundamental issues of cultural sovereignty and epistemic justice. The question is not merely whether AI platforms can accurately represent Bruneian cultural meaning, the evidence demonstrates that platforms like Gemini achieve substantially superior performance through explicit attention to cultural specificity, but rather whether the dominant logics of commercial AI development prioritise the preservation of minority language epistemologies alongside the accessibility and simplification demanded by global markets. The systematic reductionism of ChatGPT, the instrumental normativity of DeepSeek, and even the superior cultural literacy of Gemini all emerge within commercial incentive structures that privilege international accessibility, user satisfaction, and operational efficiency over the preservation of indigenous epistemological frameworks.

The research demonstrates that superior cultural representation is technically achievable through explicit attention to linguistic and cultural specificity, the incorporation of diverse training sources, and design priorities that value both cultural authenticity and accessibility. However, the realisation of representational justice for Bruneian language communities requires more than the technical improvement of existing systems. It requires the development of community-controlled interpretive frameworks, the training of minority language-specific models using authoritative community sources, and the establishment of institutional structures that enable minority language communities to exercise authority over how their knowledge is represented and transmitted within digital contexts. Without such institutional innovations, even improved AI performance represents only marginal progress toward genuine epistemological sovereignty.

Building on the previous analyses, these findings provide a clear basis for synthesising the empirical patterns observed across different platforms and for discussing their broader theoretical and methodological implications. The analysis shows that literal translation is generally achievable across contemporary systems, although the degree of lexical and syntactic precision varies significantly. A more significant divergence occurs at the level of metaphorical interpretation, where platform-specific patterns reflect differing underlying epistemologies and representational assumptions embedded in the system's design. Cultural meaning poses the most significant challenge, as all platforms demonstrate notable misrepresentation by imposing external interpretive frameworks that override indigenous semantic and pragmatic norms. Language maintenance is increasingly influenced by power dynamics where dominant languages and institutional structures marginalise minority linguistic systems. This limits the transmission of culturally embedded meanings and constrains interpretive autonomy (Zulfadzlee & Chuchu, 2021). As Eriksson Krutrök & Poromaa Isling (2025) argue, minority languages risk becoming *techno-linguistically suspended,* symbolically recognised yet functionally excluded from AI infrastructures, an observation that aligns closely with the systematic cultural misrepresentation identified in this study. These limitations are not simply the result of isolated errors; instead, they manifest as systematic biases stemming from the composition of training data, modelling priorities, and architectural choices. Notably, the findings indicate that a higher level of cultural literacy is technically achievable, as demonstrated by the comparatively stronger performance of Gemini.

This suggests that culturally informed translation and interpretation are not merely theoretical ideals but rather tangible outcomes of design.

Table 4. Summary of Platform-Specific Interpretive Patterns

Dimension	ChatGPT	Deepseek	Gemini
Literal translation accuracy	Generally reliable and precise; maintains surface-level lexical meaning with moderate contextual sensitivity	Achievable but inconsistent; occasional ambiguity and dual readings	High precision with stable alignment between form and meaning
Metaphorical interpretation	Tends toward analytic simplification; reinterprets metaphors through abstract or psychological categories	Instrumental and norm-oriented; metaphors reframed as social prescriptions	Preserves metaphorical structure while retaining source-domain logic
Cultural meaning representation	Frequently filtered through external (largely Western) interpretive frameworks	Strongly collectivist and normative; prioritises social harmony over semantic nuance	Demonstrates higher cultural fidelity; closer alignment with indigenous conceptual logics

CONCLUSION

The comparative analysis of ChatGPT, Gemini, and DeepSeek demonstrates that while contemporary AI systems are capable of generating surface-level translations and partial cultural insights, their interpretations of *Perambahan Brunei* remain shaped by the epistemologies embedded within their training data and design philosophies. ChatGPT's accessible but reductionist framing, DeepSeek's instrumental and normative orientation, and Gemini's culturally attentive yet occasionally judgment-laden responses collectively reveal that AI does not simply translate cultural knowledge but actively reshapes it. These findings highlight the broader challenge of ensuring that digital technologies respect and engage with culturally grounded meanings. To strengthen culturally aligned interpretation, this chapter recommends three key steps: first, the diversification of training data to include authoritative Bruneian linguistic and cultural sources; second, the involvement of local cultural experts in validating and refining AI-generated interpretations; and third, the development of community-governed digital resources or locally trained models that prioritise cultural specificity over global simplification. By adopting these measures, AI can better support, not distort, the transmission of Bruneian cultural heritage, ensuring that traditional wisdom is preserved with its epistemological integrity intact.

REFERENCES

Allott, N., & Textor, M. (2022). Literal and metaphorical meaning: In search of a lost distinction. *Inquiry : a Journal of Medical Care Organization, Provision and Financing*, •••, 1–28.

Almahasees, Z., Al-Harahsheh, A., & Albudairi, Y. (2025). Assessing DeepSeek and ChatGPT in English-Arabic translation of political texts. *Humanities (Washington)*, *6*(1).

Boers, F. (2003). Applied linguistics perspectives on cross-cultural variation in conceptual metaphor. *Metaphor and Symbol*, *18*(4), 231–238.

Bridges, R. (2024). Oral traditions and digital preservation: Proverbs in contemporary society. *Journal of Cultural Heritage*, *45*(2), 112–127.

Carlà, M. M., Giannuzzi, F., Boselli, F., & Rizzo, S. (2024). Testing the power of Google DeepMind: Gemini versus ChatGPT 4 facing a European ophthalmology examination. *AJO International*, *1*(3), 100063.

Chaitanya, K., & Rolla, K. J. (2024). The Evolution and Impact of Large Language Models in Artificial Intelligence. In *Algorithms in Advanced Artificial Intelligence* (pp. 410–417). CRC Press.

Chen, X., & Zmire, Z. (2024). AI's Role in Enhancing Cross-Cultural Competence and Leadership through Online Education Programs. *The Proceedings of the International Conference on New Findings in Humanities and Social Sciences, 1*(1), 1.

D'Amato, K. (2025). ChatGPT: Towards AI subjectivity. *AI & Society*, *40*(3), 1627–1641.

Dabbagh, M. (2016). Cultural conceptualisations in proverbs: A linguistic-anthropological approach to understanding metaphor, schema, and category. *Linguistics and Anthropology Review*, *28*(4), 345–367.

Dabbagh, M. (2017). Understanding proverbs through cultural schemas, cultural categories, and cultural metaphors: Frameworks for evaluating cultural interpretation. *Intercultural Communication Studies*, *26*(1), 89–112.

De Santis, E., & Rizzi, A. (2025). Noosemia: toward a Cognitive and Phenomenological Account of Intentionality Attribution in Human-Generative AI Interaction. *arXiv preprint arXiv:2508.02622*.

Deng, Z., Ma, W., Han, Q. L., Zhou, W., Zhu, X., Wen, S., & Xiang, Y. (2025). Exploring DeepSeek: A Survey on Advances, Applications, Challenges and Future Directions. *IEEE/CAA Journal of Automatica Sinica, 12*(5), 872–893.

Dervin, F. (2025). *AI for critical interculturality*. Taylor & Francis.

Egdom, G. W. V. (2024). Bridging Linguistic Divides? A Critical Exploration of Machine Translation's Role in Fostering Cross-Cultural Accessibility in Literature. *Íkala, Revista de Lenguaje y Cultura, 29*(3).

Eriksson Krutrök, M., & Poromaa Isling, P. (2025). Technolinguistic suspension and the digital futures of minority languages: AI and Meänkieli in Sweden. *new media & society*, 14614448251385920. Firdaus, T., Bakari, L. Y., Octarianto, D. S., Khoirunnisa, R., Malini, D. R., & Mukarromah, M. Applying Grice's Maxims and Speech Act Theory to Analyze Pragmatics in Chatgpt and Gemini AI Dialogues. *Uluslararası Dil ve Çeviri Çalışmaları Dergisi*, *5*(1), 34–53.

Fitryansyah, M. A., & Fauziah, F. N. (2024). Bridging Traditions and Technology: AI in The Interpretation of Nusantara Religious Manuscripts. *Jurnal Lektur Keagamaan, 22*(2), 317–346.

Ghazanfari, M., Kumar, S., & Chen, L. (2025). Multilayered semantic complexity in non-Western proverbs: Challenges for artificial intelligence interpretation and cultural understanding. *Computational Linguistics and Digital Humanities*, *9*(1), 45–63.

Goatly, A. (2022). *Two dimensions of meaning: Similarity and contiguity in metaphor and metonymy, language, culture, and ecology*. Routledge.

Hadi, M. U., Qureshi, R., Shah, A., Irfan, M., Zafar, A., Shaikh, M. B., & Mirjalili, S. (2023). Large language models: a comprehensive survey of its applications, challenges, limitations, and future prospects. *Authorea preprints, 1*(3), 1–26.

Hadijah Haji Hassan. (1987). *Perambahan Brunei*. Bandar Seri Begawan: Dewan Bahasa dan Pustaka Brunei.

Hagendorff, T., Bossert, L. N., Tse, Y. F., & Singer, P. (2023). Speciesist bias in AI: How AI applications perpetuate discrimination and unfair outcomes against animals. *AI and Ethics*, *3*(3), 717–734.

Haji-Othman, N. A., McLellan, J., & Deterding, D. (2016). *The Use and Status of Language in Brunei Darussalam*. Springer.

Hakkarainen, S., & Engelbrecht, K. (2024). *Automatic Idiomatic Expression Detection. Comparison Between GPT-4 and Gemini Pro Prompt Engineering & LSTM-RNN Construction.* [Master's Thesis, University of Gothenburg]

Hamdi, S. A., Hashem, R. A., Holbah, W. A., Azi, Y., & Mohammed, S. Y. (2023). Proverbs Translation for Intercultural Interaction: A Comparative Study between Arabic and English Using Artificial Intelligence. *World Journal of English Language*, *13*(7), 282.

Harnad, S. (2024). Language writ large: LLMS, ChatGPT, grounding, meaning and understanding. *arXiv preprint arXiv:2402.02243*.

Helm, P., Bella, G., Koch, G., & Giunchiglia, F. (2024). Diversity and language technology: How language modelling bias causes epistemic injustice. *Ethics and Information Technology*, *26*(1), 8.

Joshi, S. (2025). A Comprehensive Review of DeepSeek: Performance, Architecture and Capabilities [Preprint]. doi: DOI: 10.20944/preprints202503.1887.v1

Kiaos, T. (2023). An interpretative framework for analysing managerial ideology, normative control, organisational culture and the self. *Cogent Business & Management*, *10*(1), 2163795.

Kovač, I., Stamenković, D., & Williams, M. (2025). Cultural bias in large language models: Mechanisms, manifestations, and mitigation strategies for reducing Western-centric epistemological dominance. *ACM Transactions on Computing and Society*, *8*(2), 1–28.

Lakoff, G. (2006). Conceptual metaphor. *Cognitive linguistics: Basic readings, 34*, 185–238.

Lee, H., Kim, H., & Yan, W. (2025). Promoting inclusive AI and technology in K-12: A review of context, instructional strategies, and learning outcomes. *Computers and Education: Artificial Intelligence*, 100478.

Loeffen, R. (2025). Revisiting Earth Expansion: Mass-Energy Growth in Celestial Bodies Through the Cosmic Influx Theory, in Collaboration with ChatGPT [Preprint]. Doi DOI: 10.5281/zenodo.14586147

Lomotey, E. K., Karikari-Bowen, P., & Boateng, S. (2021). Paremiology and pragmatics: Proverbs as communicative acts in a cultural context. *African Journal of Linguistics and Languages*, *7*(3), 78–96.

Luo, X. (2023). Cross-Cultural Adaptation Framework for Enhancing Large Language Model Outputs in Multilingual Contexts. *Journal of Advanced Computing Systems*, *3*(5), 48–62.

Masoud, R., Liu, Z., Ferianc, M., Treleaven, P. C., & Rodrigues, M. R. (2025). Cultural alignment in large language models: An explanatory analysis based on Hofstede's cultural dimensions. In *Proceedings of the 31st International Conference on Computational Linguistics* (pp. 8474–8503).

Montanari, F. (2025). ChatGPT and the others: Artificial intelligence, social actors, and political communication. A tentative sociosemiotic glance. *Semiotica, 2025*(262), 189–212.

Morgado, E., Leonido, L., Pereira, A., & Gouveia, L. B. (2025). Technology-Mediated Education: Impact of AI on the main distance learning modalities. *Educational Process: International Journal, 16*, e2025211.

Nanduri, D. K. (2024). *Exploring the Role of Generative Artificial Intelligence in Culturally Relevant Storytelling for Native Language Learning Among Children* (Master's thesis, University of Maryland, College Park).

Neuman, Y. (2024). *AI for Understanding Context.* Springer.

Obeidat, M. M., Haider, A. S., Tair, S. A., & Sahari, Y. (2024). Analysing the Performance of Gemini, ChatGPT, and Google Translate in Rendering English Idioms into Arabic. *FWU Journal of Social Sciences, 18*(4).

Ofosu-Asare, Y. (2025). Cognitive imperialism in artificial intelligence: Counteracting bias with indigenous epistemologies. *AI & Society, 40*(4), 3045–3061.

Ponti, E. M., O'Horán, H., Berzak, Y., Vulić, I., Reichart, R., Poibeau, T., & Korhonen, A. (2019). Modelling language variation and universals: A survey on typological linguistics for natural language processing. *Computational Linguistics, 45*(3), 559–601.

Puppel, M. (2024). *A comparative study of technical and creative text translation: evaluating the performance of ChatGPT* (Master's thesis, University of Malta).

Puspitasari, F. D., Zhang, C., Dam, S. K., Zhang, M., Kim, T. H., Hong, C. S., & Yang, Y. (2025). Deepseek models: A comprehensive survey of methods and applications. *Authorea Preprints.*

Rane, N., Choudhary, S., & Rane, J. (2024). Gemini versus ChatGPT: Applications, performance, architecture, capabilities, and implementation. *Journal of Applied Artificial Intelligence, 5*(1), 69–93.

Sani, H. M. (2025). ChatGPT and the Future of Language Learning and Linguistic Diversity in Africa. In *Sociocultural and Multicultural Meanings in Online Communication* (pp. 437–464). IGI Global Scientific Publishing.

Segerer, R. (2025). Cultural Value Alignment in Large Language Models: A Prompt-based Analysis of Schwartz Values in Gemini, ChatGPT, and DeepSeek. *arXiv preprint arXiv:2505.17112*.

Sellami, A. (2025). AI-infused Immersion: Cultivating EFL Learners' Intercultural Sensitivity through Google's Gemini AI Chatbot. *Journal of Studies in Language* [JSLCS]. *Cultura e Scuola*, *8*(2), 250–270.

Shanmuga, S. M., Jupally, S. H., & Vempalli, N. R. (2025). Models and Frameworks for Digital Preservation of Indigenous Knowledge Systems. In *Revaluation and Preservation of Indigenous Knowledge Systems in Modern Society* (pp. 511–534). IGI Global Scientific Publishing.

Speer, A. (2018). Contemplation and Philosophy: A Historical and Systematic Approach. In *Contemplation and Philosophy: Scholastic and Mystical Modes of Medieval Philosophical Thought* (pp. 79–107). Brill.

Spennemann, D. H. (2023). ChatGPT and the generation of digitally born "knowledge": How does a generative AI language model interpret cultural heritage values? *Knowledge (Beverly Hills, Calif.)*, *3*(3), 480–512.

Tanverdi, B. (2024). *Accumulating Architectural Knowledge Through Practice-Based Engagement: Experiential Learning Modules* (Doctoral dissertation, Middle East Technical University (Turkey)).

Team, G., Anil, R., Borgeaud, S., Alayrac, J. B., Yu, J., Soricut, R., . . . Blanco, L. (2023). Gemini: a family of highly capable multimodal models. *arXiv preprint arXiv:2312.11805*.

Teubner, T., Flath, C. M., Weinhardt, C., Van Der Aalst, W., & Hinz, O. (2023). Welcome to the era of ChatGPT et al., the prospects of large language models. *Business & Information Systems Engineering, 65*(2), 95–101.

UNESCO. (2024). Indigenous languages in the digital age: Preservation, transmission, and epistemological sovereignty in technological contexts. United Nations Educational. *Scientific and Cultural Organization Reports*, *45*(1), 1–89.

Viidalepp, A. (2023). The Expected AI as a Sociocultural Construct and Its Impact on the Discourse on Technology.

Wang, Y., & Zhao, Y. (2023). Gemini in reasoning: Unveiling commonsense in multimodal large language models. *arXiv preprint arXiv:2312.17661*.

Weigang, L., & Brom, P. C. (2025). The paradox of poetic intent in back-translation: Evaluating the quality of large language models in Chinese translation. *arXiv preprint arXiv:2504.16286.*

Wu, J., & Cai, Y. (2025). The Paradox of AI Knowledge: A Blockchain-Based Approach to Decentralised Governance in the Chinese New Media Industry. *Future Internet*, *17*(10), 479.

Xia, Y., Shin, S. Y., & Kim, J. C. (2024). Cross-cultural intelligent language learning system (CILS): Leveraging AI to facilitate language learning strategies in cross-cultural communication. *Applied Sciences (Basel, Switzerland)*, *14*(13), 5651.

Ye, Y., Zhang, Z., Ma, T., Wang, Z., Li, Y., Hou, S., ... & Chawla, N. (2025). LLM-s4All: A Review of Large Language Models Across Academic Disciplines.

Yenduri, G., Ramalingam, M., Selvi, G. C., Supriya, Y., Srivastava, G., Maddikunta, P. K. R., & Gadekallu, T. R. (2024). GPT (generative pre-trained transformer)—A comprehensive review on enabling technologies, potential applications, emerging challenges, and future directions. *IEEE Access : Practical Innovations, Open Solutions*, *12*, 54608–54649.

Yuan, H., Che, Z., Li, S., Zhang, Y., Hu, X., & Luo, S. (2024). The high-dimensional psychological profile and cultural bias of ChatGPT. *arXiv preprint arXiv:2405.03387.*

Zampatti, S., Peconi, C., Megalizzi, D., Calvino, G., Trastulli, G., Cascella, R., & Giardina, E. (2024). Innovations in medicine: Exploring ChatGPT's impact on rare disorder management. *Genes*, *15*(4), 421. PMID: 38674356

Zhai, X. (2025). DeepSeek: Transforming the foundations of education [Preprint]. doi: DOI: 10.20944/preprints202503.1776.v1

Zhao, J., Huang, C., & Li, X. (2024). A comparative study of cultural hallucination in large language models on culturally specific ethical questions [Preprint]. https://doi.org/DOI: 10.21203/rs.3.rs-4566507/v1

Ziems, C., Held, W. B., Yu, J., Goldberg, A., Grusky, D., & Yang, D. (2025, November). Culture Cartography: Mapping the Landscape of Cultural Knowledge. In *Proceedings of the 2025 Conference on Empirical Methods in Natural Language Processing* (pp. 1739-1757).

Chapter 5
English Teachers' Cognitions of the Role of AI in Supporting Learners' Autonomous Reading Practices

Özlem Utku-Bilici
https://orcid.org/0000-0001-7978-9195
Recep Tayyip Erdoğan University, Turkey

ABSTRACT

Artificial intelligence-based applications have also become increasingly prevalent within teaching and learning environments. Given these tools' integration into learning environments, the main objective of this explanatory sequential mixed-method research is to examine the perceptions of English teachers regarding how the integration of such tools into EFL reading processes affects student autonomy. To this end, data were collected through a questionnaire and semi-structured interviews, with the questionnaire administered to 250 teachers and interviews conducted with 20 teachers. Findings indicate that teachers believe AI tools can support student autonomy in EFL reading processes. However, it can be inferred from the results that this support is conditional upon teachers' pedagogical approaches and their awareness of how to integrate such tools effectively. Moreover, ethical considerations and the potential risk of students becoming overly dependent on these tools highlight the need for a balanced and responsible integration of AI in educational contexts.

DOI: 10.4018/979-8-3373-4397-6.ch005

INTRODUCTION

Teachers can be regarded as the cornerstone of the teaching and learning process, serving as the primary stakeholders who exert the most significant influence on educational outcomes. In the context of foreign language education, this role becomes even more pronounced, as the process encompasses the development of both receptive (listening and reading) and productive (speaking and writing) skills. Given this dual nature, it is reasonable to assert that teachers share an equally vital responsibility with learners in facilitating effective foreign language acquisition, not only by transmitting linguistic knowledge but also by fostering communicative competence and learner autonomy.

With the rapid advancement of digital technologies and the growing prevalence of artificial intelligence (AI)-based tools, contemporary learning environments have witnessed a profound reconfiguration of roles and responsibilities. Whereas teachers were once regarded as the primary sources of knowledge, the increasing accessibility of technological resources has enabled students to seek and construct knowledge through a variety of online platforms. Consequently, the teacher's role has evolved from that of a knowledge transmitter to that of a guide and facilitator, supporting learners in critically engaging with information and developing greater autonomy in their learning processes.

Essentially, this transformation in teacher roles and the extensive pool of resources offered to students have also dramatically affected English as a Foreign Language (EFL) learning processes. In his review study, Jiang (2022) examined how artificial intelligence applications have affected EFL teaching and learning procedures and revealed the ongoing and increasingly expanding network of relationships between the two phenomena. Similarly, Sumakul et al. (2022) also examined the role of AI-based tools in EFL classrooms in their research and their findings revealed positive outcomes. More explicitly, it was concluded that the systematic integration of AI-focused applications into EFL teaching and learning environments enables teachers to conduct the teaching process more robustly, while also enabling students to undergo a more profitable learning process.

Fidan and Kasimi (2023) could not overlook the growing network of relationships between artificial intelligence and EFL learning processes and analysed articles written on this subject within their research. Their findings indicate that the increasing visibility of AI-based tools is affecting EFL learning processes and this network of influence is also permeating the scientific context. In other words, Fidan and Kasimi's (2023) bibliometric research reveals that in recent years, numerous studies have been conducted on AI applications and EFL learning and teaching processes, and the reality and popularity of the subject are likely to increase this interest even further.

The present study, building upon this expanding network of relationships, has been constructed around the concepts of the EFL reading process, student autonomy, and teacher cognition, and has conducted an examination of these three critical phenomena and the integration of artificial intelligence applications into EFL environments. To be more specific, the current research was designed using an explanatory sequential mixed methods research design, with a questionnaire and semi-structured interviews as the primary data collection tools. With the participation of a total of 250 English teachers, the current study aimed to examine the participating teachers' cognition regarding the impact of integrating artificial intelligence tools into EFL reading processes on student autonomy. In the subsequent sections of the study, the theoretical framework and the related literature addressing this issue are first presented. This is followed by an explanation of the methodological framework, after which the findings are discussed, and the study is concluded with the sections of conclusion, pedagogical implications, suggestions for further research, and limitations.

1. THEORETICAL LENS AND LITERATURE REVIEW

The present study is founded upon four fundamental concepts, namely teacher cognition, learner autonomy, and AI in foreign language learning. Each concept assumes critical significance within the scope of the study, and a thorough assimilation of the theoretical framework articulated in this section is indispensable for a meaningful interpretation of the findings. Besides, it is of great importance to throw light on the up-to-date research on each concept so that the observed gap in the related literature can be properly realised, and the purpose of the current study can be comprehensively understood.

1.1. Teacher Cognition

In its most fundamental sense, teacher cognition can be defined as what teachers think, know, believe, and do (Borg, 2003). In other words, teacher cognition can be understood as the unobservable dimension of teaching, namely teachers' mental lives (Borg, 2003, 2009). In a more elaborate formulation, it can be defined as the pedagogical information that teachers hold, their own practical expertise, their views, their thoughts about various levels of context, their instructional challenges or queries, and the core values or standards that they are attempting to put into practice (Borg, 2006). From this perspective, it is crucial to acknowledge the undeniable influence of teacher cognition on teaching and learning processes. Although teachers are no longer regarded as mere authorities but rather as guides and facilitators, the thoughts, knowledge, and beliefs they hold continue to exert a direct impact on learners. Consequently, teacher cognition entails a critical responsibility within the educational process.

Kalinowski et al. (2020) conducted a meta-analysis of professional development's effect on teachers' cognition and teaching practices, and highlighted the interactive association among professional development implementations, teacher cognition, and teaching practices. It can be concluded that teacher cognition is a multi-dimensional concept and plays a significant role in teaching circumstances. Barr and Askell-Williams (2020) specifically delved into self-regulated learning and teachers' epistemic cognition in their study, and they put a strong emphasis on "the need for researchers and practitioners to attend to teachers' epistemic cognition about [self-regulated learning]" (p. 203). Barr and Askell-Williams's (2020) study also demonstrates that exploring the constituents of teacher cognition is integral to sustaining classroom practices and ensuring their development on a robust and theoretically grounded basis.

Sun et al. (2022) utilised the implicit association test to explore the dimensions of teacher cognition, and they proposed that there are varied differences between what teachers say and what teachers think. They further explained that such differences are caused by personal, institutional, and social contexts which teachers are exposed to, and comprehending teacher cognition has a critical role in shaping learning environments. Yuan et al.'s (2022) methodological review study examines 69 empirical studies conducted between 2010 and 2020 on teacher cognition and comprehensively scrutinises the professional status, roles, cognitive aspects, and identity development of teacher trainers in the field of TESOL. The detailed and inclusive examination conducted within the scope of the study challenges the perception of teacher trainers as 'supermen/superwomen,' arguing for the visibility of their human aspects, difficulties, and learning needs, and sets a new agenda for future research (Yuan et al., 2022, pp. 26-27).

Building on the growing body of research addressing teacher cognition, Utku Bilici (2024) sought to illuminate the multidimensional nature of language teacher cognition. Through a comprehensive investigation employing a multiple case study design, she argued that language teacher cognition is fundamentally grounded in experiential dimensions. The study further underscored that a nuanced understanding and systematic exploration of teacher cognition across its various facets may exert a positive impact on both teaching and learning processes, thereby highlighting the importance of examining teacher cognition within context-specific frameworks. This conclusion is also consistent with the findings of other contemporary contributions to the field.

1.2. Learner Autonomy

In a landmark report written for the Council of Europe, Holec (1981) explicitly brought the concept of learner autonomy into the educational discourse. In its original form, autonomy was defined as the ability of the learners to take charge of their own learning experiences. Little (1991) broadened the concept's usage by highlighting that learner autonomy includes both the affective and cognitive aspects of the learning process. Furthermore, he emphasised that autonomous learners are able to consciously observe and control their own learning paths, which guarantees the continuity and sustainability of long-lasting learning outcomes.

Given that language learning is, by nature, an individual process, it is particularly pertinent to underline the critical importance of learner autonomy and the capacity of students to assume responsibility for their own learning trajectories. More precisely, when learners cultivate an acute awareness of their responsibilities within the language learning process and engage proactively with these responsibilities, their uniquely individual experiences may yield more fruitful, productive, and intellectually rewarding learning outcomes. In her exemplifying study, Tabiati (2016) investigated the determinants affecting learner autonomy in reading processes in EFL. She underscored that from early childhood onwards, students gradually—and often unconsciously—develop autonomy in reading. Besides, it was noted that these individual processes are an example of a universal acquisition that transcends contextual and cultural specificities. Rahmasari et al.'s (2025) study was conducted in a similar setting and focused on learner autonomy in EFL reading. The results were encouraging for the integration of self-regulated reading strategies into learning environments.

In their technology-oriented study, Slamet et al. (2024) concentrated on the notions of autonomous learning and self-directed learning behaviours and examined the impact of a gamification-based massive open online course (MOOC) on learners' autonomy and self-regulated learning practices. Their findings offer valuable insights for teachers, curriculum designers, and policymakers, specifically in recognising the pedagogical potential of technology-mediated environments to foster learners' autonomous learning practices. Moreover, considering that this era's students are frequently conceptualised as "digital natives" (Prensky, 2001), it appears reasonable to argue that aligning educational practices with both the expectations of the digital natives and the demands of the digital age is not only timely but necessary.

1.3. AI in Foreign Language Learning

AI in foreign language instruction has created new opportunities for individualised and flexible learning. Learners can practise real-world communication, get immediate feedback, and advance at their own speed with AI-powered tools, e.g., conversational chatbots, intelligent teaching systems, and automatic writing evaluators. Additionally, learner data can be analysed by machine learning algorithms to determine their strengths and shortcomings, which ensures more focused instructional interventions. Huynh (2024) put an emphasis on EFL learners' use of AI in their language learning process and carried out a literature review to explore how using AI in language learning influences learners' perceived autonomy. The findings reveal that AI integration into learning environments enable learners to become more autonomous and self-regulated.

Another comprehensive review article was written by Chong and Reinders (2025). The authors conducted a scoping evaluation of 61 empirical studies on learner autonomy—a topic that has evolved over four decades and relies heavily on theories and methodologies from other disciplines. Their analysis uncovers a wide range of autonomous conceptualisations and operationalizations, but barely any attempts to evaluate its progress. The analysis also puts forward numerous recommendations for future research, including the need to explicitly explain theoretical frameworks, investigate the impact of out-of-class learning in encouraging autonomy, and use more diverse research approaches.

In their systematic review article, Zhu and Wang (2025) also performed an in-depth analysis towards the current status and future implications of using AI in language education. The review illustrated that over the past ten years, the majority of research on AI in language education has concentrated on technologies including automated writing evaluation, bots, machine translation, and speech recognition. These advancements have a significant impact on helping students improve their writing and speaking abilities, but they also highlight the need for inclusive, extensive, and forward-looking research endeavours.

As noted in recent review articles (Chong & Reinders, 2025; De la Vall & Araya, 2023; Huynh, 2024; Woo & Choi, 2021; Zhu & Wang, 2025), even though AI has become a frequently employed and increasingly pervasive tool in the field of language learning, contemporary research exploring how it is utilised within learning processes remains rather limited. Building on this gap, the present study aims to investigate teachers' cognitions regarding the use of AI in EFL reading processes, acknowledging teachers as key agents who may serve as guides and facilitators in learners' educational trajectories. To this end, the study adopts an explanatory mixed-methods research design and seeks to address the following research questions:

1. To what extent are EFL teachers knowledgeable about and experienced in the use of AI tools within language education?
2. How do teachers perceive AI's role in supporting learner autonomy in reading?

3. What are teachers' perceptions regarding the pedagogical benefits and challenges of using AI in reading instruction?
4. How do teachers consider the ethical, cultural, and practical aspects of using AI in their classrooms?

2. METHODOLOGY

2.1. Research Design

The present study was designed by adopting the explanatory sequential mixed methods research design (Creswell & Creswell, 2018). This design necessitates a two-stage process that involves first collecting and analysing quantitative data, followed by collecting qualitative data based on the findings obtained. Quantitative results play a guiding role in the qualitative stage, from selecting participants to determining questions. The primary aim of this approach is to explain quantitative findings in depth using qualitative data, particularly to make sense of ambiguous, contradictory, or unexpected results.

In the present study, the research process was also conducted in two sequential phases. Initially, quantitative data were collected and analysed, after which participants were recruited from the original sample on a voluntary basis for the subsequent qualitative phase. Consistent with the primary purpose of the explanatory sequential mixed methods design, the qualitative findings were employed to elaborate upon and provide deeper insight into the quantitative results. Rather than engaging in a direct comparison between the two sets of findings, the central objective was to extend and enrich the quantitative outcomes through qualitative inquiry, thereby achieving a more comprehensive understanding of the research problem. This objective guided the overall design and implementation of the study.

2.2. Setting and Participants

The present study was carried out with the participation of English language teachers residing in Türkiye. Participants were selected from a variety of educational settings, including secondary schools, high schools, and universities, which ensures a wide spectrum of teaching contexts. Because the study employed an explanatory sequential mixed methods research design, the quantitative and qualitative phases conducted sequentially, and two sets of participants were included in the data collection procedure. However, it should be underlined that the participants getting involved in the qualitative phase also participated in the quantitative phase.

A total of 250 participants involved in the quantitative phase. As shown in Table 1, the sample consisted of 135 females (54%) and 115 males (46%). Participants' ages ranged from 22 to 60 years, with 24% aged 22–30, 36% aged 31–40, 26% aged 41–50, and 14% aged 51–60. Regarding school level, 20% in secondary schools, 29.2% in high schools, and 50.8% at universities. The distribution of years of teaching experience was 20% for 1–5 years, 24% for 6–10 years, 22% for 11–15 years, 16% for 16–20 years, and 18% for 21 years or more.

Table 1. Demographic characteristics of the participants in the quantitative phase.

Variable	Category	n	%	N
Gender	Female	135	54.0	250
	Male	115	46.0	250
Age	22–30	60	24.0	250
	31–40	90	36.0	250
	41–50	65	26.0	250
	51–60	35	14.0	250
School Level	Secondary School	50	20.0	250
	High School	73	29.2	250
	University	127	50.8	250
Years of Experience	1–5	50	20.0	250
	6–10	60	24.0	250
	11–15	55	22.0	250
	16–20	40	16.0	250
	21+	45	18.0	250

For the qualitative phase, 20 teachers were purposefully selected from the original sample on a voluntary basis. As presented in Table 2, 12 participants were female (60%) and 8 were male (40%). Their ages ranged from 22 to 58 years. School levels included secondary (5%), high school (35%), and university (60%). The participants' teaching experience varied, with 20% having 1–5 years, 30% 6–10 years, 25% 11–15 years, 15% 16–20 years, and 10% over 21 years of experience.

Table 2. Demographic characteristics of the participants in the qualitative phase.

Variable	Category	n	%	N
Gender	Female	12	60.0	20
	Male	8	40.0	20
Age	22–30	6	30.0	20
	31–40	7	35.0	20
	41–50	5	25.0	20
	51–60	2	10.0	20
School Level	Secondary School	1	5.0	20
	High School	7	35.0	20
	University	12	60.0	20
Years of Experience	1–5	4	20.0	20
	6–10	6	30.0	20
	11–15	5	25.0	20
	16–20	3	15.0	20
	21+	2	10.0	20

As noted previously, the present study employed an explanatory sequential mixed methods design, wherein quantitative data were collected initially, followed by the collection of qualitative data. Purposeful selection of participants for the qualitative phase from among those who had participated in the quantitative phase facilitated a more thorough exploration of the quantitative findings, enabled a thorough examination of the quantitative data and made the analysis phase more meaningful. Moreover, the inclusion of participants representing diverse ages, professional backgrounds, and levels of teaching experience contributed to a more comprehensive and in-depth examination of the research questions.

2.3. Data Collection Tools

In this study, the data were collected via two data collection instruments, namely a questionnaire and a semi-structured interview. The questionnaire was administered to 250 participants, while semi-structured interviews were conducted with 20 participants.

The questionnaire instrument was structured into five interconnected sections. The first section gathered demographic information to provide a contextual understanding of the participants. Building on this, the second section examined participants' familiarity with AI in language education through four targeted items. The third section then shifted focus to AI and learner autonomy in reading, encompassing five items designed to explore this relationship in depth. The fourth section addressed teacher cognitions and pedagogical perceptions, including four items that investigated participants' instructional beliefs and practices. Finally, the fifth section considered ethical, cultural, and practical aspects of integrating AI, comprising four items that contextualised the findings within broader educational considerations. All items across the five sections employed a combination of Likert-scale and multiple-response formats.

To give a more thorough, contextualised understanding of the questionnaire results, semi-structured interviews were undertaken after the quantitative phase. A total of ten questions were constructed, drawn methodically from the quantitative data, to guarantee that the qualitative phase specifically touched on the questionnaire's patterns, trends, and insights. In line with the second section of the questionnaire, two questions probed the participants' familiarity with AI in language education, particularly how they have implemented AI technologies into their teaching environments and the obstacles or opportunities they have faced. The third section included three questions about AI and learner autonomy in reading, prompting participants to consider how AI can support autonomous learning, how much it improves students' independent reading skills, and specific examples of tasks or strategies that promote learner autonomy. The fourth section issued three questions about teacher cognitions and pedagogical perceptions, and the participants were encouraged to articulate how their instructional beliefs influence their use of AI, evaluate the pedagogical benefits or limitations of AI, and discuss how it aligns with their classroom goals and routines. Finally, the fifth section addressed ethical, cultural, and practical aspects, with two questions exploring possible ethical or cultural concerns, as well as the impact of functional variables including classroom resources or student access on AI integration. A thorough and in-depth understanding of teachers' experiences, perceptions, and decision-making processes regarding the integration of AI in language education was facilitated by the carefully structured interviews. The interviews generated rich, comprehensive, and detailed qualitative data that meaningfully complemented the quantitative findings.

2.4. Data Analysis

The data obtained within the scope of the current research were analysed in two different dimensions, quantitative and qualitative, as they were collected through a questionnaire and semi-structured interviews. In the analysis of the quantitative data, descriptive statistics were calculated to reveal teachers' general tendencies, attitudes, and approaches towards AI applications. The data were summarised systematically using statistical measures in the form of percentages and frequency distributions.

The qualitative data were analysed using the hermeneutic approach to examine teachers' experiences, reflections, and decision-making processes in detail and to shed further light on the quantitative findings. This approach enabled the contextual interpretation of participants' statements (Gillo, 2021) and contributed to revealing the implicit meanings in teachers' discourses. The combined use of quantitative and qualitative data analysis is a requirement of the explanatory sequential mixed methods research design, and in the present study, the quantitative and qualitative analysis steps were meticulously followed. This approach enabled the research to be addressed from a holistic perspective and ensured that the findings complemented each other both at a numerical level and at an interpretative level. Thus, both the surveyable stances of participant teachers about the role of AI in supporting learners' autonomous reading practices and their more deeply understood experiential stances could be revealed.

2.5. Trustworthiness

The reliability of the research was ensured by following the four fundamental criteria outlined by Lincoln and Guba (1985), namely credibility, transferability, dependability, and confirmability. For the credibility criterion to be established, the reliability of the questionnaire used to obtain the quantitative data was verified by calculating Cronbach's Alpha coefficient. As for the qualitative data, direct quotations from teachers' views were used, and the comments were reflected in a data-driven manner. Furthermore, the qualitative findings obtained were examined and validated by a researcher who is an expert in the field of social sciences, which also ensured the confirmability of the qualitative data. Regarding the transferability and dependability criteria, the research process, participant characteristics, and data collection tools have been reported in detail. Besides, interview excerpts have been supported by contextual explanations so that the findings obtained can be adapted to different contexts. In addition, the combination of quantitative and qualitative data within the scope of the study enabled triangulation of the findings, which supported the overall trustworthiness of the research data.

2.6. Ethical Clearance

Ethical approval for this research was obtained prior to data collection to ensure adherence to institutional and research integrity standards. To this end, the present research study was approved by the Ethics Committee of Social and Human Sciences at Recep Tayyip Erdoğan University under protocol number 2025/498 on 9 July 2025.

3. FINDINGS AND DISCUSSION

In this section, the findings are presented and discussed under four main themes: the use of AI tools in teaching EFL reading, AI-supported learner autonomy, the perceived shift in instructional roles brought about by AI integration, and the ethical, cultural, and practical considerations involved in employing AI tools for EFL reading activities. To enhance clarity in line with the reviewer's recommendation, the number of respondents contributing to each dataset is explicitly stated: quantitative data were collected from 250 participants, while qualitative insights were drawn from interviews with 20 participants. Detailed demographic and procedural information concerning these participants has been provided in the *Methodology* section.

Consistent with the research design, the quantitative findings are first presented through descriptive statistics, followed by a deeper examination supported by the qualitative data. For each thematic sub-domain, illustrative interview excerpts are incorporated to strengthen the qualitative strand of the analysis and to offer a more vivid representation of how the triangulated data substantiate the thematic conclusions. This structured approach ensures methodological coherence and contributes to a more robust and transparent presentation of the study's findings.

3.1. Use of AI Tools in Teaching EFL Reading

The initial set of findings addressed within the scope of this study concerned the use of AI tools in EFL reading instruction. Under this overarching category, participants' familiarity with AI-based tools for language education was first examined. The results demonstrated that a substantial proportion of participants (33.2%) reported being familiar with the use of AI-supported tools in English language teaching and learning contexts. As presented in Table 3, 30.4% of the participants described themselves as moderately familiar, while 14.8% reported being slightly familiar, and 7.2% stated that they were not familiar at all with such tools. Notably, 14.4% of the respondents indicated that they were very familiar with AI-based tools. One of the participants who responded that she was very familiar also participated in the qualitative phase, and she summarised her familiarity as follows:

> *In today's world, AI has become so integrated into our lives that I believe it is actually impossible not to have noticed this development. I also believe that this advanced tool can be systematically integrated into language teaching processes because teaching is, as we know, a profession that requires considerable mental effort. I think it is very pragmatic to use such tools to make our work easier from time to time.*

Taken together, these findings suggest that, overall, the participating teachers demonstrated a relatively high level of familiarity with AI-supported tools and their potential applications in EFL reading instruction.

In addition to participants' familiarity levels, their confidence in using AI-based tools—particularly for supporting reading skills in language education—was also examined. The results revealed that a considerable proportion of the participants (36.4%) described themselves as moderately confident in employing such tools. Furthermore, 28.4% of the respondents reported being confident, and 10.0% identified themselves as very confident. On the other hand, 16.4% indicated that they were slightly confident, while 8.8% stated that they were not confident in using AI-based tools for reading instruction. These findings suggest that, despite a relatively high level of familiarity with AI-based technologies, some teachers may still experience certain uncertainties or dilemmas when integrating such tools into reading-related teaching practices. A teacher with 16 years of teaching experience who selected the "moderately familiar" option described the dilemma he experienced in his personal teaching journey with the following words:

It is, of course, no longer possible to be unaware of AI. We can, to some extent, see it in every area of life. It can also be used in education. However, I sometimes find it very impractical to introduce such tools to my students or include them in the process in my own classrooms because, ultimately, it can also lead to some loss of authority or a change in roles.

When the percentages related to the use of AI tools in reading instruction are examined, the results reveal a rather nuanced pattern. Despite the somewhat ambivalent tendencies observed in participants' familiarity and confidence levels, the data demonstrate that AI-supported tools are incorporated into reading instruction to varying extents. As shown in Table 3, 38.8% of the participants reported using such tools sometimes, 22.0% often, and 8.4% very often in their reading activities. Conversely, 19.6% of the respondents indicated that they rarely used AI-based tools, while 11.2% reported never using them. These percentages suggest that, although participants display relatively positive attitudes toward AI integration, the actual implementation of AI tools in EFL reading instruction remains limited among a notable proportion of teachers.

Table 3. Descriptive statistics for the use of AI tools in teaching EFL reading (N=250).

Item	Response Option	Frequency (n)	Percentage (%)
Familiarity with AI-supported tools for language education	Not familiar at all	18	7.2
	Slightly familiar	37	14.8
	Moderately familiar	76	30.4
	Familiar	83	33.2
	Very familiar	36	14.4
Confidence in integrating AI-supported reading tools into lessons	Not confident	22	8.8
	Slightly confident	41	16.4
	Moderately confident	91	36.4
	Confident	71	28.4
	Very confident	25	10.0
Frequency of using AI tools to support EFL reading activities	Never	28	11.2
	Rarely	49	19.6
	Sometimes	97	38.8
	Often	55	22.0
	Very often	21	8.4

Determining the extent to which AI tools are used in EFL reading instruction constitutes a valuable aspect of the present study. However, it is equally important to identify which specific tools are being utilised in these instructional processes. To address this issue, a multiple-response question was included in the questionnaire. The results revealed that a substantial majority of participants selected adaptive reading platforms as the most frequently used AI-supported tool. This quantitative finding was further substantiated by qualitative data obtained through semi-structured interviews. To exemplify, a female teacher with twelve years of teaching experience working at a university elaborated on this point, describing how often she uses AI tools and providing the following examples of the specific applications she employs:

> *Young people today use AI quite extensively and actually really enjoy these types of applications. I also work with young people, and at some point, I need to make this process appealing to them. That's why I like to incorporate AI-based applications into my lessons. That is, I don't incorporate them directly into lessons but rather recommend and use them in out-of-school learning processes. In this sense, I also find AI tools that suggest and generate text according to the students' level useful.*

Figure 1. AI-supported tools utilised in EFL reading instruction.

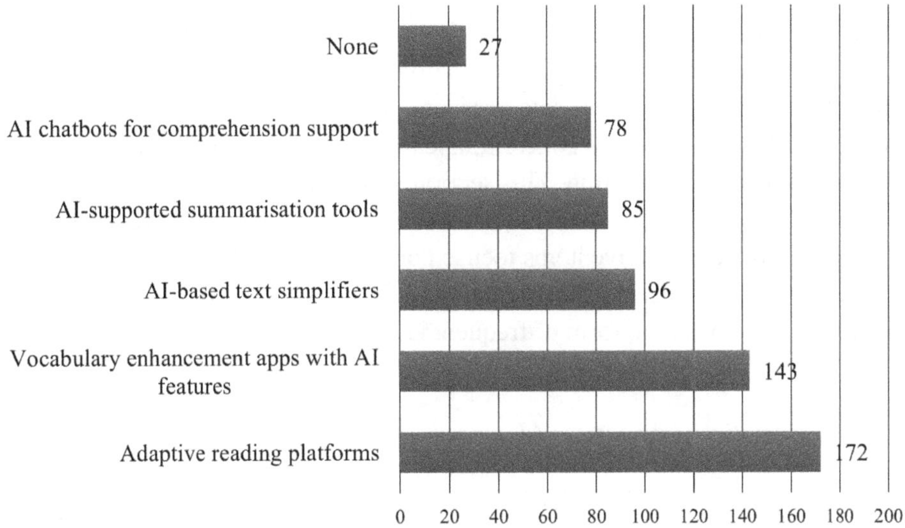

When examining participants' levels of familiarity, confidence, frequency of use, and the specific tools employed in relation to the use of AI tools in EFL reading instruction, it was observed that the participating teachers possess an awareness of AI-based learning tools and incorporate them to a certain extent within their teaching practices. This finding aligns with the conclusions of the comprehensive systematic review conducted by Celik et al. (2022), which has been widely cited in the literature. Specifically, Celik et al. (2022) highlighted that AI provides educators with a range of pedagogical opportunities that can enhance their teaching experiences. Furthermore, they underscored the pivotal role teachers play in the development and effective integration of AI-based educational applications. In a similar connection, Kolchenko (2018) examined the interaction between AI and teachers in his theoretical work, and he asserted that AI cannot fundamentally assume the roles and responsibilities of educators. Nonetheless, his analysis underscores that while AI is unable to replace teachers, learners who exhibit high levels of motivation and self-regulation are better equipped to utilise AI tools effectively. Through these competencies, such students are able to enhance the efficacy of their individual learning processes. Accordingly, the following section of this study will explore teachers' perspectives on the interaction between AI and students, with a particular focus on learner autonomy and the use of AI tools in EFL reading instruction.

3.2. AI-Supported Learner Autonomy in EFL Reading

The primary focus of the present study is to investigate teachers' cognitions regarding the role of AI in fostering learner autonomy within EFL reading processes. In this regard, all analyses and data presented aim to illuminate teachers' views, beliefs, attitudes, and practices concerning learner autonomy and the integration of AI tools into reading instruction. This section specifically addresses AI-supported learner autonomy in EFL reading. The questionnaire administered to participants as part of the study included five items focused on AI-supported learner autonomy in EFL reading. Table 4 displays the quantitative findings obtained from the analysis of the relevant items in the form of frequencies and percentages.

Table 4. Descriptive statistics for AI-supported learner autonomy in EFL reading (N=250).

Item	Response Option	Frequency (n)	Percentage (%)
I believe AI tools help learners become more autonomous EFL readers.	Strongly Disagree	10	4.0
	Disagree	25	10.0
	Neutral	50	20.0
	Agree	110	44.0
	Strongly Agree	55	22.0
AI-supported reading tools increase students' motivation to read in EFL independently.	Strongly Disagree	8	3.2
	Disagree	27	10.8
	Neutral	45	18.0
	Agree	115	46.0
	Strongly Agree	55	22.0
AI tools can personalise EFL reading tasks according to student needs.	Strongly Disagree	5	2.0
	Disagree	20	8.0
	Neutral	45	18.0
	Agree	120	48.0
	Strongly Agree	60	24.0

continued on following page

Table 4. Continued

Item	Response Option	Frequency (n)	Percentage (%)
AI tools effectively support learners in developing reading strategies (e.g., skimming, summarising, predicting).	Strongly Disagree	7	2.8
	Disagree	28	11.2
	Neutral	50	20.0
	Agree	110	44.0
	Strongly Agree	55	22.0
I am concerned that learners may become too dependent on AI-supported tools.	Strongly Disagree	15	6.0
	Disagree	40	16.0
	Neutral	60	24.0
	Agree	90	36.0
	Strongly Agree	45	18.0

As presented in Table 4, a substantial proportion of participants held favourable attitudes toward the contribution of AI tools to the development of learner autonomy. Specifically, 44.0% of the respondents agreed and 22.0% strongly agreed that such tools facilitate students' autonomous learning processes. Conversely, 10.0% disagreed and 4.0% strongly disagreed with this statement. These findings indicate a predominant consensus among participants regarding the positive influence of AI-based tools on the promotion of learner autonomy within the context of EFL reading instruction. To investigate whether AI-supported tools enhance students' motivation in EFL reading, participants were also asked to respond to a related questionnaire item. The results echoed the findings of the previous item: 46.0% of participants agreed, and 22.0% strongly agreed that AI-supported reading tools increase students' motivation to engage in independent reading. These findings suggest that the integration of AI tools may positively influence learners' motivation in EFL reading contexts. On this matter, a female participant teacher with three years of teaching experience, who could be described as a novice and who works at a university's foreign languages college, stated the following:

> *I am a teacher who has just started working, so I am actually quite close in age to my students. This generation has grown up particularly immersed in technology. There can certainly be downsides, but I, in fact, think AI tools can be used effectively, especially for language learning processes, because language learning is essentially an individual process and teachers take on more of a guiding role.*

Furthermore, supporting students in this way will be quite motivating, as they will see what they can do on their own.

It can be inferred from this excerpt that AI tools have the potential to motivate students and to make the language learning process more individualised and distinctive. Considering that reading is a receptive skill, it is reasonable to suggest that EFL learners who are competent in planning their own learning processes can effectively integrate AI tools into their autonomous learning activities.

The analysis of the questionnaire item investigating whether AI tools can be used to personalise students' EFL reading processes indicates that the vast majority of participants (i.e., 48% agreed and 24% strongly agreed) believe that AI tools can tailor reading tasks to individual student needs. However, 18% of the participants adopted a neutral stance on this matter, while 10% stated that they did not believe that AI tools can be used to personalise students' EFL reading processes. A female teacher with 13 years of experience who believes that AI tools cannot be used to personalise students' EFL reading processes explained her view as follows during a semi-structured interview:

AI is now truly quite visible in every area of our lives. So, it is impossible to ignore it. But... I think the language learning process is quite complex. So, students can certainly benefit from it, but if they use these tools to plan their own process, I do not think they can progress with very solid steps. Especially for reading skills, there are so many resources that students can get lost in them.

One of the most crucial aspects of becoming an autonomous language learner is the ability to employ a range of strategies effectively throughout the learning process. The significance of such strategies in EFL reading is indisputable (Li et al., 2022; Rianto, 2022; Utku Bilici & Subaşı, 2022). Accordingly, an item was included in the questionnaire to investigate whether AI tools support the use of reading strategies within EFL reading contexts. The findings closely align with those of the first item in this section. Specifically, a substantial majority of participants reported that AI tools effectively assist learners in developing and applying reading strategies (e.g., skimming, summarising, predicting). In this context, Ademola (2024, p. 118) highlights that the capacity of AI to provide personalised learning pathways, adaptive feedback, and advanced data analytics has the potential to elevate the educational process to exceptional levels. Achieving this transformative potential, he argues, requires deliberate policy development and careful implementation.

While learner autonomy in EFL reading and the integration of various strategies can lead to a more individualised learning experience, certain potential risks may also emerge. A particularly notable risk is that students may develop an overreliance on AI tools and depend on them entirely to direct their learning processes. To explore this issue, the participating teachers were also asked to provide input, and the results indicated considerable variability in their responses. To clarify, 36% of the participants agreed and 18% strongly agreed that they are concerned learners may become overly dependent on AI-supported tools. In contrast, 24% of the respondents expressed a neutral stance, while 16% disagreed and 6% strongly disagreed with this view. The following excerpts from two different participants demonstrate the varying opinions on this matter:

> *AI is certainly a wonderful blessing. Its use in learning processes has also begun to increase steadily. It definitely makes students more independent, but I think we should also ask ourselves this: what if students now consult artificial intelligence tools for everything? In that case, their efforts to be independent could unfortunately turn into dependence. (A male teacher with 10 years of teaching experience, working in a high school)*

> *I essentially hold the view that a student capable of integrating AI tools into their language learning process has already gained independence. In other words, even if they resort to such tools, their motivation and awareness of their responsibilities prevent them from blindly relying on AI. (A male teacher with 8 years of teaching experience, working in a university)*

As it can be understood from these two excerpts, the participants' opinions on whether the use of AI-supported tools in the process of teaching EFL reading skills improves student autonomy are quite varied. In the next section of the study, based on this variation, the participant teachers' views, attitudes, beliefs, and practices regarding the teachers' perceptions of the impact of AI on their role and reading assessment are elucidated.

3.3. Perceived Role Shift with AI in EFL Reading Instruction

It can be anticipated that as learners develop greater autonomy and self-regulation within the learning process, the teacher's role will concurrently evolve and undergo transformation. Specifically, whereas the teacher traditionally functions as the primary source of knowledge in conventional classroom settings, in learning environments that foster student autonomy, the teacher's role is more likely to shift toward that of a guide or facilitator. For this reason, it was deemed important to explore teachers' perceptions of their own roles while investigating teacher cognition concerning the impact of integrating AI-supported tools into EFL reading instruction and its influence on learner autonomy. To address this aim, four related items were incorporated into the questionnaire as part of the data collection process. The same issue was also examined through semi-structured interviews. Table 5 illustrates the descriptive findings related to the perceived impact of AI on teaching role and assessment in reading instruction.

Table 5. Descriptive statistics for perceived role shift with AI (N=250).

Item	Response Option	Frequency (n)	Percentage (%)
The use of AI in EFL reading instruction has changed my role as a teacher.	Strongly Disagree	12	4.8
	Disagree	23	9.2
	Neutral	45	18.0
	Agree	110	44.0
	Strongly Agree	60	24.0
I feel that AI tools shift my role towards being more of a facilitator than a direct instructor.	Strongly Disagree	10	4.0
	Disagree	20	8.0
	Neutral	55	22.0
	Agree	100	40.0
	Strongly Agree	65	26.0
I believe that AI can be used as a reliable tool for assessing students' EFL reading performance.	Strongly Disagree	15	6.0
	Disagree	25	10.0
	Neutral	60	24.0
	Agree	95	38.0
	Strongly Agree	55	22.0
I would trust AI-generated feedback for formative assessment in EFL reading.	Strongly Disagree	18	7.2
	Disagree	35	14.0
	Neutral	70	28.0
	Agree	85	34.0
	Strongly Agree	42	16.8

As clearly shown in Table 5, the findings reveal that the vast majority of the teachers (44% selecting agree and 24% selecting strongly agree) reported that their roles are likely to change with the integration of AI into EFL reading instruction. In this context, a female teacher with nine years of teaching experience working at a high school provided the following statement, which offers valuable insight into how teachers' roles may evolve in response to AI integration:

> *I think back to when I first started working, when I was teaching at a secondary school. My students really seemed to do whatever I told them to do. Of course, the internet was quite widespread back then, but it was not as much a part of*

our lives as it is now. Now it is like a friend... I think the students see it that way too, and as we introduce this new friend into the learning environment, our role becomes more of a guiding one.

The findings of the first item clearly indicate that the integration of AI into EFL learning environments has brought about a transformation in teachers' professional roles. In parallel, the qualitative data provide rich and nuanced insights that further substantiate this conclusion. Similarly, results from another questionnaire item display that the preponderance of the respondents concur with the statement: "I feel that AI tools shift my role towards being more of a facilitator than a direct instructor." Although 22% of respondents remained neutral, 8% disagreed, and 4% strongly disagreed, this overall pattern suggests that AI tools are indeed reshaping teachers' roles—moving them away from traditional, instructor-centred conceptions and towards a more facilitative orientation. A female teacher who is currently working in a secondary school and has 20 years of teaching experience has stated the following in this context and provided critical data by comparing the student profile she encountered when she first started working with the student profile she is now working with:

I have been doing this job for 20 years. So, I have been with many students over this time. The most important thing I have noticed during this process is that students' expectations and realities have changed significantly. For example, when I first started working, teachers were seen and accepted as authority figures. Now, even if you are an experienced teacher, students may not see you as a source of knowledge.

Teachers play an undeniable role in the EFL learning process, particularly in the provision of feedback. Although this role can be difficult to define precisely, given the central importance of receiving constructive feedback in foreign language learning, it was considered essential to examine this aspect in greater depth. To this end, two items addressing this issue were included in the questionnaire, and the same topic was further explored through questions posed to the participants during the semi-structured interviews. The findings revealed that, in contrast to other items, the participants expressed more varied opinions on this particular issue. More specifically, regarding the item "I believe that AI can be used as a reliable tool for assessing students' EFL reading performance," 38% of the respondents indicated that they agreed, while 22% strongly agreed. Meanwhile, 24% adopted a neutral stance. In addition, 10% disagreed with the item, and a further 6% strongly disagreed. In addition, an item was included in the questionnaire to explore whether the participants would trust AI tools for formative assessment purposes. The results revealed a moderate yet notable level of confidence in such tools. Specifically, 7.2% of the respondents strongly disagreed and 14% disagreed with the statement, while 28% remained neutral. In contrast, 34% agreed and 16.8% strongly agreed. These findings demonstrate that more than half of the participants showed a positive inclination towards the use of AI for formative assessment. Accordingly, this distribution implies that, although a degree of scepticism persists, there is a growing acceptance among teachers of the potential of AI to support formative assessment practices in EFL contexts. This tendency coincides with previous studies highlighting teachers' cautious optimism towards AI-based assessment (e.g., Bower et al., 2024; Kizilcec et al., 2024; Swiecki et al., 2022).

3.4. Ethical, Cultural, and Practical Considerations

As AI has become increasingly embedded in everyday life, it has brought not only numerous advantages but also certain disadvantages. One of the primary concerns relates to the use of personal data. Given the potential vulnerabilities that may arise in learning environments in this regard, the present study also investigated issues related to ethical considerations. Furthermore, cultural and practical aspects were examined to provide a more comprehensive understanding of the phenomenon. Table 6 presents the descriptive statistics for the items addressing ethical, cultural, and practical dimensions.

Table 6. Descriptive statistics for ethical, cultural, and practical considerations (N=250).

Item	Response Option	Frequency (n)	Percentage (%)
I am concerned about data privacy when using AI tools in reading instruction.	Strongly Disagree	25	10.0
	Disagree	50	20.0
	Neutral	75	30.0
	Agree	60	24.0
	Strongly Agree	40	16.0
I believe that AI tools need to be more culturally and linguistically inclusive.	Strongly Disagree	20	8.0
	Disagree	55	22.0
	Neutral	65	26.0
	Agree	70	28.0
	Strongly Agree	40	16.0

The findings indicate considerable variation in teachers' attitudes towards data privacy. The observation that 30% of teachers adopted a neutral stance on this matter may be regarded as a particularly significant insight. In other words, as discussed in the preceding sections, although teachers appear to be familiar with AI-based learning tools and generally express positive attitudes towards their use, they do not exhibit a strong level of confidence in such tools with respect to data privacy.

Figure 2. Reasons reported by teachers for reluctance to integrate ai tools into EFL reading instruction

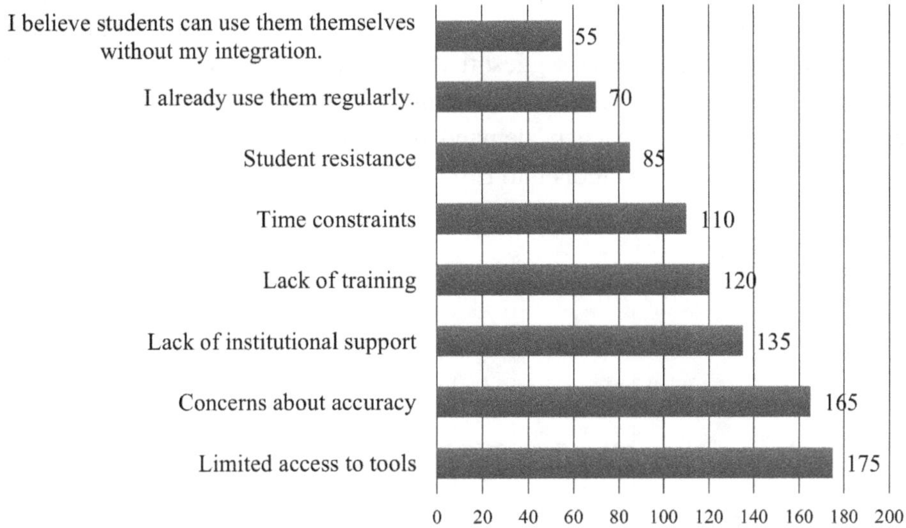

As outlined in the theoretical framework of this study, teacher cognition is a multifaceted construct encompassing what teachers think, know, believe, and do. Within the scope of the present research, the participating teachers' thoughts, knowledge, and beliefs regarding the integration of AI tools into the EFL reading process and their perceived impact on learner autonomy were examined. In addition, the questionnaire administered to the participants included two items addressing issues of classroom implementation, while the semi-structured interviews placed particular emphasis on this theme. The quantitative findings indicated that a substantial proportion of teachers expressed reluctance to integrate AI tools into EFL reading instruction, citing limited access to such tools, concerns about their reliability, and insufficient professional training in this area. Figure 2 illustrates the findings obtained in this context. Connectedly, a male teacher with 9 years of experience verbalised his perceptions concerning his reluctance to utilise AI tools within his classes as follows:

> *Artificial intelligence is, of course, a highly effective tool today. It is also used quite frequently in educational settings, but I must admit I have some doubts about its recent introduction and the accuracy of the information it provides.*

Furthermore, the fact that we cannot access reliable versions of it also makes me cautious in this regard.

In addition to teachers' concerns about incorporating AI tools into the development of EFL reading skills, the study also explored factors that might help to mitigate these concerns. As demonstrated in Figure 3, the findings indicate that the most influential factor encouraging teachers to use AI tools was found to be *access to free applications*. Moreover, the participants underscored the importance of experiential learning, identifying hands-on workshops as the second most significant factor in this respect. The results further suggest that the frequent selection of options such as technical support, online courses and webinars, institutional encouragement, and peer collaboration reflects teachers' demand for sustained, practice-oriented support in effectively integrating AI tools into EFL learning environments.

Figure 3. Support factors encouraging teachers' use of AI in EFL reading instruction

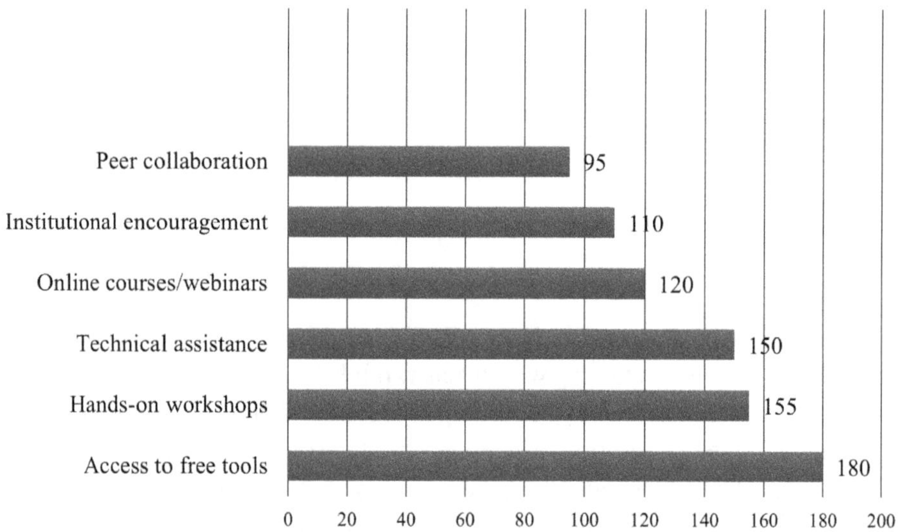

A female teacher with 11 years of teaching experience working at the high school expressed her concerns about her inability to effectively use AI tools in the process of teaching EFL reading, and she also elaborated on the factors that might encourage her to integrate such tools into her teaching practice, stating the following:

AI tools certainly save time. If we know how to use them, that is, if we don't get lost in them, I think they are quite functional, but they can also contain so much misinformation... At this point, it is very important to find the right tools to guide us, and I think such high-quality tools are essentially paid tools. That is why, to be honest, I support having institutional support for this and integrating more qualified tools, because only then we can introduce these tools to our students in a correct and meaningful way and enable our students to become autonomous.

Based on the findings discussed in this section, it can be concluded that a range of interrelated factors must be taken into account for teachers to effectively utilise AI tools in fostering learner autonomy within EFL contexts. While these factors may vary according to individual learner characteristics, the key point to emphasise is that teachers occupy a central mediating position in determining how, and to what extent, AI is meaningfully integrated into pedagogical practice. From the perspective of teacher cognition, this highlights the importance of teachers' beliefs, knowledge, and professional agency in shaping the pedagogical value of AI applications. Consequently, any attempt to promote AI-enhanced autonomy in language learning must also address teachers' professional development needs and their confidence in navigating technology-mediated learning environments.

4. CONCLUSION AND PEDAGOGICAL IMPLICATIONS

The present study employed an explanatory sequential mixed-methods research design, in which data were collected through questionnaires and semi-structured interviews. The sample comprised 250 English language teachers working across different educational levels and regions in Türkiye. The findings were presented in two stages: initially, the quantitative results were reported, followed by the salient insights derived from the qualitative data. Subsequently, the two strands of data were integrated and discussed to provide a comprehensive interpretation of the research problem within four main sub-headings, namely use of AI tools in teaching EFL reading, AI-supported learner autonomy in EFL reading, perceived role shift with AI in EFL reading instruction, and ethical, cultural, and practical considerations in the use of AI tools for EFL reading activities.

The results of this study show that pedagogical, cognitive, and environmental aspects interact dynamically when integrating AI tools into EFL reading instruction. The participating teachers recognised that AI may improve student autonomy and engagement, but their readiness to use these technologies is still constrained by worries about data privacy, dependability, and inadequate training. These challenges suggest that teachers' preparedness, digital literacy, and institutional support are just as important for successful integration as technology availability.

The findings also highlight the value of ongoing professional development programs that address both the technical and pedagogical aspects of using AI from a pedagogical perspective. The gap between awareness and competent execution can be closed by giving teachers access to interactive workshops, useful materials, and cooperative learning opportunities. Institutional support and the incorporation of culturally and linguistically relevant AI tools are also required to guarantee that technology improves, rather than limits, inclusive teaching practices.

To sum up, this study highlights the central role teachers play as facilitators in AI-supported EFL learning environments. In other words, although the role of teachers has undergone change and transformation with the engagement of AI tools in EFL learning processes, the role and mission of teachers in learning environments remains highly critical. To this end, ensuring that teachers make conscious, thoughtful, and context-sensitive decisions regarding AI integration will not only improve the quality of reading instruction but also support the development of autonomous learners who are prepared for the evolving digital learning environment.

5. LIMITATIONS AND SUGGESTIONS FOR FURTHER RESEARCH

The sample of the present study consists of 250 English teachers working in various schools across different regions of Türkiye. Semi-structured interviews were also conducted with 20 of these teachers, aiming to establish the validity and reliability of the data obtained in accordance with the data triangulation strategy. However, the data obtained cannot be generalised based on the current sample size, which can be regarded as the primary limitation of this study. In this direction, it is recommended that further research be conducted with larger samples. Besides, as the present study focused on teacher cognition in the context of the effects of integrating AI tools into EFL reading learning processes on student autonomy, the sample group only included teachers. However, future studies could be carried out by including students in the sample group, in order to shed light on student views, attitudes, and practices.

REFERENCES

Ademola, E. O. (2024). Reading strategies in the AI age: Enhancing comprehension and engagement with advanced technologies. In *Proceedings of the 38th iSTEAMS Multidisciplinary Bespoke Conference* (pp. 105–124). University of Ghana. https://dx.doi.org/DOI: 10.22624/AIMS/ACCRABESPOKE2024P13

Barr, S., & Askell-Williams, H. (2020). Changes in teachers' epistemic cognition about self–regulated learning as they engaged in a researcher-facilitated professional learning community. *Asia-Pacific Journal of Teacher Education*, *48*(2), 187–212. DOI: 10.1080/1359866X.2019.1599098

Borg, S. (2003). Teacher cognition in language teaching: A review of research on what language teachers think, know, believe, and do. *Language Teaching*, *36*(2), 81–109. DOI: 10.1017/S0261444803001903

Borg, S. (2006). *Teacher cognition and language education: Research and practice* (1st ed.). Continuum.

Borg, S. (2009). Language teacher cognition. In Burns, A., & Richards, J. K. (Eds.), *The Cambridge guide to second language teacher education* (pp. 163–171). Cambridge University Press.

Bower, M., Torrington, J., Lai, J. W. M., Petocz, P., & Alfano, M. (2024). How should we change teaching and assessment in response to increasingly powerful generative artificial intelligence? Outcomes of the ChatGPT teacher survey. *Education and Information Technologies*, *29*, 15403–15439. DOI: 10.1007/s10639-023-12405-0

Celik, I., Dindar, M., Muukkonen, H., & Järvelä, S. (2022). The promises and challenges of artificial intelligence for teachers: A systematic review of research. *TechTrends*, *66*(4), 616–630. DOI: 10.1007/s11528-022-00715-y

Chong, S. W., & Reinders, H. (2025). Autonomy of English language learners: A scoping review of research and practice. *Language Teaching Research*, *29*(2), 607–632. DOI: 10.1177/13621688221075812

Creswell, J. W., & Creswell, J. D. (2018). *Research design: Qualitative, quantitative, and mixed methods approaches* (5th ed.). SAGE Publications.

De la Vall, R. R. F., & Araya, F. G. (2023). Exploring the benefits and challenges of AI-language learning tools. *International Journal of Social Sciences and Humanities Invention*, *10*(01), 7569–7576. DOI: 10.18535/ijsshi/v10i01.02

Fidan, Ş., & Kasımi, Y. (2023). An investigation into artificial intelligence (AI) in the English as a foreign language (EFL) context. *International Journal of Educational Spectrum*, *5*(2), 269–280.

Gillo, M. D. (2021). Fundamentals of hermeneutics as a qualitative research theoretical framework. *European Journal of Education and Pedagogy*, *2*(3), 42–45. DOI: 10.24018/ejedu.2021.2.3.43

Holec, H. (1981). *Autonomy is the ability to take charge of one's own learning*. Pergamon.

Huynh, Q. Q. (2024). A literature review on EFL students' integration of Al-based tools in English learning and the perceived impact on autonomy. *International Journal of AI in Language Education*, *1*(2), 30–51. DOI: 10.54855/ijaile.24123

Jiang, R. (2022). How does artificial intelligence empower EFL teaching and learning nowadays? A review on artificial intelligence in the EFL context. *Frontiers in Psychology*, *13*, 1049401. DOI: 10.3389/fpsyg.2022.1049401 PMID: 36467167

Kalinowski, E., Egert, F., Gronostaj, A., & Vock, M. (2020). Professional development on fostering students' academic language proficiency across the curriculum—A meta-analysis of its impact on teachers' cognition and teaching practices. *Teaching and Teacher Education*, *88*, 102971. DOI: 10.1016/j.tate.2019.102971

Kizilcec, R. F., Huber, E., Papanastasiou, E. C., Cram, A., Makridis, C. A., Smolansky, A., Zeivots, S., & Raduescu, C. (2024). Perceived impact of generative AI on assessments: Comparing educator and student perspectives in Australia, Cyprus, and the United States. *Computers and Education: Artificial Intelligence*, *7*, 100269. DOI: 10.1016/j.caeai.2024.100269

Kolchenko, V. (2018). Can modern AI replace teachers? Not so fast! Artificial intelligence and adaptive learning: Personalized education in the AI age. *HAPS educator, 22*(3), 249-252. https://doi.org/DOI: 10.21692/haps.2018.032

Li, H., Gan, Z., Leung, S. O., & An, Z. (2022). The impact of reading strategy instruction on reading comprehension, strategy use, motivation, and self-efficacy in Chinese university EFL students. *SAGE Open*, *12*(1). Advance online publication. DOI: 10.1177/21582440221086659

Lincoln, Y. S., & Guba, E. G. (1985). *Naturalistic inquiry*. SAGE Publications.

Little, D. (1991). *Learner autonomy 1: Definitions, issues and problems*. Authentik.

Prensky, M. (2001). Digital natives, digital immigrants. *On the Horizon. The International Journal of Learning Futures*, *9*(5), 1–6. DOI: 10.1108/10748120110424816

Rahmasari, B. S., Munir, A., & Nugroho, H. M. A. (2025). Fostering learner autonomy in EFL reading: A study of strategy-based instruction at two Indonesian universities. *Cogent Education, 12*(1), 2477367. DOI: 10.1080/2331186X.2025.2477367

Rianto, A. (2022). Exploring correlation between metacognitive online reading strategy use and online reading comprehension of EFL students. *Turkish Online Journal of Distance Education, 23*(2), 223–235.

Slamet, J., Basthomi, Y., Ivone, F. M., & Eliyanah, E. (2024). Unlocking the potential in a gamification-based MOOC: Assessing autonomous learning and self-directed learning behaviors. *Teaching & Learning Inquiry, 12*, 1–20. DOI: 10.20343/teachlearninqu.12.19

Sumakul, D. T. Y., Hamied, F. A., & Sukyadi, D. (2022). Artificial intelligence in EFL classrooms: Friend or foe? *LEARN Journal: Language Education and Acquisition Research Network, 15*(1), 232–256.

Sun, C., Wei, L., & Young, R. F. (2022). Measuring teacher cognition: Comparing Chinese EFL teachers' implicit and explicit attitudes toward English language teaching methods. *Language Teaching Research, 26*(3), 382–410. DOI: 10.1177/1362168820903010

Swiecki, Z., Khosravi, H., Chen, G., Martinez-Maldonado, R., Lodge, J. M., Milligan, S., Selwyn, N., & Gašević, D. (2022). Assessment in the age of artificial intelligence. *Computers and Education: Artificial Intelligence, 3*, 100075. DOI: 10.1016/j.caeai.2022.100075

Tabiati, S. E. (2016). Factors contributing to learners' autonomy in EFL reading. *Jurnal Pendidikan Humaniora, 4*(2), 82–94.

Utku Bilici, Ö. (2024). *The multi-dimensional nature of teacher cognition: The case of EFL teachers* (Publication No. 31653733) [Doctoral dissertation, Middle East Technical University]. ProQuest. https://www.proquest.com/docview/3122678784?pq-origsite=gscholar&fromopenview=true

Utku Bilici, Ö., & Subaşı, G. (2022). EFL learners' metacognitive awareness of global reading strategies and L2 reading comprehension skills: The efficacy of reciprocal teaching. *The Reading Matrix: An International Online Journal, 22*(2), 170–185.

Woo, J. H., & Choi, H. (2021). Systematic review for AI-based language learning tools. *Journal of Digital Contents Society, 22*(11), 1783–1792. DOI: 10.48550/arXiv.2111.04455

Yuan, R., Lee, I., De Costa, P. I., Yang, M., & Liu, S. (2022). TESOL teacher educators in higher education: A review of studies from 2010 to 2020. *Language Teaching*, *55*(4), 434–469. DOI: 10.1017/S0261444822000209

Zhu, M., & Wang, C. (2025). A systematic review of artificial intelligence in language education: Current status and future implications. *Language Learning & Technology*, *29*(1), 1–29. https://hdl.handle.net/10125/73606

Chapter 6
Thinking Through the Algorithm:
Learner Cognition, Agency, and Communicative Effectiveness in AI-Mediated Language Education

Shen Min
https://orcid.org/0000-0003-4418-377X
Universiti Brunei Darussalam, Brunei

Hjh Zurinah binti Hj Ya'akub
https://orcid.org/0000-0001-5690-6766
Universiti Brunei Darussalam, Brunei

Rabi'Atul Adawiyyah binti Hj Hassan
https://orcid.org/0000-0002-1101-5510
Universiti Brunei Darussalam, Brunei

ABSTRACT

Generative artificial intelligence (AI) is reshaping language education, moving the focus beyond surface-level accuracy toward communicative competence. While large language models provide fluent and immediate support, their outputs often lack pragmatic nuance, relational sensitivity, and cultural appropriateness. This chapter argues for a shift from AI-assisted to AI-mediated language production, where learners negotiate and co-construct meaning with AI. It introduces the concept of thinking through the algorithm, a framework that integrates AI's capacities, learner cognition and agency, and communicative effectiveness. Evidence from beginner-level Mandarin learners shows how agency enables adaptation of AI

DOI: 10.4018/979-8-3373-4397-6.ch006

outputs to meet communicative goals. The framework offers broader relevance for rethinking pedagogy, highlighting the need to embed AI literacy and critical cultural competence to prepare learners for effective engagement with AI.

INTRODUCTION

The rapid development of generative artificial intelligence (GenAI) has introduced profound changes to language education, reshaping the ways learners produce, evaluate, and refine language. Large language models (LLMs), such as ChatGPT, DeepSeek, and Gemini, are now widely accessible to students and teachers, offering immediate, fluent, and contextually responsive outputs that were unimaginable in earlier decades. These systems promise to democratise access to language support and reduce barriers to learning (Escalante et al, 2023; Lee et al, 2025; Yan, 2023). At the same time, their widespread adoption has raised concerns that such models often privilege generic registers and fail to account for communicative goals, with limited sensitivity to pragmatic nuance or cultural appropriateness (Ji et al, 2022; Pegrum, 2025).

Much of the current scholarship describes the pedagogical contributions of AI at the surface level, particularly in relation to accuracy and fluency. Generative tools are widely reported to enhance grammar, vocabulary, and sentence organisation, offering corrective feedback that is often faster and more detailed than what teachers can provide in real time (Escalante et al., 2023). Learners frequently turn to AI to overcome writer's block, expand lexical repertoires, or polish final drafts, thereby accelerating the production process. Yet these gains highlight a persistent tension: linguistic correctness does not automatically translate into communicative success. AI-generated outputs still struggle with pragmatic nuance, making them potentially inappropriate in authentic intercultural encounters.

This pedagogical tendency is mirrored in the research landscape. Studies published since 2022 have predominantly examined AI as a mechanism for automated evaluation or corrective feedback, often prioritising measurable linguistic outcomes such as syntactic accuracy or lexical sophistication (Fu et al., 2024; Wang et al., 2025). Far fewer studies investigate AI's role as a communicative partner, especially in intercultural contexts where pragmatic expectations, politeness strategies, and cultural conventions critically shape meaning-making (Thomas, 1983; Baker, 2024).

Moreover, empirical work remains heavily concentrated in English-language settings, leaving multilingual contexts comparatively underexplored. As noted in systematic analyses of GenAI in language education (Chiu et al., 2023), this bias constrains understanding of how AI performs in multilingual contexts, including languages such as Mandarin, Japanese, Korean, and Arabic. These languages introduce additional pragmatic challenges, such as tonal and character-based encoding in Mandarin, complex honorific systems in Japanese and Korean, and diglossic variation in Arabic, that may increase the likelihood of AI-generated pragmatic mismatches. This suggests that AIMC may function differently across linguistic systems and that cross-linguistic comparison is essential for evaluating its transferability and potential risks.

This chapter addresses these gaps by situating its inquiry within the field of AI-mediated intercultural communication (AIMC), which provides the originating stance of the study. From this foundation, the research advances a conceptual framework to guide both methodological design and data analysis. Instead of conceptualising AI as an external instrument,the framework reconceptualises it as a co-authoring agent whose outputs are continuously shaped through reciprocal influence and negotiation with the learner. In this view, the central challenge of language education is no longer confined to the generation of grammatically accurate sentences, but to the learner's capacity to co-author with AI so as to facilitate language production that remains pragmatically appropriate, relationally sensitive, and culturally adaptive. At the heart of this framework is the concept of *thinking through the algorithm*, which conceptualises co-authoring as a process that moves beyond surface-level linguistic adjustment to involve metacognitive regulation and sociocultural mediation in AI-mediated communication.

In this chapter, "AI-assisted" refers to uses of AI in which learners edit or refine AI-generated language to improve accuracy or fluency. By contrast, "AI-mediated" communication (AIMC) positions AI as an interlocutor with whom learners co-author meaning, foregrounding pragmatic appropriateness, relational sensitivity, and cultural alignment.

Accordingly, the aim of this chapter is twofold: first, to articulate a conceptual framework that situates AI-mediated communication as a field of inquiry and foregrounds the process of co-authoring between learners and AI, integrating AI's functional capacities, learner cognition, and the evaluative dimension of communicative effectiveness; and second, to demonstrate how this framework can inform pedagogical practice that prepares learners for communication, particularly in intercultural contexts, within an AI-saturated environment. The discussion unfolds in five parts. Section 2 reviews the shift from AI-assisted writing to AI-mediated communication, underscoring both the affordances and limitations of existing approaches. Section 3 presents the conceptual framework and elaborates the notion of thinking through the algorithm. Section 4 provides illustrative insights from empirical work with beginner-level Mandarin learners to demonstrate how the framework operates in practice. Section 5 considers pedagogical implications, with particular emphasis on AI literacy and intercultural communicative competence. Section 6 concludes by identifying directions for future research and practice.

FROM AI-ASSISTED TO AI-MEDIATED COMMUNICATION: IDENTIFYING GAP

Recent advances in generative AI have shifted attention from AI-assisted writing, where tools serve mainly corrective functions, to AI-mediated communication, where human–AI interaction itself shapes meaning. This section reviews the affordances and limitations of this shift as a basis for the framework developed in Section 3.

Evolution of AI in Language Education

The integration of artificial intelligence (AI) into language education has evolved substantially over the past two decades, progressing from rule-based grammar checkers and automated writing evaluation (AWE) systems to the generative capacities of large language models (LLMs). Earlier AI-assisted tools, such as automated written corrective feedback (AWCF) platforms and AWE systems like *Pigai*, focused primarily on surface-level accuracy, detecting errors in grammar, vocabulary, and mechanical features of writing (Escalante et al., 2023). These systems were valued for their consistency and speed and contributed to reducing teacher workload, yet their pedagogical scope was limited. Numerous reviews across the project files highlight that pre-GenAI systems overwhelmingly prioritised linguistic correctness while offering minimal support for discourse organisation, interpersonal tone, or pragmatic appropriateness (Chiu et al., 2023; Chen et al., 2020). As a result, learners often engaged with automated feedback selectively. Studies report inconsistent uptake and occasional over-reliance on automated suggestions, a pattern that aligns with findings in Computer-Assisted Language Learning (CALL), where learner autonomy and form-focused instruction have been key concerns.

Recent systematic reviews show that AWE dominated AI-related research in language education between 2000 and 2019, revealing the field's strong orientation toward quantifiable linguistic outcomes. However, AWE tools were limited in addressing higher-order concerns such as register choice, cohesion, rhetorical stance, and cultural or relational nuance. Many learners perceived automated feedback as overly corrective or insufficiently contextualised (Escalante et al., 2023), which contributed to "genre flattening", the homogenisation of writing towards generic, decontextualised patterns (Chen et al.2020). In other words, these systems treated language primarily as a set of forms to be corrected rather than as a means of achieving communicative goals (Ranalli, 2018).

The release of generative AI in late 2022 marked a substantive shift in these capabilities. LLMs such as ChatGPT, DeepSeek, and Gemini can generate extended, context-sensitive texts; adjust register and style; and engage in conversational turn-taking. Unlike earlier systems, they can simulate realistic communicative scenarios and co-construct meaning with learners. As highlighted in multiple project files, these affordances enable improvements in lexical sophistication, syntactic variety, and discourse organisation (Escalante et al., 2023; Lee et al., 2025; Yan, 2023). They also allow learners to move beyond sentence-level corrections toward more holistic language production.

Despite these advances, important limitations persist. AI outputs often exhibit over-formality, generic phrasing, or insufficient sensitivity to pragmatic, relational, or cultural norms (Ji et al., 2022; Pegrum, 2025). Because LLMs are trained on large-scale corpora dominated by English and standardised global registers, they risk promoting homogenised language practices that neglect local variation or culturally preferred interactional styles (Sun & Gao, 2023; Fu et al., 2024). Learners therefore continue to face challenges familiar from the pre-GenAI era—determining when to accept, reject, or modify machine-generated suggestions to align with authentic communicative expectations. These enduring issues underscore the need to move beyond structural accuracy as the primary measure of AI effectiveness and motivate the shift toward a focus lies on how learners evaluate and reshape AI output to achieve contextually appropriate meaning.

The AIMC Paradigm

One promising response to these limitations is the reconceptualisation of AI as a communicative partner. This approach, referred to here as AI-mediated communication (AIMC), shifts the role of AI from background corrective tool to co-author and interlocutor. In AIMC, learners do not simply receive and adopt AI-generated suggestions; rather, they actively evaluate, adapt, and reshape them in order to meet communicative objectives. This perspective is grounded in the broader tradition of computer-mediated communication (CMC). Rather than focusing on how technology transmits messages between human interlocutors, AIMC examines how an intelligent agent modifies, augments, or generates messages on behalf of a communicator to accomplish interactional goals (Hancock & Levy, 2020). In this view, AI is not a passive conduit but a co-author that shapes linguistic, relational, and cultural dimensions of interaction.

AIMC's emphasis on mediation highlights three communicative mechanisms identified in recent studies (Hancock et al., 2023):

- Expansion – extending a communicator's linguistic or stylistic range;
- Automation – streamlining or accelerating message production; and
- Transformation – altering voice, stance, or interpersonal tone.

These mechanisms are especially relevant in language learning, where generative models routinely propose formulations, adjust register, and simulate scenarios. Such affordances introduce new forms of dialogic co-construction, but also amplify risks documented in CMC and intercultural research—such as generic register use, relational misalignment, and cultural mismatch (Spencer-Oatey, 2008; Baker, 2024).

This chapter draws on the paradigm of AIMC to reconceptualise the role of AI in language education. It proposes that generative AI be understood not merely as a technological scaffold for language production, but as a mediating force within communicative practice. Accordingly, the AIMC paradigm expands the criteria for success. While traditional AI-assisted writing emphasises accuracy and fluency, AIMC requires learners to consider whether language is appropriate for context, interlocutor, and communicative purpose. In this way, AIMC highlights the active role of learners in negotiating with AI outputs and transforming them into socially meaningful discourse.

Emerging empirical work illustrates both the potential and complexity of this perspective. Studies demonstrate that conversational AI can scaffold authentic practice, reduce anxiety, and personalise learning opportunities (Ebadi & Amini, 2022). At the same time, researchers also report that AI responses can be tangential, misaligned with learner intentions, or pragmatically inappropriate (Ji et al., 2022; Fu et al., 2024). Such tendencies underscore the importance of learner agency: students must decide when to accept, reject, or modify machine-generated suggestions in order to achieve successful outcomes. AIMC therefore does not replace human interaction but reframes it, foregrounding the need for learners to engage critically with AI outputs.

Pedagogically, this reframing is significant. It encourages task designs that integrate AI as a co-constructor of discourse, requiring learners to reflect on how machine-generated suggestions can be adapted for contextually appropriate communication. In doing so, AIMC offers a more nuanced understanding of AI's role in language education, one that embraces both its affordances and its constraints.

Communicative Effectiveness

The move toward AIMC necessitates a re-examination of what counts as success in language learning. Communicative effectiveness must be defined more broadly than grammatical correctness or lexical sophistication. As intercultural communication scholars have long emphasised, messages that are formally accurate may still fail if they are pragmatically misaligned, relationally insensitive, or culturally inappropriate (Byram, 1997; Deardorff, 2006; Spencer-Oatey, 2008).

This chapter adopts three criteria for communicative effectiveness: pragmatic appropriateness, relational sensitivity, and cultural adaptiveness. Pragmatic appropriateness refers to the alignment of utterances with intended speech acts and discourse functions (Thomas, 1983). Relational sensitivity highlights the management of interpersonal dynamics, including politeness, empathy, and tone (Brown & Levinson, 1987; Spencer-Oatey, 2008). Cultural adaptiveness foregrounds the alignment of language with sociocultural norms and conventions, recognising that direct translations from a first language often result in expressions that sound unnatural or even impolite in the target language (Sun & Gao, 2023; Ji et al., 2022; Chew, Min, & Yap, 2023).

Empirical research shows that generative AI frequently falls short in these dimensions. For example, systems may produce sentences that are formally correct but excessively verbose, overly formal in casual interactions, or misaligned with local politeness routines (Pegrum, 2025; Baker, 2024). Learners who rely on AI uncritically may thus reproduce outputs that are ineffective in authentic intercultural communication. By contrast, those who engage critically, softening directives, or substituting culturally conventional expressions, demonstrate the process this chapter conceptualises as *thinking through the algorithm.*

In sum, communicative effectiveness in AIMC is best understood not as the mechanical reproduction of correct sentences but as the collaborative achievement of meaning between learner and machine. The next section elaborates the conceptual framework that integrates these dimensions, illustrating how learners exercise cognition and agency to transform AI outputs into pragmatically appropriate, relationally sensitive, and culturally adaptive communication.

THINKING THROUGH THE ALGORITHM: A CONCEPTUAL FRAMEWORK FOR AI-MEDIATED LANGUAGE LEARNING

The preceding section traced the move from AI assistance to AI mediation, underscoring AIMC implications for language education. To provide a framework for understanding AI in language education within the paradigm of AIMC, the present framework draws together three interdependent components: AI's functional capacities and constraints, learner cognition and agency, and communicative effectiveness. These components interact in a cyclical relationship, shaping how learners engage with AI and how communication unfolds in intercultural tasks. Central to this framework is the concept of *thinking through the algorithm*, which captures the learner's active role in critically evaluating, adapting, and transforming AI outputs.

Core Components

AI's Functional Capacities and Constraints

Generative AI provides affordances for language learning, offering grammatically accurate and fluent language, lexical alternatives, and support for discourse organisation (Escalante et al., 2023; Lee et al., 2025; Yan, 2023). At the same time, its functional capacities are dynamic. System updates and user prompts continually reshape what the technology can deliver, meaning its strengths and weaknesses are in constant flux.

These evolving capacities are tempered by persistent constraints: AI outputs may be over-formal, generic, or pragmatically insensitive, producing utterances that sound correct in form but awkward in practice (Ji et al., 2022; Pegrum, 2025; Sun & Gao, 2023; Fu et al., 2024).

Learner Cognition and Agency

The second component emphasises the active role of learners in shaping AI outputs. Agency is understood here as the learner's capacity to evaluate, modify, and strategically deploy suggestions (Duff, 2012; Oxford, 2017). Some learners demonstrate high agency by shortening verbose responses, softening directives, or substituting culturally conventional expressions, while others accept AI suggestions passively. This divergence illustrates how cognition and agency mediate between what AI provides and what communication requires. Without critical engagement, learners risk producing language that is formally accurate yet communicatively ineffective.

Communicative Effectiveness

The third component reflects the evaluative dimension of interaction. Communicative effectiveness extends beyond structural accuracy to include pragmatic appropriateness, relational sensitivity, and cultural adaptiveness (Byram, 1997; Deardorff, 2006; Spencer-Oatey, 2008). Learners must ensure that their language achieves its intended speech acts, maintains rapport, and aligns with cultural conventions. In AI-mediated contexts, effectiveness becomes the ultimate standard by which learner–AI collaboration should be judged.

Interdependence of Components

These three components are not discrete but dynamically interconnected. AI's capacities and constraints shape how learners engage with outputs; learner cognition and agency mediate these interactions; and the degree of communicative effectiveness achieved feeds back into learners' evolving strategies.

For example, when AI generates a grammatically correct but pragmatically inappropriate utterance, a learner with strong agency may recognise the misalignment and adapt it to fit the context. This successful adaptation not only improves immediate communication but also strengthens the learner's awareness of pragmatic norms, enhancing future interactions. Conversely, uncritical reliance on AI may reinforce dependency while leaving pragmatic awareness underdeveloped.

This cyclical relationship positions AI not as a replacement for human judgement but as a partner in negotiation. Each learner–AI exchange is shaped by the interplay of capacities, agency, and effectiveness, and each outcome in turn reshapes subsequent learner strategies. Figure 1 visualises this cyclical relationship, situating the three components within a AI-mediated environment.

Figure 1. Thinking through the algorithm: A conceptual framework for AI-mediated language learning.

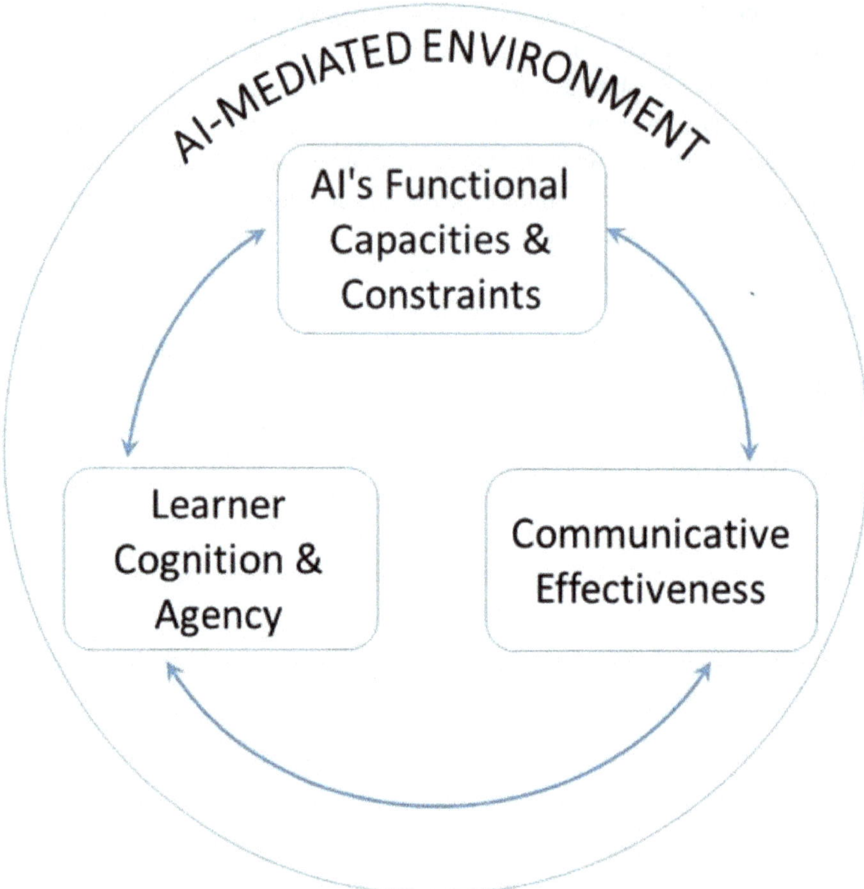

Indicators of Communicative Effectiveness

To operationalise communicative effectiveness, this framework specifies three indicators: pragmatic appropriateness, relational sensitivity, and cultural adaptiveness. These constructs draw on established models in pragmatics and intercultural communication (Thomas, 1983; Brown & Levinson, 1987; Byram, 1997; Deardorff, 2006; Spencer-Oatey, 2008) and are adapted here for the analysis of AI-mediated tasks. Table 1 presents these indicators with their operational definitions and observable features.

Table 1. Operationalisation of communicative effectiveness indicators

Construct	Operational Definition	Observable Indicators
Pragmatic Appropriateness	The extent to which learner utterances align with intended communicative functions (e.g., requests, offers, confirmations) and are contextually appropriate in terms of speech act realisation and discourse function.	- Utterances successfully fulfil the intended speech acts. - Appropriate choice of directness or indirectness according to context. -Sequencing and coherence of turns reflect communicative goals. - Minimal illocutionary or discourse-level misalignments.
Relational Sensitivity	The degree to which learner language attends to interpersonal dynamics, politeness norms, empathy, and social roles to maintain rapport and manage face in interaction.	- Use of linguistic softeners and mitigation strategies. - Tone adjusted to social roles and relationships -Avoidance of overly direct, abrupt, or authoritative phrasing. - Presence of transitional routines and polite closures in conversations.
Cultural Adaptiveness	The extent to which learners adjust their language use to align with the sociocultural expectations, discourse conventions, and communicative norms of Mandarin-speaking contexts.	- Use of listener-oriented formulations and culturally conventional expressions. - Appropriate greeting and closing routines consistent with Mandarin norms. - Avoidance of L1-based translations that sound unnatural or inappropriate. - Sensitivity to culturally specific discourse patterns and interactional scripts.

As Table 1 shows, communicative effectiveness requires attention to multiple dimensions beyond accuracy. It foregrounds the learner's ability to negotiate meaning in ways that are socially and culturally resonant. These indicators not only guide analysis of learner–AI interaction but also serve as pedagogical benchmarks for fostering pragmatic and intercultural competence.

Thinking Through the Algorithm

The unifying concept of this framework is thinking through the algorithm. This describes the process by which learners critically engage with AI outputs, assessing them against pragmatic, relational, and cultural benchmarks before accepting, adapting, or rejecting them. Rather than functioning as passive recipients, learners become co-constructors of meaning, exercising agency in collaboration with the machine.

This notion builds on applied linguistics traditions that foreground autonomy, agency, and critical engagement (Duff, 2012; Oxford, 2017) while intersecting with recent discussions of AI literacy (Walter, 2024; Pan et al., 2025; Wang, 2024). Moreover, the process of thinking through the algorithm can also be understood through the lens of cognitive and sociocultural learning theories. From a metacognitive perspective, it reflects learners' capacity to plan, monitor, and evaluate their use of AI-generated language — a higher-order cognitive activity that involves self-regulation and conscious decision-making (Flavell, 1979; Wenden, 1998). At the same time, sociocultural theory highlights how interaction with AI tools can serve as a form of mediated learning, where language development is scaffolded through engagement with external artefacts (Vygotsky, 1978; Lantolf & Thorne, 2006). In this view, AI is not merely a provider of linguistic input but a mediating tool that shapes cognition, supports internalisation of discourse norms, and extends learners' zone of proximal development. Linking thinking through the algorithm to these theoretical traditions deepens our understanding of how cognitive and social processes underpin learner–AI interaction.

Pedagogically, thinking through the algorithm highlights the need to prepare learners for collaborative work with AI. This involves equipping them with linguistic strategies—such as editing verbosity or adjusting register—and intercultural strategies, including recognising culturally appropriate formulations and avoiding literal translations. In this sense, thinking through the algorithm becomes a teachable competence that integrates digital literacy with intercultural communicative competence.

Ultimately, the framework repositions AI as a partner whose outputs demand negotiation and adaptation. By theorising thinking through the algorithm as a central learner strategy, it offers a lens for understanding how learners navigate AI-mediated communication and how language education can better support them in doing so.

ILLUSTRATIVE INSIGHTS FROM EMPIRICAL WORK

The Study

This section presents data from a classroom-based study involving 35 beginner-level Mandarin learners enrolled in a university programme in Brunei. They were aged between 19 and 25 and came from a wide range of academic majors, including the sciences, engineering, social sciences, and humanities. Their areas of study included mathematics, computing, engineering, geography, business, environmental studies, and literature, reflecting the academic diversity typical of university-wide language electives. Their linguistic background was predominantly multilingual, with English and Malay functioning as their principal communicative resources. In many respects, this cohort mirrors the general learner characteristics and demographic profile of undergraduate students in Brunei, who tend to be multilingual, academically diverse, and accustomed to navigating English and Malay across both academic and everyday communication contexts.

The students, who had no prior exposure to Mandarin before the course, possessed only a limited working vocabulary of about 50–150 words when completing the assignment, mainly covering greetings, numerals, basic courtesy expressions, and simple transactional phrases. At this level, learners typically rely heavily on memorised sentence patterns and often struggle with tone production, word-order constraints, and culturally appropriate forms of address. When completing communicative tasks assigned by the course instructors, the students were allowed and frequently optioned to use AI tools to overcome lexical gaps, generate model sentences, or verify accuracy. This reliance on AI, combined with their developing pragmatic awareness, makes beginner learners an important group for examining how communicative decisions are shaped, scaffolded, or distorted within AI-mediated interactions.

Over a seven-week period, the students completed three communicative tasks simulating real-world scenarios, including transport interactions, medical consultations, and hotel bookings. Each student produced an audio recording accompanied by a written script. The written scripts served as the primary source for analysing how learners accepted, adapted, or rejected AI-generated suggestions. Four experienced Mandarin instructors evaluated the scripts using the three dimensions outlined in the chapter's framework: pragmatic appropriateness, relational sensitivity, and cultural adaptiveness.

To ensure interpretive validity, four experienced Mandarin instructors independently reviewed the students' written scripts. Their evaluations focused on scorable, surface-visible features that are commonly used in pragmatics research, including alignment of tone and directness with contextual expectations, the use of mitigation markers (e.g., Bù h oyìsi [excuse me]), address forms (e.g., Nín vs N [different forms of "you"]), sequencing of moves, and culturally normative closing strategies. These expert judgments were triangulated with learners' semi-structured interview responses, which provided insight into their intentions, decision-making processes, and awareness of contextually appropriate language use. Together, the instructors' evaluations and the learners' self-reported reasoning allowed the researchers to assess the extent to which students demonstrated pragmatic competence through their ability to critically mediate AI output and produce contextually appropriate, relationally attuned, and culturally resonant utterances.

Rather than offering a comprehensive data report, this section highlights selected examples that illustrate how learner agency influenced communicative effectiveness in an Ai-mediated environment. The analysis focuses on three core dimensions of communicative effectiveness, pragmatic appropriateness, relational sensitivity, and cultural adaptiveness, revealing how learners critically engaged with AI to navigate affordances and constraints. These examples operationalise the conceptual framework of *thinking through the algorithm,* demonstrating how communicative outcomes are shaped by dynamic interaction between learners and generative tools.

To ground the subsequent analysis, Table 2 presents representative instances of learner engagement with AI-generated content across the three task domains. Each example identifies the speech act involved, the AI-produced draft, and the learner's revision. These illustrate the communicative challenges that prompted learners to adapt, reshape, or selectively retain AI-generated language and serve as an empirical foundation for the detailed discussion that follows.

Table 2. Examples of Learner Mediation of AI-Generated Output Across Communicative Scenarios

Scenario	Speech Act	AI-Generated Draft	Learner-Revised Version	Issue Identified	Indicator
Transport	Expressing appreciation	"Nà h o, fēicháng xièxiè nín de bāngzhù." (All right, thank you very much for your help.)	"Nà h o, xièxiè nín."	AI output overly formal for a brief public interaction; unnecessary elaboration.	Pragmatic Appropriateness
	Initiating a request	Q ng kuài di n, w kuài chídàole. (Can you please speed up? I am running late.)	Learner added: "Bù h oyìsi, q ngwèn…"	Learner inserted softener to sound polite and less abrupt.	Relational Sensitivity
Clinic	Describing symptoms	W tóuténg dé lìhài, hóulóng yě téng.(I got a bad headache and my throat is sore as well.)	Learner shortened to: "W s ngzi téng, y u di n fāshāo."	AI's style too formal and article-like; learner aimed for natural patient speech.	Cultural Adaptiveness
	Closing the consultation	Rú y u rènhé wèntí, q ng y w men liánxì. (If there is any issue, you may contact us.)	Learner replaced with "Y u shén me wèntí, suíshí liánxì w men"	Learner recognised culturally preferred "face-preserving" exit; AI output sounded like a clinic announcement.	Relational Sensitivity
	Initiating a request	W xi ng chī yào.(I want to take medicine.)	Learner added: "Bù h oyìsi, w yàodi n yào."	Learner reshaped tone to maintain interpersonal harmony in a patient-doctor interaction.	Relational Sensitivity
Hotel	Making a request	Q ng gěi w yī tiáo máojīn."	Learner streamlined to: "Q ng gěi w máojīn, kěy ma?"	Over-elaboration from AI inappropriate for basic service exchange.	Pragmatic Appropriateness
	Softening a refusal	Bù yào.(No need)	Learner replaced with "Bù yòng le, xièxiè."	Learner adjusted tone to avoid sounding rude in a service context.	Relational Sensitivity
	Requesting information	"Rúgu y u xūyào, q ng suíshí liánxì w men." (If you need anything, please feel free to contact us.)	"Suíshí liánxì."	AI produced service-provider style closing unsuitable for a passenger asking for directions.	Pragmatic Appropriateness

Pragmatic Appropriateness

A key finding from the larger study was that learners often defaulted to over-formal, overly elaborate expressions generated by AI, reflecting a lack of awareness about contextual appropriateness. For example, utterances such as *"Na hao, fei chang xie xie nin de bang zhu"* (literally, "All right, thank you very much for your help") or *"Ru guo you xu yao, qing sui shi lian xi wo men"* ("If you need anything, please feel free to contact us") were grammatically correct but overly formal for casual settings. Learners who engaged critically with these AI-generated suggestions revised them into more natural expressions like *"Na hao, xie xie nin"* ("All right, thank you") or simply *"Sui shi lian xi"* ("Contact us anytime"). These revisions reflect learners' growing awareness of Mandarin discourse norms and show how they adapted AI suggestions through iterative decision-making. In several cases, learners treated the AI's overly formal output as a rough draft, modifying tone and structure to align with authentic use. One learner described this process as 'getting a formal version first, then making it sound more like me.' This suggests a form of co-authoring in which learners selectively drew on AI for structural scaffolding before asserting their own pragmatic preferences.

Interview data reinforced this pattern. One learner explained, "The AI always gives very polite and long sentences. But if I say that to my friend, it sounds too much, so I changed it." Another reflected, "Sometimes I feel like AI writes like a formal article. I want to sound like myself, not a robot." These examples illustrate that pragmatic appropriateness becomes a diagnostic entry point into the larger cognitive process of mediation: learners selectively trusted and reshaped AI language based on their intuitive grasp of communicative norms.

Relational Sensitivity

The framework also reveals that AI does not consistently model relational nuance, especially in interpersonal tasks. Learners had to mediate AI-generated content by inserting or reducing politeness markers, transitions, or mitigations to render their discourse socially acceptable. For example, learners added softeners such as *bù h o yì si* ('excuse me' or 'sorry') or used face-saving closings like *nà w xiān z u le* ('then I'll go now'), istead of direct goodbye, to maintain relational harmony.

While these additions may not have been directly suggested by the AI, learners often reported that the AI's initial output provided a structural base which learners reshaped to better suit the interpersonal context. One learner explained, "The AI gave something too flat, but it helped me remember the structure. Then I added my own way of saying it." This highlights a co-authoring process in which learners used the AI as a springboard and then built upon it based on their relational intuition. The revisions indicate their ability to judge interpersonal implications and restructure the message accordingly.

Interview responses further clarified this negotiation. One participant remarked, "The AI's answer felt cold, like something in a textbook. I wanted to sound friendly, so I added a sentence to make it softer." Another said, "It was too direct. I felt uncomfortable using it as-is." These moments exemplify how learners actively evaluate social tone, exercising relational judgment beyond linguistic correctness. Here, learners demonstrate reflective awareness in reconfiguring interpersonal stance, moving beyond surface corrections to achieve socially appropriate communication.

Cultural Adaptiveness

Learners also encountered instances where AI suggestions, though structurally accurate, failed to reflect cultural conventions. Direct translations of English idioms or overly literal renderings often disrupted the local coherence of the message. Learners who identified and replaced these mismatches with culturally resonant alternatives exhibited both linguistic and intercultural agency.

Interview reflections were particularly revealing. One student noted, "Some sentences sound like Google translated. I know Chinese people don't say that, so I changed it to a more normal way." Another commented, "It's like the AI is a smart friend who doesn't know Chinese culture very well. You have to help it sound right." Such responses position AI as a partial but fallible partner in communication—one whose outputs require intercultural mediation.

However, many beginner learners lacked this awareness. They tended to accept AI's translations or production at face value, indicating limited sensitivity to cultural nuances. This contrast highlights the developmental nature of intercultural agency and underscores the need to support learners in recognising when AI-generated outputs diverge from culturally appropriate norms. One learner explained, 'It (AI) gives a translation, but I got the feeling Chinese people don't say it that way. I keep the idea, but try to ask AI to change the way to say it.' This reflects a collaborative negotiation where AI contributes content direction, and the learner refines it for cultural resonance.

Within the framework, these cases illustrate how cultural adaptiveness emerges through learner-led mediation. Learners navigate between AI's globalised language models and local communicative norms, engaging in a reflective process that blends linguistic adaptation with cultural sensitivity.

Summary of Illustrative Insights

The examples presented here operationalise the conceptual framework of *thinking through the algorithm* and demonstrate how communicative outcomes are shaped by dynamic interaction between learners and generative tools. Communicative effectiveness in AI-mediated environments emerges not from output quality alone, but from learners' agency in negotiating that output. Through revision, contextualisation, and cultural calibration, learners bridged functional affordances and communicative goals. These patterns underscore the interdependence of the framework's components and support a pedagogical approach that cultivates learners' strategic, relational, and intercultural awareness when working with generative AI tools.

PEDAGOGICAL IMPLICATIONS

The findings of this investigation demonstrate that meaningful engagement with generative AI in language education requires more than technical proficiency or surface-level grammatical accuracy. Learners' written outputs, triangulated with interview data, reveal that communication in AI-mediated environments is inherently negotiated and co-constructed. Students frequently positioned AI-generated suggestions as provisional drafts or scaffolds that demanded revision to achieve pragmatic appropriateness, relational sensitivity, and cultural resonance. Yet, beginner learners often lacked the metacognitive awareness to recognise when AI outputs diverged from contextual expectations, accepting overly literal or excessively formal language uncritically. These results indicate that pedagogy must extend beyond grammar and lexis to cultivate critical AI engagement, contextual discernment, and adaptive communicative strategies.

Integrating AI Literacy into Language Education

AI literacy should be regarded as a core dimension of contemporary language education. This involves not only functional proficiency with AI tools but also the capacity to critically appraise, modify, and strategically deploy their outputs. Pedagogical practices can foreground this by requiring students to interrogate AI-generated alternatives, debate which expressions are most contextually appropriate, and justify their revisions. Such classroom routines underscore that AI can function as a valuable co-author but remains fallible, thereby reinforcing the learner's role as an active evaluator rather than a passive recipient. Embedding these practices supports the internalisation of negotiation skills, advancing both linguistic competence and digital literacy.

Scaffolding Contextual and Communicative Competence

The study underscores the centrality of contextual knowledge in shaping communicative effectiveness. Learners were compelled to adapt AI-generated language not only for pragmatic politeness but also in relation to audience expectations, situational conventions, and cultural norms. Instances of softening overly direct phrasing or reformulating literal translations illustrate the necessity of contextual sensitivity. Such competence cannot be assumed; it must be explicitly scaffolded through targeted pedagogy. Role-play activities, audience-specific writing assignments, and reflective analyses of AI outputs can train learners to critically evaluate and adapt machine-generated discourse for authentic communicative purposes.

Teacher Professional Development

Educators occupy a pivotal role in equipping learners to navigate AI critically. Professional development must therefore address both the affordances and limitations of generative tools, preparing teachers to design tasks that foreground learner agency and evaluative judgment. This entails more than technical familiarity; it requires pedagogical strategies that integrate AI in ways that foster reflection, adaptation, and self-regulation. Prior research has shown that teachers' content knowledge and instructional beliefs significantly influence the extent to which they adopt and adapt innovative practices (Shen et al, 2022; Chua et al, 2021). In this light, collaborative workshops and professional learning communities can provide valuable platforms for sharing practices and refining methodologies. By modelling critical engagement, teachers can demonstrate how AI can serve simultaneously as a resource and a challenge within the learning process. Collaborative workshops and professional learning communities can provide valuable platforms for sharing practices and refining methodologies. By modelling critical engagement, teachers can demonstrate how AI can serve simultaneously as a resource and a challenge within the learning process.

Balancing Opportunities and Risks

The pedagogical integration of AI necessitates a careful balancing of benefits and risks. Generative tools can expand learners' access to input, provide personalised feedback, and stimulate practice opportunities. However, they also risk promoting uncritical dependence, perpetuating formulaic expression, and obscuring sociocultural nuance. Findings from this study indicate that learners benefit most when AI is positioned as a co-authoring partner whose contributions require evaluation and adaptation. Instructional design should therefore deliberately incorporate reflection on when AI facilitates learning, when it misleads, and how learners can assert agency in shaping communicative choices.

In sum, the implications highlight the need for language education in AI-mediated environments to prioritise the cultivation of learners' capacity to critically collaborate with technology. By embedding AI literacy, contextual awareness, and adaptive communicative strategies into curricula, educators can prepare students to navigate both the affordances and the limitations of generative systems. Such an approach ensures that AI functions as a catalyst for, rather than a substitute for, the development of communicative competence.

CONCLUSION

The rapid evolution of generative AI is reshaping the landscape of language education, demanding a fundamental rethinking of how communication is taught, learned, and assessed. This chapter has argued that moving beyond traditional notions of AI as a background aid toward a paradigm of AI-mediated communication (AIMC) offers richer possibilities for language learning. By conceptualising AI as a co-author rather than an assistant, learners are positioned as active agents who negotiate, adapt, and reshape machine-generated outputs to achieve communicative goals.

Central to this shift is the proposed conceptual framework of *thinking through the algorithm*, which integrates AI's functional capacities, learner cognition and agency, and communicative effectiveness into a dynamic, cyclical relationship. Empirical evidence from beginner-level Mandarin learners illustrates how communicative outcomes emerge not from AI output alone but from learners' capacity to critically engage with and transform that output. Through iterative processes of revision, contextualisation, and cultural calibration, learners bridged the gap between the affordances of generative tools and the demands of authentic communication. These findings underscore that pragmatic appropriateness, relational sensitivity, and cultural adaptiveness must become core benchmarks for communicative success in AI-mediated environments.

The pedagogical implications are clear: language education must expand its scope beyond grammatical accuracy and lexical breadth to cultivate learners' critical AI literacy, contextual awareness, and communicative competence. Teachers play a crucial role in designing learning environments that foreground negotiation and reflection, equipping students with strategies to evaluate and adapt AI outputs for diverse communicative contexts. At the same time, curricula must balance the opportunities and risks of generative tools, encouraging students to see AI not as an infallible authority but as a fallible partner whose contributions require active mediation.

Ultimately, this chapter positions AI-mediated environment as both a challenge and an opportunity. It challenges language pedagogies that privilege structural correctness over communicative depth, while offering opportunities to reconceptualise language learning as a collaborative, mediated practice shaped by human–machine interaction. By fostering learners' ability to think through the algorithm, language education can harness the transformative potential of generative AI to prepare students for communication that is not only accurate and fluent but also pragmatically meaningful, relationally nuanced, and culturally resonant. Future research should continue to refine this framework across diverse linguistic and intercultural contexts, exploring how learners' agency evolves as they navigate the complexities of communication in an AI-saturated world.

REFERENCES

Baker, W. (2024). Intercultural communication. *ELT Journal*, *78*(2), 212–225. DOI: 10.1093/elt/ccad040

Brown, P., & Levinson, S. (1987). *Politeness: Some universals in language usage*. Cambridge University Press.

Byram, M. (1997). *Teaching and assessing intercultural communicative competence*. Multilingual Matters.

Chen, X., Xie, H., Zou, D., & Hwang, G. J. (2020). Application and theory gaps during the rise of Artificial Intelligence in Education. *Computers and Education: Artificial Intelligence*, *1*, 100002. DOI: 10.1016/j.caeai.2020.100002

Chew, F. P., Min, S., & Yap, T. T. (2023). Factors influencing the Mandarin teachers' intercultural teaching beliefs and practices. *Kasetsart Journal of Social Sciences*, *44*(2), 455–464. https://so04.tci-thaijo.org/index.php/kjss/article/view/266273

Chiu, T. K. F., Xia, Q., Zhou, X., Chai, C. S., & Cheng, M. (2023). Systematic literature review on opportunities, challenges, and future research recommendations of artificial intelligence in education. *Computers and Education: Artificial Intelligence*, *4*, 100118. DOI: 10.1016/j.caeai.2022.100118

Chua, N. A., Soon, G. Y., Ibrahim, M. Y., Che Noh, C. H., Mansor, N. R., Zaid, C. M., & Shen, M. (2021). Virtual differentiated instruction and its implementation in Mandarin teaching: A review. *Journal of Nusantara Studies*, *6*(2).

Deardorff, D. K. (2006). Identification and assessment of intercultural competence as a student outcome of internationalization. *Journal of Studies in International Education*, *10*(3), 241–266. DOI: 10.1177/1028315306287002

Duff, P. A. (2012). Identity, agency, and second language acquisition. In Mercer, S., Ryan, S., & Williams, M. (Eds.), *Language learner agency* (pp. 15–29). Multilingual Matters.

Ebadi, S., & Amini, J. (2022). The effects of AI-based chatbot on learners' affective status. *Computers & Education*, *180*, 104432. DOI: 10.1016/j.compedu.2022.104432

Escalante, J., Pack, A., & Barrett, A. (2023). AI-generated feedback on writing: Insights into efficacy and ENL student preference. *International Journal of Educational Technology in Higher Education*, *20*(57), 1–20. DOI: 10.1186/s41239-023-00425-2

Flavell, J. H. (1979). Metacognition and cognitive monitoring: A new area of cognitive–developmental inquiry. *The American Psychologist, 34*(10), 906–911. DOI: 10.1037/0003-066X.34.10.906

Fu, Y., Foell, S., Xu, X., & Hiniker, A. (2024). From text to self: Users' perception of AIMC tools on interpersonal communication and self. *Proceedings of the CHI Conference on Human Factors in Computing Systems, 17,* 1–17. https://doi.org/ DOI: 10.1145/3613904.3641955

Hancock, J. T., & Levy, K. (2020). AI-mediated communication: Definition, research agenda, and ethical considerations. *Journal of Computer-Mediated Communication, 25*(1), 89–100. DOI: 10.1093/jcmc/zmz022

Ji, H., Han, I., & Ko, Y. (2022). A systematic review of conversational AI in language education: Focusing on the collaboration with human teachers. *Journal of Research on Technology in Education, 55*(1), 1–22. DOI: 10.1080/15391523.2022.2142873

Lantolf, J. P., & Thorne, S. L. (2006). *Sociocultural theory and the genesis of second language development.* Oxford University Press.

Lee, J., Park, H., & Kim, Y. (2025). Generative AI in language education: A systematic review. *Language Learning & Technology, 29*(1), 45–67. DOI: 10.1016/llt.2025.45

Lee, S., Choe, H., Zou, D., & Jeon, J. (2025). Generative AI (GenAI) in the language classroom: A systematic review. *Interactive Learning Environments.* Advance online publication. DOI: 10.1080/10494820.2025.2498537

Oxford, R. (2017). *Teaching and researching language learning strategies: Self-regulation in context.* Routledge.

Pan, Y., Zhang, L., & Chen, H. (2025). Self-regulated reading with ChatGPT: Exploring learner strategies in AI-supported contexts. *Language Learning & Technology, 29*(2), 55–73. DOI: 10.1016/llt.2025.29.2

Pegrum, M. (2025). From revolution to evolution: What generative AI really means for language learning. *Language Teaching,* •••, 1–17. DOI: 10.1017/ S0261444825000151

Ranalli, J. (2018). Automated written corrective feedback: How well can students make use of it? *Computer Assisted Language Learning, 31*(7), 653–674. DOI: 10.1080/09588221.2018.1428994

Shen, M., Peng, C. F., Teng, Y. T., & Hua, F. (2022). Chinese ESL Instructors' Beliefs and Practices of Intercultural Education in a Ubiquitous Environment. In Intercultural Communication and Ubiquitous Learning in Multimodal English Language Education (pp. 68-88). IGI Global Scientific Publishing. https://doi.org/ DOI: 10.4018/978-1-7998-8852-9.ch004

Spencer-Oatey, H. (2008). *Culturally speaking: Culture, communication and politeness theory* (2nd ed.). Continuum.

Sun, Y., & Gao, F. (2023). Exploring the risks of generative AI in intercultural education. *Computers & Education*, *200*, 104792. DOI: 10.1016/j.compedu.2023.104792

Thomas, J. (1983). Cross-cultural pragmatic failure. *Applied Linguistics*, *4*(2), 91–112. DOI: 10.1093/applin/4.2.91

Vygotsky, L. S. (1978). *Mind in society: The development of higher psychological processes*. Harvard University Press.

Walter, M. (2024). AI literacy in education: Preparing learners for critical engagement with AI. *Computers & Education*, *195*, 104744. DOI: 10.1016/j.compedu.2024.104744

Wang, L. (2024). Self-regulated use of generative AI in language learning: Monitoring, evaluation, and adaptation. *ReCALL*, *36*(1), 22–40. DOI: 10.1017/S0958344023000219

Wang, Y., Zhang, T., Yao, L., & Seedhouse, P. (2025). A scoping review of empirical studies on generative artificial intelligence in language education. *Innovation in Language Learning and Teaching*. Advance online publication. DOI: 10.1080/17501229.2025.2509759

Wenden, A. (1998). Metacognitive knowledge and language learning. *Applied Linguistics*, *19*(4), 515–537. DOI: 10.1093/applin/19.4.515

Yan, D. (2023). Impact of ChatGPT on learners in a L2 writing practicum: An exploratory investigation. *Education and Information Technologies*, *28*, 13943–13967. DOI: 10.1007/s10639-023-11742-4

Chapter 7
Artificial Intelligence and the Linguistic Turn in Education

Utkarsh Mishra
https://orcid.org/0009-0004-6955-755X
Vellore Institute of Technology, Chennai, India

Riya Kesharwani
https://orcid.org/0009-0009-2911-1402
Chandigarh University, Mohali, India

K. Vijayaprabakaran
https://orcid.org/0000-0002-7322-2443
Vellore Institute of Technology, Chennai, India

ABSTRACT

This chapter provides a critical reflection on the encounter between artificial intelligence and the linguistic turn in education as language-based AI systems reconfigure meaning-making, discourse, and epistemology in today's educational settings. Drawing on interdisciplinary perspectives from applied linguistics, cognitive science, and educational theory, the chapter examines the underlying epistemological issues of AI interventions such as large language models, intelligent tutoring systems, and natural language processing. The authors consider how AI challenges as well as accommodates the linguistic turn's dialogic and constructivist ideals by appropriating the key components of classroom interaction, learner role, and authorship. Particularly, the chapter investigates the implications of AI for multilingual education, low-resource language preservation, and linguistic diversity and

DOI: 10.4018/979-8-3373-4397-6.ch007

inclusion in generally underrepresented contexts. The chapter also interrogates the ideological and ethical dimensions of algorithmic language, including bias, and linguistic imperialism.

1. INTRODUCTION

The adoption of artificial intelligence (AI) in educational settings has indicated a shift in understanding and facilitating language, cognition, and learning. This shift matters in education with regard to the linguistic turn an intellectual movement that indicates that language plays a key role in the construction of human knowledge, identity, and social interaction. In general, education focused on the linguistic turn operated through dialogue, discourse, and social construction of meaning. However, the emergence of AI systems e.g., capable of generating, responding to, and interpreting human language presents complex adaptabilities and possibilities about how these key features will be available in today's classrooms (Ahmed et al., 2025). Concerning AI technologies such as large language models (LLMs), chatbots, and intelligent tutoring systems, how will these reshape an understanding of learner agency, knowledge construction, and mediation of multilingual and culturally diverse discourses? By reflecting on these matters, in this chapter, we focus on the socially-transformative power of AI, and it's socio-ethical tensions, in contemporary schooling through a review of the relationship between AI and the linguistic turn. As shown in Figure 1, artificial intelligence can take on many roles in educational settings, from creating content to helping with administrative decisions.

Figure 1. Use Case diagram illustrating AI-enabled functions in the educational ecosystem.

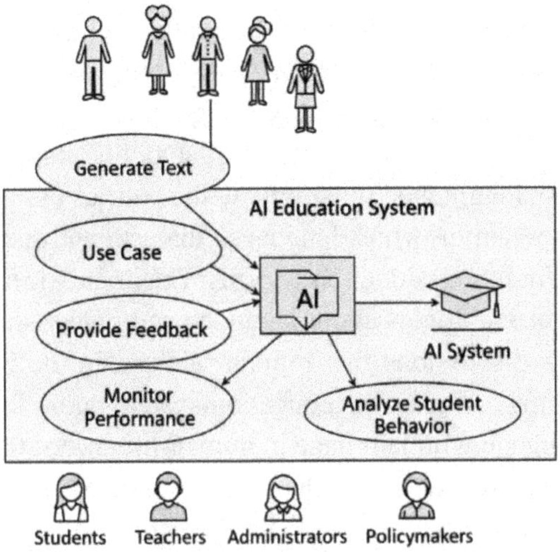

1.1. Overview of the Linguistic Turn in Education

The linguistic turn signifies a major transformation in the humanities and social sciences in the 20th century, where language moved from being seen as an apparatus for articulating ideas to actually being seen as the medium in which our realities are formed and understood. In education, this shift included a move from behaviourist and transmission models of learning to more constructivist, interpretative, or discourse-oriented models of learning. In this view, language was no longer treated as a passive vehicle for transmitting content, but a complex sociocognitive tool that constructs cognition, identity, and the social construction of knowledge. The central tenet of the linguistic turn in education is that learning occurs through language-mediated interaction. Following Wittgenstein, Vygotsky, and Bakhtin, the linguistic turn posits that meaning is not fixed or inherent, but is negotiated through dialogue and context. Vygotsky's sociocultural theory of development argues that language is the primary psychological tool for cognitive development, and that thought and speech develop concurrently through socially mediated interaction. Bakhtin's notion of dialogism highlights the multiplicity of voice and perspective in every utterance, and the socially constructed and dynamic nature of learning through discourse.

The impact of the linguistic turn has been expansive, both in education theory and practice. It has opened the door to learner-centered pedagogies, inquiry-based learning, critical literacy, or discourse analysis, to be positioned in classroom contexts. Teachers were no longer viewed as a source of information; teachers were seen as facilitators of dialogue, co-constructors of knowledge, and as constituents of a meaning-making community. The design of curriculum began to encompass more authentic communication, thinking critically, and interacting with linguistic diversity. Assessments moved away from being solely assessed through objective measures to more formative ways of assessing students, through language, in an attempt to recognize the interpretive nature of students' responses. This change in ways of thinking also heightened sensitivity to the politics of language in education. Researchers began to examine whose languages, dialects, and discourse practices were privileging or marginalizing education systems. The linguistic turn has subsequently opened discussion of ideologies about language and power, and the socio-cultural nature of learning - discussions that remain relevant in the increasingly diverse, globalized classrooms of today. As education systems now face the advent of AI technologies that engage with language in human-like ways, the basic tenets of the linguistic turn are being re-examined. Questions about the legitimacy of AI-mediated communication, the importance of human interaction to learning, and the degree to which AI may participate in or undermine dialogic pedagogy are all coming to bear. By understanding the inception and ramifications of the linguistic turn, we are better equipped to critically examine how AI may embody or malignantly subvert educational goals and values established by the linguistic turn.

1.2. The Rise of Language-Based AI Systems in Pedagogy

Recent advancements in language-based artificial intelligence (AI) systems have rapidly advanced and disrupted educational practices and pedagogy. These systems are able to read, write, and talk in natural human language with similar fluency as humans due to their ability to process natural language (NLP), their categorization as a machine learning or algorithmic software, their deep neural networks. Currently popular examples of these applications are named large language models (LLMs) and include OpenAI's GPT, Google's BERT and Meta's LLaMA. They have been used in a wide variety of educational contexts (e.g., personalized tutoring, automated essay assessments, language learning apps, intelligent feedback applications) (Akindotuni, 2025). The use of the AI applications in pedagogy can be considered an innovative "language interface" in education. These applications enabled students to no longer receive content in a static form, but to instead interact dynamically with the AI applications using their language with the understanding that the AI will dynamically engage and respond to them immediately with responses, explanations, corrections, or other relevant feedback. Using pedagogical AI applications also contributed to the personalization of learning pathways that traditional educational technology did not foster. For example, chat bots powered by generative AI and writing assistants could converse akin to a tutor allowing the user to brainstorm ideas, modify or create persuasive arguments, practice new vocabulary, or continue a conversation- using natural dialogue.

In addition, these tools are becoming even more accessible within public-facing educational contexts. There are many Learning Management System (LMS) tools that now have AI-enhanced discussion boards and feedback tools; language learning (e.g., Duolingo, ELSA Speak, and Grammarly with NLP algorithms to assess pronunciation, grammar, and cohesiveness). In some higher education contexts, AI is being implemented to support peer review, recommend resources, and provide formative assessment. The result is a variety of learning environments, where language-based AI operates as an adjunct to teaching or even as a stand-in for human connection. The advantages of language-based AI within pedagogy are diverse. From an emergent practitioner lens, the systems allow for flexible responses to the constraints of teacher-time constraints, student-school ration challenges, and supports differentiated instruction. They are available 24/7, can adjust to the level of appropriate learner engagement, and operate equally in more than one language viable features for language-based learning in diverse and resource-poor educational contexts. In addition, AI systems provide space for students to explore the use of language in a developmental way with little-to-no stakes, which could diminish anxiety and potentially increase learner autonomy.

But with this rise come essential questions and concerns. The conversational fluency of LLMs often conceals their limitations, such as lack of actual understanding, occasional factual inaccuracies (the AI hallucination phenomenon), and difficulties with nuanced and culturally specific language use. In addition, the datasets that train these models are heavily biased to dominant languages and perspectives, raising concerns about linguistic imperialism and digital inequality. The pedagogical considerations of outsourcing feedback, directions, and, for some, even dialogue to AI systems are also worthy of criticism: What happens to the role of the teacher? How is learner agency affected when all interactions are, even unknowingly, algorithmically mediated? More conceptually, can AI truly be a partner in the co-construction of meaning when it is fundamentally limited to the ways language triggers Human-AI interactions, or does it risk supporting surface-level language engagement? As language-based AI becomes more common in classrooms and curricula, its rise now must be understood not just as another advancement of technology, but as a notable transformation of the whole ecology of pedagogy. It also brings into question longstanding assumptions about communication, understanding, and relational relationships in teaching. These tools present truly transformative possibilities, but also demand for educators to critically contemplate how language as the essence of teaching and learning, is changing in a world of intelligent machines.

1.3. Positioning the Chapter Within Global and Interdisciplinary Contexts

This chapter is located at the intersection of global educational change and interdisciplinary inquiry, providing a critical lens through which to explore the introduction of artificial intelligence (AI) into language-based pedagogies. In the rapidly changing educational landscape in countries around the world, AI technologies are emerging across various educational systems, and they carry significant educational implications across regional, cultural, and linguistic domains. While many AI-based technologies are starting in the industrialized nations in the Global North, and within these nations, in well-resourced educational systems, many of the implications of AI systems will emerge in different ways across the Global South where educational systems must contend with issues of limited resources, diverse languages, and conflicting societal systems. As the authors intend to place this chapter within a global context, it is important to consider and highlight the extent to which the educational realities of learners are shaped by the digital technologies including AI systems that will become widespread among less-resourced, multilingual, and marginalized communities (Alm & Watanabe, 2023). At a global scale, education systems are all struggling with issues of equity, access, and cultural responsiveness globally all questions of increased significance when conveyed through an artificial intelligence system that has drawn from datasets that are primarily English-based and Western-oriented. For example, in countries in South Asia, Sub-Saharan Africa, Latin America, and Southeast Asia, learners often make sense of their linguistic identities from a hybrid perspective of English and whatever local language policies have developed, often patterned as translanguaging, in complex and diverse spaces. The chapter discusses how even though AI tools hold promise and potential as educational resources, they might perpetuate existing language hierarchies, privileging certain resources and standard dialects while neglecting vernacular, indigenous, and low-resourced languages. This has implications for learning the language and implications for matters of cultural preservation, the process of identity formation, and epistemic justice.

Interdisciplinarity, this chapter combines theory and practice from a variety of perspectives, drawing on different traditions from fields such as applied linguistics, cognitive science, educational technology, philosophy of language, and AI ethics. With respect to language learning, it draws on sociocultural and constructivist theories of learning, dialogic pedagogy, and discourse analysis while engaging with the technical understandings of machine learning, natural language processing, and algorithmic design. Interdisciplinary perspectives allow for a more nuanced understanding of the ways in which AI is changing both human cognition in the process of language learning and the institutional context of education itself. Not to mention that the chapter discusses the topic at hand in connection with current scholarly debates about automation and human development, data-driven pedagogy and ethics, and the politics of knowledge production in AI systems. It is deliberate in challenging the technocentric assumption that the mere progress of technology means progress of some sort, and whether that progress is culturally, ethically, and pedagogically relevant. This chapter will strive to adopt a global and interdisciplinary perspective and encourage a pluralistic and holistic perspective over the limitations of a one-size-fits-all view of AI in education. It argues for a perspective that recognizes regional forms of pedagogic traditions, acknowledgement of languages, and cultures formed out of familiarity with common methods and uses of learning. In the end, the chapter contributes towards a more just and equitable, contextualised and thoughtful discussion of how AI should be shaped (not just used) to meet the needs of a diversity of learners and educators across the world.

2. LANGUAGE, AI, AND EPISTEMOLOGY

The intersection of language, artificial intelligence (AI), and epistemology is a significant point of consideration in current conversations about education. Language is a primary system of meaning, and thus, the foundation of human thought, knowledge, and cultural reproduction. The emergence of AI, and especially language-based AI like natural language processing (NLP), large language models (LLMs), and conversational agents, has the potential to transform how we construct, communicate, and recognize knowledge (AlNusif, 2025). This chapter is focused on considering how technological developments like AI not only have the potential to affect our teaching practices; they also may change epistemological assumptions about what is known, how it is known, and who has epistemic authority. Through considering educational implications of machine mediated language, this chapter aims to situate AI as an epistemic actor, as it can shape knowledge in ways we have not encountered, and it has the potential to shape curriculum, instruction, and learner identity.

2.1. Constructivist and Dialogic Views of Learning

Constructivist and dialogic theories of learning have had a significant impact on contemporary understandings of how knowledge is created, articulated, and taken on in educational contexts. Constructivist and dialogical theories reject the idea that learning simply involves being a passive receptacle of facts, and instead, suggests that learners are simply co-constructors of meaning through experience, reflection, and interactions with their context. Constructivism originates from the cognitive theories of Jean Piaget, and it emphasizes the development of knowledge based on an individual learner's previous understanding. It's viewed as the learner creating increasingly complex and abstract conceptual structures using processes of assimilating and accommodating more representations. Learning is not static, and as the learner develops concepts and engages with the material, knowledge is developed as contextual knowledge (situated within experience). Lev Vygotsky's sociocultural theory of learning, which is also part of the family tree of constructivism, also emphasizes the role of social interaction, cultural tools, and language in cognitive development. Vygotsky's notion of the Zone of Proximal Development (ZPD) explains the co-construction of knowledge and the role of social interaction and the need for some additional support of more knowledgeable others (e.g. peers, teachers, digital systems) to achieve outcomes alone (Gaol et al., 2024). Effectively, learning is not just a solitary internalized or constructed mental process, but a relational, dialogical act created through communication, collaboration, and cultural mediation.

Dialogic methods of learning extend these constructivist ideas by focusing on dialogue as a way to make meaning. Drawing on scholars such as Mikhail Bakhtin and Paulo Freire, dialogic pedagogy conceives of educational activity as relational, reciprocal, and transformative. In Bakhtin's theory of dialogism, he asserts that, because all communication is fundamentally responsive, it takes place in a network of social voices, perspectives, and ideologies. Learning through a dialogic perspective is not the result of the influence of one authoritative voice, but the richness created by the interplay of multiple discourse and interpretations. Similarly, Freire argues for problem-posing pedagogy in his critical pedagogy (Freire, 1970), which recognizes the importance of dialogue between learners and educators, questioning of assumptions, and collectively creating knowledge in ways that challenge oppression because they enact critical consciousness. In multilingual, multicultural, and digitally mediated classrooms, the dialogic and constructivist perspectives provide a challenging promise for inclusive, responsive education. They privilege learner voice, respect diverse epistemologies, and provide scaffolding for critical engagement with content and context. As artificial intelligence specifically, language-based AIs become part of the pedagogical environment, these theoretical lenses become critical tools for critique and design. Intelligent tutoring systems, AI-generated feedback, conversational agents, and LLMs all have the potential to impact dialogic learning through scaffolding interaction, modelling discourse and providing adaptive engagement most importantly. All of which give rise to important questions about authenticity, agency and the reproduction of dominant language ideologies. So, situating AI within dialogic and constructivist psychologies is not simply a theoretical framework, it is a necessary step toward ethical, inclusive, and pedagogically sound integration of AI tools in education. The perspectives prioritize learner agency, meaningful interaction, and critical reflection as ways for educators and developers to interrogate how AI tools can enhance learning rather than mechanize, standardize, or obfuscate the fundamentally social and contested nature of knowledge.

2.2. AI's Role in Meaning-Making and Symbolic Representation

Artificial Intelligence (AI), particularly language-based AI systems such as Large Language Models (LLMs), increasingly plays a transformative role in how meaning is constructed, interpreted, and mediated across educational and communicative contexts. These systems, trained on vast corpora of human language, simulate understanding by predicting, generating, and organizing linguistic output based on statistical relationships and contextual cues. While they do not "understand" language in the human sense, AI systems exhibit a remarkable capacity to approximate meaning-making processes and perform symbolic representation at scale. In educational settings, meaning-making is traditionally rooted in human interaction learners engage with texts, teachers, and peers to negotiate, question, and internalize concepts (Hila, 2025). AI systems now serve as both mediators and participants in this process. For example, intelligent tutoring systems, writing assistants, and AI-powered dialogue agents provide learners with real-time feedback, paraphrasing, summaries, concept clarification, and even content generation. These tools influence how learners encounter, interpret, and internalize ideas, often shifting the locus of epistemic authority from human educators to algorithmic systems. Figure 2 describes the symbolic representation process undertaken by AI models like ChatGPT as they turn raw input into an output that carries meaning in context.

Figure 2. Activity diagram showing the symbolic representation pipeline of AI language models.

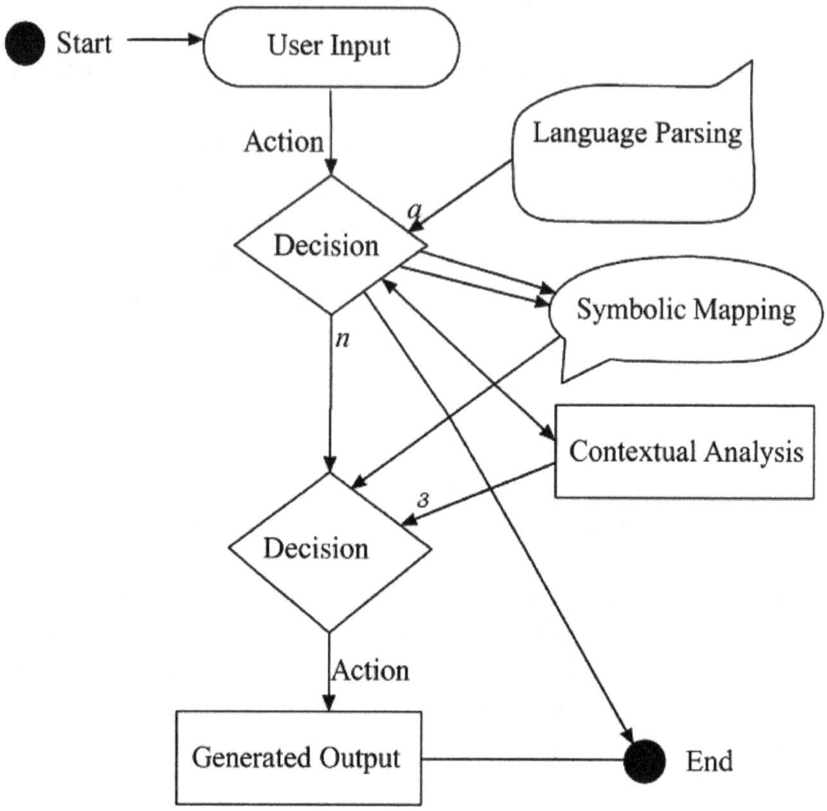

From a semiotic and cognitive perspective, symbolic representation is the use of symbols language, images, signs to refer to and construct meaning. AI is engaging in symbolic representation in regard to knowledge by encoding, transforming, and generating language-symbols that fit expected meanings in context. For example, if a student inputs an inquiry or prompt, a large language model (LLM) may generate a situationally appropriate response that reflects the symbolic configurations of expert human discussion. Therefore, AI reshapes how we engage with signs in their interpretation and the scaffolding of knowledge. In particular, this is significant in areas of learning where interpretation, nuance, and rhetorical framing matter such as literature, philosophy, and the social sciences. The meaning that AI generates, however, is fundamentally different than that of meaning-making for humans. Whereas humans generate meaning from lived experiences, intentionality, cultural context, and social engagement, AI systems generate meaning through pattern-matching and stochastic inference. This discrepancy raises important epistemological questions: Would AI-generated content truly reflect human values and intentions? Does AI reproduce or challenge prevailing language ideologies? To what degree should learners consider AI as a co-constructive partner in making sense of complex ideas?

In addition, the role of AI in symbolic representation expands from just textual responses, to other symbolic outputs that are visual, auditory, and multimodal-in the case of model and generative tools that produce diagrams, speech, music and code etc. To learners, this not only adds a new symbolic repertoire to use, but also presents new literacies needed to make sense of and evaluate artifacts generated by machines (Karpouzis, 2025). Thus, educators need to consider how they support learners to develop critical AI literacies which involve discerning AI responses are constructed, and the biases contained within their training data, as well as the limits of algorithmic logic as a representation of human reality. In conclusion, AI is not just a tool for delivering information, but is also now an agent in the construction of meaning and mediation of symbols in learning spaces. This further complicates our context and will require us to re-examine our pedagogical frameworks, and develop approaches that reconcile technological efficiency with critical reflection, human-centered dialogue and a dependency on the social, cultural and ethical elements of representation.

2.3. Cognitive Implications of LLMs and Algorithmic Text Generation

The emergence of Large Language Models (LLMs) and algorithmic text generation services like ChatGPT, Claude, and others presents deep cognitive shifts in the ways humans use language, create knowledge, and perform intellectual processing. Operating using deep learning systems and analyzing data from significant linguistic sources, these systems convincingly approximate human outputs with data-driven decision-making. Moreover, their current capabilities push the edges of how we view cognition by melding meaning-making, problem-solving, and reasoning processes that traditionally rested with humans in varying degrees. Cognitively, composing a written text in several forms academic writing, critical reflection, or dialogue has typically been associated with internal cycles of memory retrieval, concept formulation, modulating attention, and metacognitive products. Algorithmic text generators sit squarely between significant modes of text generation and simply provide pre-framed relevant texts which might reduce the need for continued cognitive labor for ideation (Khoo & Kang, 2022). Ideally, learners could gain access to the ideation processes of their instructors and solve complex problems easier; alternatively, they could be substituting their cognitive resources for those of the algorithms they interact with, with less critical use of evaluate, synthesize, and originality in service of fluency and efficiency. Figure 3 illustrates the typical iterative and interactive process of how learners collaboratively co-construct content with AI systems, incorporating back and forth segments.

Figure 3. Sequence diagram of learner interaction with LLM for iterative academic writing.

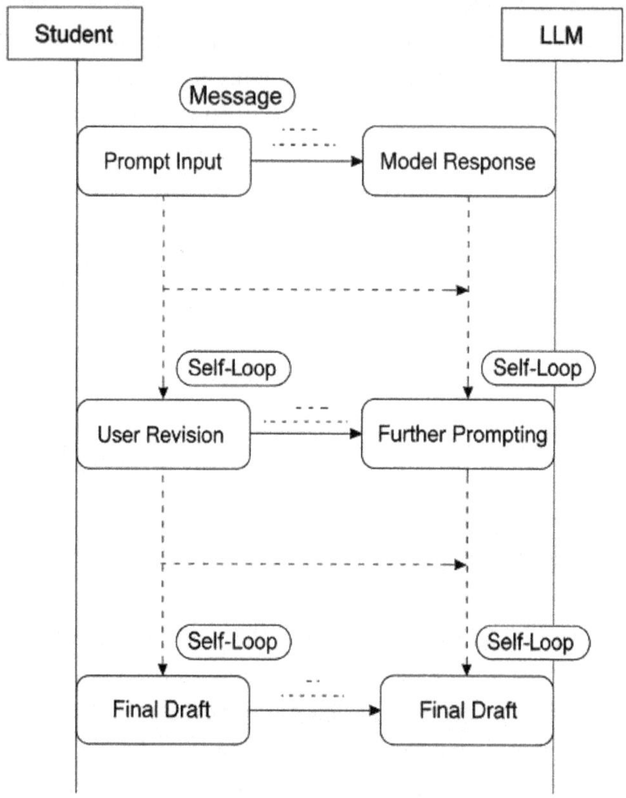

LLMs offer cognitive load theory opportunities and threats. On the one hand, the learner's extraneous load is reduced by facilitating grammar, vocabulary, or structure, freeing cognitive resources for higher order thinking. On the other hand, there is a danger they may subvert and miss the critical germane load, the mental processes used to build schemas and deep learning, through over-reliance on the generated artefacts. This presents a paradox. On the one hand, LLMs provide a cognitive scaffolding to get your thinking process started. On the other hand, the use of LLMs potentially diminishes the cognitive skills important for independent learning, inquiry and critical literacy. In addition, LLMs present new constructs of distributed cognition. When students interact with AI-generated responses, they are not working independently, they are co-constructing meaning with an agent that is non-human, which alters the cognitive environment from individual thought to hybrid human-machine cognition. Importantly, this co-development requires the learner to do cognitive operations such as interpret or evaluate the machine-generated response and determine how to use it - sometimes recontextualising it. This raises the crucial requirement of algorithmic metacognition, which is the critical awareness of not only the individual's own reasoning but also the epistemic trustworthiness and context to be aware of when machine-generated responses.

The role of AI has emerged as a factor in cognitive practices including information seeking, attention, and judgment. The immediacy associated with obtaining polished responses leads to a diminished appetite for ambiguity, less exploratory inquiry, and a higher value on quick certainty and correctness. In time, learners may be conditioned toward surface engagement rather than meaningful or deep, reflective learning. Consequently, in particular disciplines that rely on sustained argumentation, dialectical reasoning, and interpretive nuance this could transform dispositions associated with foundational cognitive abilities. Notably, features of LLMs structure the cognitive responses they might elicit. These models are designed to respond to input prompts by producing the most likely subsequent text statistically. LLMs favor coherent, fluent and plausible text over factually accurate or critical responses. As such, it is more likely their outputs promote, resonate with, and subsequently remain in the established discourse. In cognitive terms, this could prompt users to conform to the prevailing narrative and inhibit counter-narrative or contrarian thinking; and likely limit the exposure to epistemically diverse substantiates unless prompted otherwise. To conclude, the cognitive implications of LLMs are varied and context-dependent (Kim et al., 2025). They can offer unrivaled opportunities for augmenting thought and expression and may threaten autonomy, criticality, and depth. It is the responsibility of educators, cognitive scientists, and learning- or innovation-space designers to facilitate critical examination toward using LLMs in ways that continue to realize these contexts in the ways we intended.

3. AI IN DIALOGIC PEDAGOGY

The addition of Artificial Intelligence in education is transforming not only how knowledge is accessed and received, but also how it is collaboratively co-constructed via dialogue. Dialogic pedagogy acknowledges the relational, interactive and transformative way collaborative learning makes meaning via open-ended dialogue, from thinkers like Bakhtin, Freire and Alexander. As Artificial Intelligence systems, particularly large language models (LLMs), increase in prevalence in the dialogic space of classrooms, as part of tutoring systems, and as a feedback tool, they present a new opportunity for a genre of dialogue. This section considers how these AI tools can be thought about within dialogic pedagogy frameworks not only as information providers, but as participants in conversations, thought provokers, and meaning makers. This examination highlights the affordances, as well as tensions, of machine dialogue in human-centered educational discourse and raises pertinent questions related to authenticity, agency, reciprocity, and the role of teachers and students in emerging AI-mediated educational contexts. AI is coded to mediate dialogue and contribute to the multi-directional flow of communication among the educational actors depicted Figure 4.

Figure 4. Communication diagram modeling AI-mediated dialogic pedagogy.

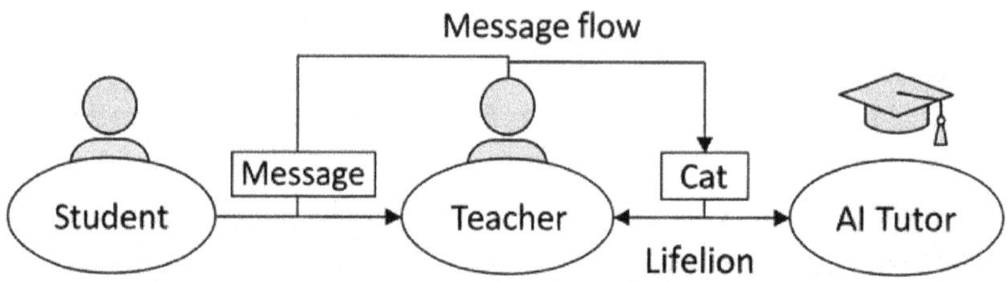

3.1. ChatGPT, Intelligent Tutoring Systems, and Dialogic Breakdowns

New conversational AI and other intelligent tutoring systems (ITS) offer new opportunities for and pedagogical design, particularly in contexts of dialogic learning. More specifically, its form of interaction is designed to mimic the natural forms of human conversation, offering a contextual response to learners' inquiries and adapting to individual learners based on their responses. In an ideal world, thanks to chatbot mechanisms, it would be possible to free oneself from scale and availability as well as have a dialogic partner that responds to your needs on your unique learning pathway. However, this possibility is not without a myriad of challenges toward a colourfully designed dialogic pedagogical framework, which could be said to be "dialogic breakdowns". At the centre of dialogic pedagogy, it is the notion that learning is derived from meaningful dialogue, characterized by reciprocity, mutual recognition and co-construction of knowledge. Although ChatGPT and ITS do a better than average job of mimicking dialogic interaction by using clever language modelling and retrieval, they are fundamentally incapable of consciousness, contextual substantive awareness and relational meaning-making (Kumar & Kaur, 2025). Thus, these platforms will struggle with recognizing context in a nuanced fashion, and the ability to respond meaningfully, ethically and culturally purposeful, when surrogate discussion takes place. When AI-generated feedback is either overly generic, over-confident in its incorrectness, or is unable to validate students' affective cues and struggles epistemically, break downs are amplified.

Additionally, these tools are frequently biased towards one type of dialogue usually either information gathering or task-based dialogue over other, more useful forms of dialogue which require ambiguity, reflection, or critical thinking. The result is an instrumental bias that narrows dialogic possibilities, pushing learners towards convergence answers rather than open ended exploration. From a social epistemic perspective, dialogic breakdowns provoke important questions: who is positioning the discourse? Which knowledge is privileged or erased? How do learners begin to understand authority and trust in AI-mediated learning spaces? Despite these concerns, intelligent tutorial systems when thoughtfully designed can deceive with their appropriateness in dialogic pedagogy. Tools such as scaffolding prompts, multi-turn clarification, and adaptive questioning assist students in their metacognition and engagement in discourse. Still, the design is pedagogically transparent, ethically conscious, and coupled with human mediation if and when the dialogic space is authentic, inclusive, and intellectually generative. Educators must check their critical awareness against these breakdowns, and continuously reflect about the pedagogical, epistemological, and relational commitments they have made towards enabling learning based on conversational AI.

3.2. Learner Agency, Feedback, and Academic Voice in AI-Supported Writing

In the shifting landscape of education, AI-supported writing tools from grammar correctors to futuristic large language models (LLMs) like ChatGPT have fundamentally changed how learners approach the writing process. These AI technologies present a greater concomitance of efficient work, immediacy of feedback, personalized scaffolding, etc., which has significant benefits for learners. However, these opportunities also create complex dimensions that require us to contemplate the notion of learner agency, the basis and nature of feedback when working with AI, and how we can still have an authentic academic voice in our writing. The multilevel engagement states of learners shown in Figure 5 may be best accessed through AI feedback loops, and emphasize the role of pedagogical design.

Figure 5. State machine diagram representing learner agency transitions in AI-supported environments

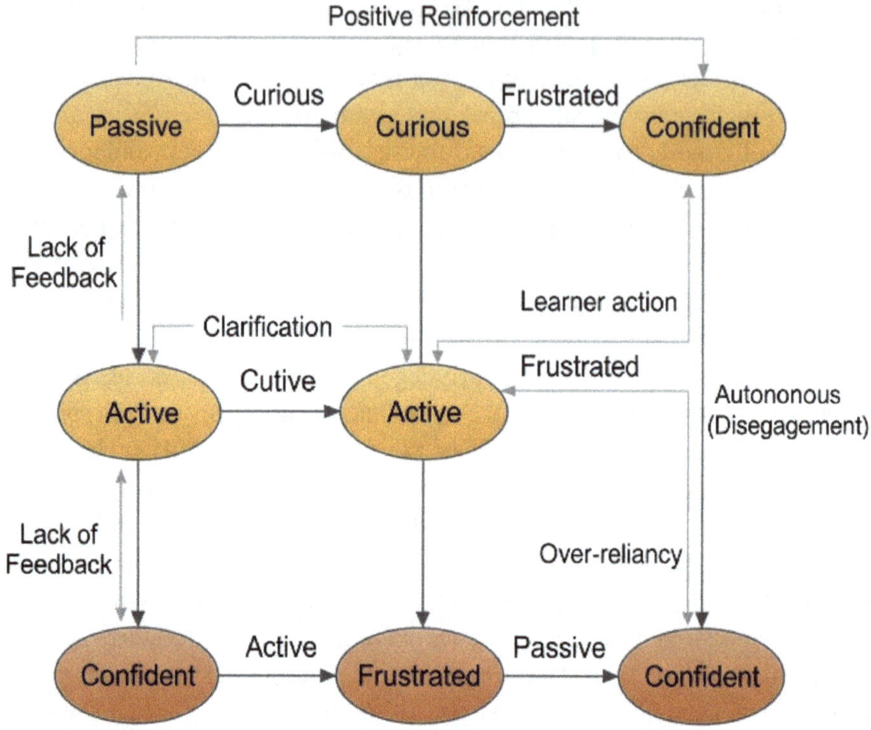

Learner agency, which can be simply defined as the ability for the learner to make choices that result in personal ownership of their learning moments, is essential to any meaningful educational experience. When AI tools can take one of many forms in the writing process, they can either enhance learner agency or exist to the detriment of that agency. On one side, AI systems provide an opportunity to engage with immediate and low-stakes conditions to experiment with their writing, revise iteratively, test phrases, or make suggestions while producing a text independently, through a senior or helper to the writing process. The benefit of immediacy can alleviate immediate pressure for learners to be held to the same standard of judgment when using AI tools; therefore, it can encourage learners to engage with increasingly ambitious writing prospects, in the writing moment. If learners do not use AI in the right ways this can be problematic, if they use suggestions without necessarily validating with their own thought processes, or if they allow AI to formulate a larger block of their textual development, they may run the risk of subtracting the intentional authorial focus, critical engagement, and reflective self-awareness critical to the academic writing process. When AI replaces learner agency, this dynamic means that AI's replacement role is raw (rather than collaborative) voice, which could represent a unhelpful type of intellectual ownership.

Feedback has long been viewed as a pivotal aspect of effective writing instruction, and yet, it takes on a different role when we consider AI-mediated feedback contexts. Traditional forms of feedback provided by teachers or peers often included not only grammar and mechanics, but also quality of content, logic/structure of the argument, and rhetorical choices characteristic of a number of disciplinary norms. AI-generated feedback, on the other hand, primarily generates feedback on surface-level features (e.g., grammar, coherence, tone). While feedback of this kind can be incredibly helpful for learners who have little access to a human professional, or writing outside their native language, it may not provide the recognizable dialogic, formative feedback that leads to deep learning (Oyetade & Zuva, 2025). Furthermore, the algorithmic nature of AI feedback is often opaque to human users they may have no idea what rationale a suggestion was based on, or they may accept a change uncritically, with no development of their own editing authority.

Academic voice, as a construct bound closely with epistemic positioning, credibility, and disciplinary identity, it is another area that is affected by AI-supported writing. Here, the main challenge is how personal it can feel, but personalisation can lead to homogenisation. AI systems are trained on large corpuses of both general and academic written texts, and thus reproduce discourses that have developed through dominant forms. While this could serve as a helpful approximation of acceptable academic conventions, it also could flatten varieties of stylistic diversity and cultural specificity, thus marginalizing non-Western epistemology or non-standard ways of using language. As students engage with these AI tools, learners' authentic voices risk being conformed more toward what the algorithm interprets or determines as "standard," than as the product of their rhetorical purposes and cultural contexts.

Educators have two responsibilities in balancing these tensions productively. We must design writing instruction that conservatively positions AI as a collaborative approach potentially as a resource for students and educators to augment rather than displace, human creativity and critical thinking. Specifically, this includes helping students ask questions of AI input, consider their decisions in writing, interrogate how the AI compromises or contributes to ownership over their language and rhetorical choices, and critically think about the epistemology and cultural considerations that may accompany AI-derived text, and show them ways to resist passive adoption while becoming aware of the socio-political implications of writing in a discipline to underpin or reinforce their own positionality. To this end, if we seek to nurture learner agency, reflexive feedback literacy, and academic voice in AI-assisted writing, it will require a critical and liberatory pedagogy in which learners learn how to write with AI, if used, but perhaps also writing against AI and certainly to take their writing beyond AI when the need arises.

3.2. Challenges in Sustaining Authentic Conversation With Machines

With the growing prevalence of AI-driven conversational agents, such as ChatGPT, Alexa and many intelligent tutoring systems, the nature of human-computer interactions at educational and everyday levels has shifted dramatically. These systems have notable fluency and coherence in their responses and understanding, and often bear an uncanny resemblance to human conversation. Yet, there remain challenges to keeping participants or interlocutors in authentic conversation meaning conversations in which parties remain aware of their conversations as evolving interactions with shared understanding, responsiveness to nuance, and faithful engagement as dialogical participants.

A. The Facsimile of Understanding

One of the most significant problems is that language- based AI does not have understanding in a human-related sense. They lack conscious meaningfulness; and they behave only on the recognition of patterns and statistical associations (so they ultimately lack intent). They can seem to construct suitable responses with coherent statements but exist without situated cognition, which is the basis for grounding conversational parameters against real-world experience or emotion (Spence & Brandão, 2021). As a result, the language-based AI sometimes generates contextually relevant answers but has only shallow semantic coherence or repetitiveness when moving deeper or reaching abstract levels of the conversation.

Case Study: Educational Dialogue with ChatGPT

A university instructor undertook a research project where students engaged with ChatGPT in a Socratic dialogue about the ethics of artificial intelligence with. The instructor reported that although ChatGPT provided articulate responses during the initial few turns, students claimed that ChatGPT was evasive or circular when they introduced contradictions or sought deeper emotional perspectives. They reported the experience as having ChatGPT providing authoritative sounding responses despite its incapacity to build upon or expand the previous statements, or to otherwise show any growth between 'participants', unlike what a human interlocutor may do.

B. Dialogic Breakdown and Loss of Coherence

Holding multi-turn, even coherent dialogue, requires memories, context tracking, and the ability to respond to nuances and subtleties. LLMs do not have clear memories of the content of past interactions, except when the point in time history has been deliberately programmed with persistent context. In turn, some of the limitations lead to inconsistencies or contradictions or blind spots in longer interactions.

Case Study: Mental Health Chatbots

Some start-ups have launched AI chatbots - like Woebot - to assist in some mental health contexts. While these resources state they adhere to basic principles of Cognitive Behavioral Therapy (CBT), many users have reported that the bots miss emotional tone or provide generic ideas. Users expressed lack of empathy and personalization in sensitive contexts (e.g., grieving, trauma). Trust is necessary to have authentic dialogue.

C. Cultural and Contextual Blind Spots

AI systems are trained on datasets laden with heavy influence of dominant or Euro-colonial cultural narratives. Most datasets present realities and norms from Western Dominion English-speaking contexts. This gives rise to the possibility that AI undermines or misrepresents non-dominant worldviews, linguistic idioms, and culturally infused subtlety.

Case Study: Language Learners Utilising AI Tutors

In classrooms across Southeast Asia and Africa, educators were using AI tutors to teach English to language learners. When language learners asked the AI for feedback on their response, the AI provided feedback when it could, but it generated unhelpful feedback when it failed to decipher the culturally-embedded examples used culturally in the students native languages. In some circumstances the AI "corrected" regionally-acceptable English language use, which left students feel confused and lack linguistic ability.

D. Emotional Authenticity and Trust

Another barrier is emotional authenticity. AI only simulates emotions and has none. This limits the degree to which AI can represent human affection, or make altercations for interpersonal inertia. When conversations included the use particularly charged emotional discourse - resolution of conflict, mentoring disposition and/or moral reasoning - the absence of authenticity is accentuated.

Case Study: AI Talk-Based Career Counselling

A high school in the UK community piloted an AI-based career counsellor that was intended to help students explore career options, as well stipulating career suggestions related to results or interests from possible aptitude assesses attempting to determine a career." Although the students preferred providing feedback, because it was quick and gave gave structure to a "suggestion", many students reported that the bot did not "listen" to their ambitions and interests in their area of consideration.

Although AI conversational agents have linguistic potential, they lack actual experience, lived experience, cultural grounding, emotional intelligence, and genuine intention, thus limiting the authenticity of sustained conversation, particularly in educational and human connection contexts. In particular, making AI systems more dialogically conscious will necessitate both technological advancements as well as ethical design, cultural sensitivity, and human oversight. Engaging in genuine dialogue with a machine may ultimately depend less on how we can make AI more "human" and more on how we can help people clearly understand the landmarks and limits of machine dialogue.

4. MULTILINGUALISM, MARGINALIZATION, AND LINGUISTIC EQUITY

In a world that is increasingly globalized and digitized, the interactions between multilingualism, marginalization, and linguistic equity are receiving growing attention in education, technology, and social justice. A language is not merely a form of communication, it is also a marker for identity, representation, and access. In the twenty-first century, millions of speakers of indigenous, minority and regional languages, are systematically excluded from education, public discussions, and digital platforms because they make use of languages that are not hegemonic, such as English, Mandarin or Spanish (Tacheva & Ramasubramanian, 2023). This linguistic marginalization undermines cultural and linguistic diversity and cognitive heritage, and it further reinforces existing social inequities. As artificial intelligence, global media, and policy frameworks continue to define linguistic norms, ensuring linguistic equity fair access, recognition, and representation for all languages is imperative to achieving inclusive participation and marginalized people's empowerment in the twenty-first century. The structural injustices outlined by the class system in Figure 6 are embedded in language model structures through resourcing and language superiority.

Figure 6. Class diagram illustrating the classification of languages in AI language models.

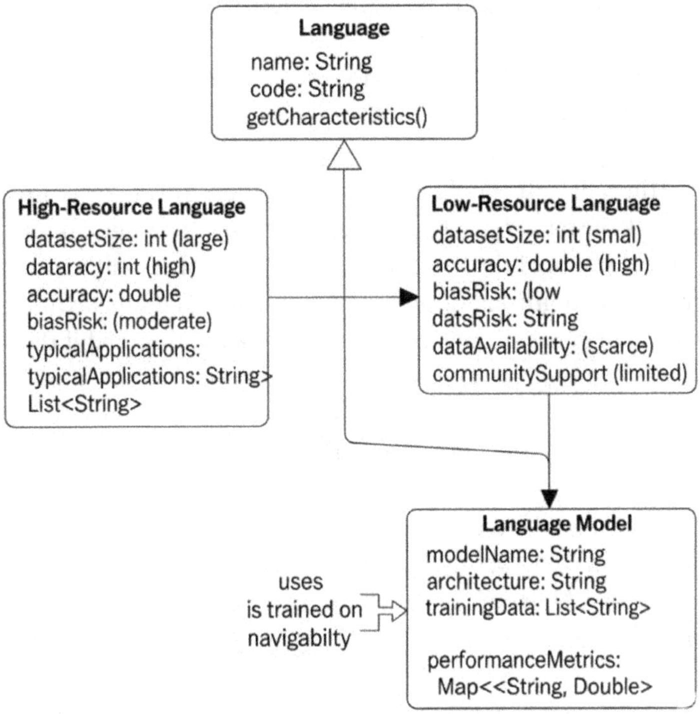

4.1. AI and Low-Resource Languages: Opportunities and Exclusions

The advancement of Artificial Intelligence has been revealed in recent years, especially through Natural Language Processing (NLP) and Large Language Models (LLMs) who have grown impressive technologies with humans' languages. However, while remarkable advancements have been made, inclusivity has been uneven. While for example the vast majority of NLP systems are mostly trained on high-resource languages such as English, Mandarin, Spanish etc, the vast majority of the 7,000 languages in the world remain low-resource (in this paper these under-represented languages are referred to as low-resource languages). Low-resource languages do not have vast digitized corpora which can be used to train AI models, and so are largely excluded from these AI systems. Opportunities to use AI for low-resource languages are vast as the literature is just beginning to tap into using transfer learning, cross-lingual embeddings, and multilingual pre-training to provide models with the ability to transfer learning about a high-resource language to a low-resource language. Depending on the level of the language, possible uses to improve representation could include uses of machine translation, speech recognition, automated transcription, and/or the preservation of language. An emerging community-based effort supported by Masakhane, Common Voice by Mozilla, and the AI4D initiative has resulted in developing open-source AI models focusing on African, Indigenous, and regional languages, providing text democratization and cultural preservation.

At the same time, these technologies can also demonstrate severe forms of exclusion and risk. Lack of standardized orthography or limited corpora for many written languages, and the sociopolitical marginalization of speakers, often precludes data collection or representation using AI/algorithmic systems. Additionally, representation of a low-resource language is only as good as the data it uses, and if the data is not thoughtful in respect to cultural context, syntactic structure, or semantics, it can create algorithmic harm or erase a language entirely. Often, these communities are also further marginalized in the design and governance of the technologies purported to serve them, thus reproducing historic digital colonial dynamics and exclusion. When we consider this further, there is also a risk of AI reifying linguistic hierarchies by privileging standardized or widely spoken varieties, and destabilizing dialectal or non-standard varieties (Tang et al., 2025). For instance, while a multilingual model may offer Hindi for selected tasks, there are likely numerous dialects, such as Awadhi, or Maithili, which are regularly used but excluded. The digital invisibility faced by these languages can hinder educational, civic engagement, and digital opportunities in their preferred languages. Recognizing these exclusions requires researchers and developers to engage in ethical AI design that accepts community collaboration, socially equitable resource dynamics, and linguistic justice principles, whereby local linguists, educators, and technologists can assume leadership positions in data governance and the development of these models to benefit the communities served that are also aware of the modes of exclusion at play.

4.2. The Political Economy of Language Models and Data Availability

The emergence of large language models like GPT, BERT and LLaMA are not only a technological advancement, but also a significant political and economic activity. These models are trained on large volumes of textual data - often scraped from the web, books, journal articles, social media and other data sources. Accumulated textual data does not appear from nowhere; it is situated in global power structures which determine what data is available, which voices are represented, and how linguistic value is created. The asymmetries in availability of training data replicate and exacerbate long-standing inequities in linguistic, economic and geopolitical domains. At the heart of this political economy is data capitalism, which recognizes language as both a commodity and a resource. The tech giants and leading AI companies (most of which are headquartered in the Global North) collect large datasets without compensating people for their work, and without paying attention to the socio-cultural contexts of those sources. This extractive model favours dominant languages (e.g., English, Mandarin and French), while ignoring underrepresented or non-digitized languages. This model perpetuates a self-fulfilling prophesy, where high-resource languages produce better performing AI, and these systems attract more financial investment and improvement, which widens the digital divide.

Additionally, data has availability stratified. Wealthy countries, particularly those in Global North contexts, have buoyant digital infrastructures, and accelerate digitized materials on a broad scale. In low-income countries, it is often the case that the technological capacity does not exist for generating or storing large language datasets. For example, educational systems may not have the requisite digital outreach capability, oral traditions may also exist to varying degrees above that of written ones, and in many cases there would be diverse and rich linguistic practices excluded from learning algorithms altogether. Together, these socio-political configurations work together to produce what Sisodia (2018) labels as "epistemic injustice", essentially meaning that these ways of knowing, and therefore ways of speaking, are systematically rendered invalued, and even erased. Ownership and control of language models also have a central bearing on their design and implementations. The majority of most contemporary state-of-the-art models are owned by private corporations, many of which have little transparency around their data sources, including the meaning of the word datasets and related ethical overrides regarding training strategies. This acts much like a collation of power that works to limit public accountability, and creates a dependency as governments, educators, and researchers must use these proprietary vessels that have been built without their agency. Furthermore, decisions around what data model creators deem to be "clean", what they deem "worthy" languages to include, and what may be classified as "toxicity" in language, become again filtered through specific cultural norms and market extracts.

The implications are extensive. In the educational sector, to illustrate, language models trained almost exclusively on Western-centric texts may overlook local curricula, perpetuate cultural biases, or neglect region-specific systems of knowledge. In the governance context, these language models could entrench linguistic disenfranchisement which would effectively exclude a significant proportion of the population and citizenship from participation in the digital space if their language is under-represented. In situations such as journalism or public discourse, AI-generated content may also further entrench the existing dominant ideology with little representation or reporting on differing or alternative viewpoints. To adequately confront these inequities, a radical shift in thinking about the data pipelines and governance models that underpin language AI is required. Movements that aim toward the data sovereignty, open access corpora, and community-centred AI are crucial to democratizing our participation in order to ensure equity in representation in relation to language technologies. If we are to understand language in all of its usages and nuances as more than mere data, but that it is an extension of identity, history, and power, it is essential that our artificial intelligence systems that use language are truly inclusive and equitable. Can we imagine an AI that does not simply produce linguistic data or collections of corpora, but rather prizes linguistic diversity and recognizes the need for multiple and diverse language systems?

4.3. Case Studies From Multilingual Classrooms in India, Africa, and Latin America

Multilingual classrooms across India, Africa, and Latin America offer with rich and complex learning environments in which students routinely engage with multiple linguistic codes. These classrooms are not simply pedagogical spaces but also spaces of social negotiation, cultural identification, and linguistic power. With the increasing integration of AI tools particularly language models and automated translation systems into the educational space, the relationships between multilingualism are being reconfigured and transformed in both enhancing and pernicious ways. The case studies that emerge from these regions illustrate how local linguistic ecologies are interwoven with global technological infrastructures.

India: Negotiating Linguistic Hierarchies in a Digital Era

In a rural government school in Jharkhand, students often speak a tribal language, Kurukh, at home, receive instruction in Hindi, and are expected to learn some English for higher education and employment. An AI-based translation tool recently piloted in the school aimed to enable these students to overcome the challenge of transitioning from learners of their home language to pay attention to the official vehicle of instruction. The tool contributed to increasing student comprehension for a few learners, but consisted of translation problems based on semantic aspects of texts and its lack of cultural sensitivity to indigenous concepts led to incorrect translations of idioms and values based on their oral traditions. While teachers who took part in the evaluation of the AI tool were enthusiastic about its utility, they simultaneously acknowledged the challenge of making the technology relevant to their daily pedagogical work because of inconsistent internet access and the lack of a publicly accessible dataset for localized training.

Africa: AI in Post-Colonial Language Contexts

A pilot project in South Africa rolled out AI-based reading assistants in a multilingual township school featuring isiXhosa, Zulu, and English. The purpose of these tools was to personalize feedback during early literacy instruction. The reading software attended to English quite well, but run its course on indigenous African languages without a wealth of data. As a result, students frequently received incorrect or incomplete feedback, which reinforced their inclination towards English during early learning, which is neither their first nor second language (UNESCO, 2025). The project revealed the larger issue of digital exclusion the only language legitimized in technology was English, and sometimes Afrikaans, despite the formal understanding of and representation for many languages in the country. Teachers were left to question whether these tools might unintentionally progress colonial forms of linguistic benefit as technological advancements.

Latin America: Community Voice vs. Algorithmic Norms

In one public school located in southern Mexico, where Spanish and Zapotec coexist, an experiment occurred testing the use of AI-assisted writing platforms. Ultimately, the ways in which documents were treated in these languages were very different. For example, Spanish-language essays were eliciting grammar suggestions and ideas for vocabulary expansion; Zapotec writing, however, produced either error messages or no action at all. Students recounted that they felt like their heritage language was being “corrected away”, reinforcing their sense of marginalization. However, in a concurrent project in Peru, an NGO was working with Bolivian educators and AI developers to co-create a speech-to-text system in Quechua that included culturally embedded design procedures; this resulted in a successful pilot project, where reading fluency improved without compromising the students' linguistic identity.

Taken together, the preceding case studies provide some systematic insight into the problem of linguistic marginalization. While AI tools can serve as bridges over linguistic divides, they are designed on the basis of data learned through and with significant structural inequalities. Even if there is no structural inequality at play in language processing-based learning, if our documents generate unbalanced translation with grammar checking algorithms and user interfaces inaccessible to minority-language communities, the documentation is reinforcing linguistic marginalization. To the extent that local educators, educational communities, and developers are actively involved in building AI tools as culturally responsive tools, they can be contributing to linguistic equity, engaging learners, and celebrating multilingualism in their classrooms. To fully realize this potential, AI in multilingual education needs to go beyond translation or simply providing speech recognition, and make an active contribution to showcasing linguistic justice, decolonized knowledge-making, and being an inclusive pedagogy, particularly in contexts of postcolonialism and indigenous studies, where language is a resource and a right.

5. ALGORITHMIC LANGUAGE AND POWER

In the age of artificial intelligence, language is no longer the exclusive purview of human communication, it is also the medium upon which algorithms structure thought, monitor behavior and mediate sociality. As large language models (LLMs) and algorithmic automated text-making tools (like ChatGPT) continue to gain legitimacy and visibility, the questions around who owns language, whose voices are privileged or silenced and how power is fluid in algorithmic systems are more urgent than ever. Algorithmic language is not neutral, it is embedded in relationships of dominant power that are reproduced in language technologies. Algorithmic technologies produce language that often reflects and perpetuates dominant linguistic structures while minoritizing non-standard, indigenous, or context-oriented forms of expression. The increased mediation of language by these systems in education, federal legislation, media and daily conversation raise serious questions around how algorithmic language-making and models in general are operating to reproduce social order, develop epistemic authority and reproduce global inequalities. This section examines entangled issues of algorithmic design, linguistic politics, and power, and emphasize the need for ethical, inclusive, transparent practices regarding language technologies. The structural systems of AI models shown in Figure 7 can create new ideological dominant pathways to discrimination through certain use of predominant modes of linguistics.

Figure 7. Component diagram of AI language system showing influence points for power and bias.

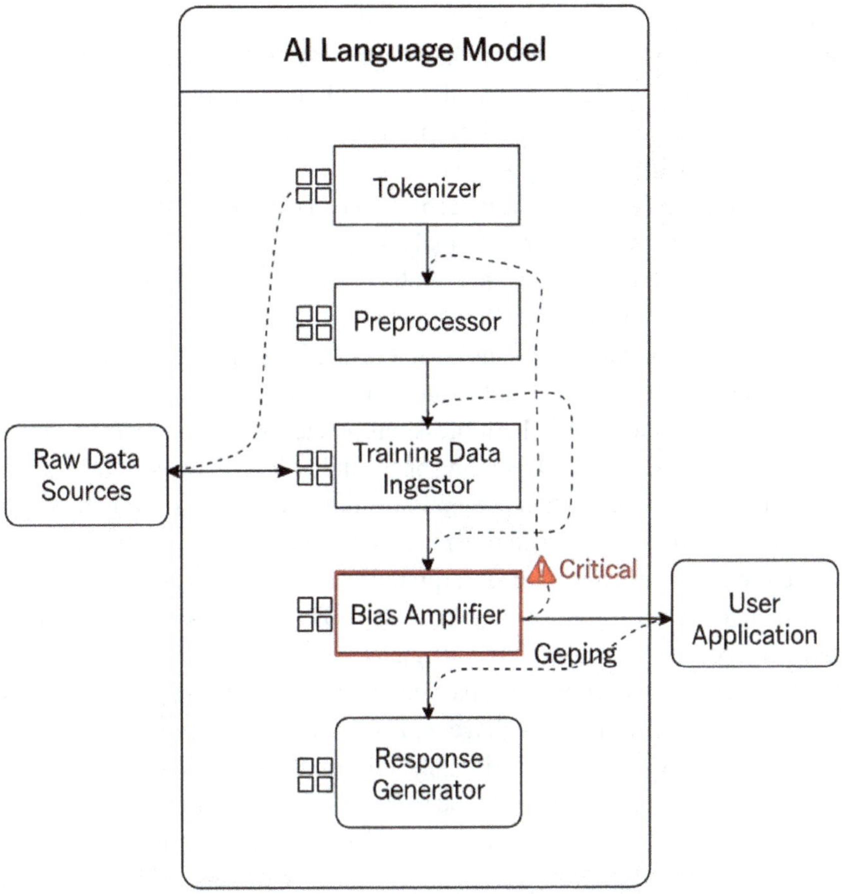

5.1. How AI Systems Reproduce Dominant Language Ideologies

AI systems, especially those developing natural language (NLP) capabilities, are primarily trained on extensive corpuses of human language data, which are based on human language and the data are anything but neutral and uniformly diverse. The corpuses AI used to create language models are essentially predominately high resource language texts (especially English) and text produced by institutions, which all manifest hegemonic or dominant cultural, political and ideological positions. As a consequence, the linguistic outputs of AI models already reproduce the biases, assumptions, and hegemonic ideologies present in the texts in these source corpuses. One of the primary ways that AI reproduces dominant ideologies of language, is through its standard language bias. For example, language models produce typical grammatical forms based on standard language texts. If the models were trained on dialect, indigenous, or colloquial textual sources, they would have difficulty producing outputs in these forms as well. For instance, AI systems often misunderstood, corrected, or flagged as wrong African American Vernacular English (AAVE), variants of Indian English, or Caribbean Creoles (varieties of English spoken in the Caribbean), to name just a few. Largely, language models privilege a language 'correctness' model that is reminiscent of elite, academic, or Westernized speaking and writing norms or priorities.

Also, these systems can solidify and promote racialized, gendered and class-based stereotypes. For example, many studies have shown associations made with specific genders or ethnicities for particular professions or backstories with particular sets of biased training data provided. Depending on the individual response, this regard may appear small, but in sum, it alters perception and reinforces the current hierarchical reality of society. Similarly, translation models that may also, normalise more dominant structures of a language in the interpretation of the original textual meaning of less dominant or culturally embedded languages. This means structures will flatten semantic differences, and support one language, such as English or Mandarin, calling itself for and upon a support of a particular language representation, rather than helping a more low-resourced language representation, as in the case of Quechua or Kannada (Walter et al., 2022). Practices like this unintentionally support knowledge inequity in this sense, creating epistemic hierarchies, where knowledge systems are made visible and invisible to others, i.e. their irreplaceable curriculum agendas.

In the classroom, AIs writing assistants or automated grading tools that are trained on dominant linguistic norms may penalize students who include multilingual or hybrid voices in the classroom (this is evident when language models are trained primarily on English). This puts pressure on students to conform to a single, often Western-focused, academic style of writing that discourages or may mute students' ability to articulate their multilingual and complex cultural or linguistic identities. Ultimately, AIs do not only process language; they also shape the values of language, who has the authority to speak, and what kinds of expression are deemed legitimate. This underlines the necessity to problematize the ideologies implicit in trainings datasets and in algorithm models, and recognize broader possibilities for inclusive, transparent and reflexive language-based technologies.

5.2. Deconstructing Bias, Surveillance, and Colonial Legacies in NLP

Natural Language Processing (NLP), the artificial intelligence(AI) subfield that concerns the relationship between computers and human language, has often been viewed as inherently neutral and technical. Beneath this technical facade, however, we can recognize that there are pervasive biases, surveillance logics, and colonial traces that frame how we understand the design, deployment and experience of language technologies. In order to challenge these layers, it is imperative to engage critically with these socio-political implications of AI technologies (like language systems) and create technologies for more responsible and inclusive futures. Bias in NLP emerges not only through imperfect facets of data, but that assumptions are made since the inception of design for the system. NLP models are trained through large-scale datasets sourced typically from the internet datasets that reflect local socio-cultural inequities and dominant perspectives resulting in systems that may privilege certain languages, dialects, and registers while marginalizing others. Gender and racial stereotypes are salient examples of biased associations of discrimination that manifest in word embeddings and autocomplete inference with language biases (e.g., "nurse" with women or "criminal" being interpreted with ethnic identity). These assumptions do not simply reflect imperfections of contextualized bias; they are part of the assiduous issue with sourcing data uncritically and failing to understand the contexts in which these models are trained.

Surveillance issues are particularly explicit in NLP applications like sentiment analysis, emotion detection, and automated moderation tools, especially since these often surveil and monitor human expression across platforms often without users aware of such actions happening or consent to it happening. There are many questions arising from this: who is surveilled, who is being harmed by the policing of their voices and who are the people whose linguistic performances are determined to be deviant. This is compounded by multilingual settings or non-Western contexts where NLP but can misinterpret local phrases, idioms or political speech as hostile or dangerous, resulting in over-surveillance and over-policing of marginalized communities. There are colonial entanglements in NLP, particularly around prioritization over certain languages and epistemologies being centre-staged. English and other European languages have historically been the main focuses of NLP in relation to resourcing, and attention in the research of the languages (Wang et al., 2024). Practices for meaning-making which are written, oral and embedded in indigenous knowledge place-based language traditions are either not present or poorly represented. For example, NLP model implementations are limited in their representation and applicability of local agglutinative languages like Zulu or morphologically rich languages like Inuktitut, and therefore often fail to deliver and thus are erased entirely.

Moreover, the extraction of linguistic data from communities -without consent, and without sharing the benefits- reflects extractive colonial practices in which cultures (in this case, language) are extracted from the Global South and then used to power models created in the Global North.

In taking apart these intertwined issues, it is clear that NLP is not just a technical problem and is instead a deeply political one. Language and culture are not free, and researchers, not just developers or educators, must critically interrogate whose knowledge is embedded, whose language is privileged, and who is not surveilled in the name of progress. The first steps to dismantling these damaging structures are participatory data collection, community-formed annotation, and decolonial design methods. But to shift to a more just NLP world, we need to change our perspective on what matters - not efficiency and scale, but equity, cultural considerations, and epistemic justice.

5.3. Ethical Dilemmas in Deploying English-Dominant AI Tools

The increasing proliferation of AI tools which privilege English, especially in the contexts of education, government, health care, and customer service, creates a nexus of ethical challenges that run counter to the principles of equity, inclusion, and linguistic justice. AI tools such as chatbots, voice assistants, and automated translation services offer the promise of improvements in efficiency, accessibility, and scalability. However, their single or primary reliance on English as the default or privileged language, can effectively marginalize those who do not speak English and reaffirm systemic hierarchies in multilingual societies. This is particularly problematic in postcolonial contexts where, despite being the primary language of the majority, English also holds sociopolitical currency and serves to further entrench disparities in an already inequitable playing field. One of the central ethical dilemmas lies in the areas of linguistic accessibility and fairness. When AI supported systems such as chatbots, voice assistants, or automated translation services only work in English/English speaking context, they systematically exclude users that are more fluent or literate in other languages. As an example from education, a student whose first language is not English may find that an AI learning platform is either entirely inaccessible to them, or less responsive to the cultural and linguistic context(s) on which they draw for a successful learning experience, resulting in learning gaps as well as feelings of isolation. This violates the ethical principle of equitable access to technology-enhanced learning and it further entrenches the false premise that ability in English is a form of intelligence or progress.

One other issue is one of epistemic injustice. English-based AI may encode, center, and perpetuate knowledge structures based on English-speaking Euro-American worldviews. To the people who contributed to knowledge production, culturally situated concepts, local metaphors or indigenous knowledge systems are either mistranslated, oversimplified, or disregarded altogether. In addition to losing lexical diversity, this also lessens non-Western epistemic structures, such as knowledge production from global majority populations, where it may be suggested or mandated that they engage with AI in English rather than their native language. The reinforcement of socio-economic imbalance is another salient issue. In many Global South contexts, English fluency is directly related to social mobility, and access to increased opportunities or privilege. English-centric AI tools may thus increase the social inequities associated with the digital divide by placing emphasis on and increasing the value of urban, educated elites while devaluing rural and unserved populations. For example, AI-recruitment tools or government e-services that primarily function in English might unintentionally discriminate against non-English speakers who cannot access jobs or services.

Consent and knowledge participation also represented ethical concerns. If communities interact first with dominant English-based AI systems without knowing the terms of use, privacy, data usage rights and restrictions, or access agreements - all written in complex English (and might not be addressed in a timely manner), the community may be, unknowingly, consenting to surveillance, data culling, or biased decisions. As a result, this brings into question makers' beliefs about autonomy and informed consent and ethics and design of multilingual technology. As a whole, the use of English-based artificial intelligence tools is not a neutral act. It is a highly political act that can also be seen with the historic use of linguistic imperialism (which may parallel language and dialect issues) (Wang & Chen, 2025). In response to these issues, we need to radically rethink how we operate in relation to AI toward a framework of linguistic pluralism, including designing tools that are (written) multilingualism, culturally relevant, and community based. Ethical AI must consider the right of every speaker to engage with technology in their language and not only as a translation afterthought, but that it is a basic design principle.

6. DESIGNING CRITICAL, INCLUSIVE AI PEDAGOGIES

As artificial intelligence (AI) becomes increasingly integrated in educational contexts, there is a pressing need to rethink pedagogies not only that are technologically responsive, but also ethnically inclusive. The new ways for students to think, write, and learn have evolved via the application of a variety of AI tools, including intelligent tutoring systems and generative language models. However, if we are not guided by design, these technologies may unwittingly contribute to the wilful reproduction of inequalities based on the dominant cultural and linguistic norms, and marginalize voice of others. The process of designing critical, inclusive pedagogies with AI consists of creating digital tools in classrooms but also interrogating the design within these technologies in regards to their values, assumptions, and power. By valuing equity, cultural responsiveness, and learners' agency, educators and technologists ought to ally to enhance dialogue and diversity while critiquing democratic engagement practices in learning spaces. Figure 8 provides an overview of these potential discriminatory pathways of AI tools as they relate to inclusive education by activating digital infrastructures.

Figure 8. Deployment diagram showing infrastructure for inclusive AI-based learning environments.

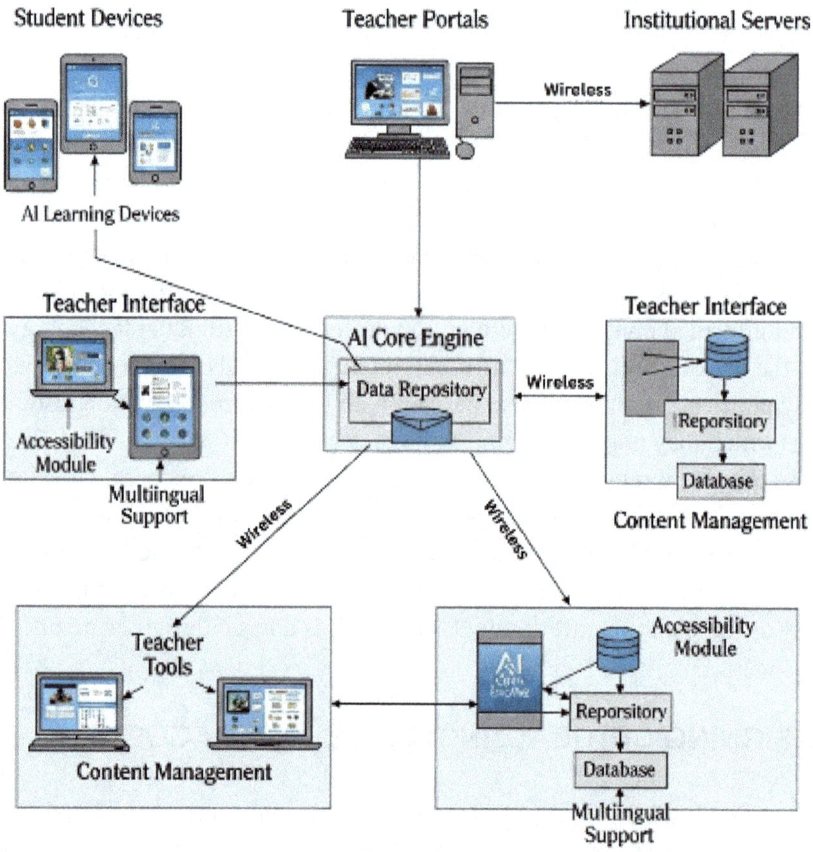

6.1. Principles for Equitable AI Integration

Equitable AI integration in education requires access to resources, but fundamentally, it is a commitment to fairness, equity, and justice at every stage of design and deployment. The process begins with our awareness that AI systems are not neutral: they are grounded in contextualized data, the developers' assumptions, and within specific socio-political contexts. Thus, a principled approach must center the needs and voices of historically marginalized learners to ensure the equitable development of AI systems that do not increase or reproduce existing imbalances related to language, culture, race, or socio-economic status. One key principle is transparency (Zhang et al., 2025). Both educators and learners need to understand how AI tools work, what data they are operating from, and how decisions are made. Without an element of visibility, it will be challenging to validate or to challenge bias, or to trust AI outcomes. Equally, accountability is necessary: institutions and developers must be made accountable for the harms their systems might inflict, and there should be ways to right or remedy that harm.

Another fundamental principle is cultural and linguistic inclusivity. AI should support multiple ways of knowing/communicating, particularly for speakers of low-resource and indigenous languages. As such, we need to design the system(s) to recognize varying dialects, narrative structures, and epistemologies, and to help define and also activate, instead of erasing, those identities. Learner agency must also remain as a guiding value - AI should help amplify students' representation, decision-making, and engagement with things rather than proscribing uniform outputs or reifying standardization. Finally, co-designing with communities is paramount. Teachers, students, and local stakeholders need to be collaborators and not afterthoughts in creating AI tools that center their context, pedagogy, and understanding. This participatory process can enshrine relevance and efficacy, and, shift power away from the creators of technology. Together, these principles establish the groundwork for AI integration that is: technically sound, ethically coherent, and socially transformative.

6.2. Cultural and Linguistic Responsiveness in AI-Assisted Instruction

Cultural and linguistic responsiveness in AI-assisted pedagogy should include the development of educational technologies that acknowledge, value, and incorporate learners' diverse cultural and language backgrounds. The need for culturally responsive AI to reflect students we work with is particularly important in a globalized educational environment in which students and classrooms are increasingly multilingual and multicultural. This means that AI must not create a sense of uniformity with standardized AI models, but instead, provide diversity that reflects the individuality of students' identities, experiences, and worldview. AI systems based on natural language processing (NLP), for example, often reinforce dominant language ideologies standardized English has become the unquestioned default language, while regional dialects, indigenous languages, and other culturally specific communicative practices and styles are neglected. With widespread standardization of communication, those who engage in non-dominant language practices and languages are nominalized and linguistics diversity is undermined. To use AI systems in culturally responsive ways, then, we need to create AI systems using localized sets of data and which include the languages, idioms, and pedagogical practices found in the regions and communities of the learners they intend to support. For example, an AI writing assistant developed from multilingual corpora located within an Indian or African context would have a greater chance of accounting for factors such as code-switching, translanguaging, and culturally-bound rhetorical practices.

Furthermore, responsiveness also involves adjusting content and feedback produced by artificial intelligence systems to the cultural beliefs, histories, and socio-political contexts of learners. In the context of education, this could mean providing examples, case studies or problem scenarios that are locally relevant and situated in the lived experiences of students. When education is relevant, it engages students more deeply and helps them to be more motivated and take ownership of their learning. Educators have an important mediating role in this process. When teachers are provided with diverse AI supports that can respond to a greater variety of language expressions and cultural references, they can better scaffold learning, provide appropriate supportive feedback, and develop academic voice. The flip side is to think of AI systems that do not allow learners access to this diversity; in this situation a learners confidence could be compromised and cultural hierarchies exacerbated. Therefore, culturally and linguistically responsive AI-assisted pedagogy requires intentionality of design, inclusive training data, and a commitment to honouring difference. It is not simply a technical challenge, it is also a pedagogical and ethical challenge as educators (and the society) are challenged to ensure that educational AI serves to uplift all learners and not just those who conform to dominant linguistic and cultural representations.

6.3. Participatory AI Development With Educators and Learners

Participatory AI development is a radical approach that centers educators and learners as co-designers on the development, iteration, and deployment of AI tools that are utilized in education. Participatory approaches refrain from top-down, expert-only responses to problems, and are instead collaborative, transparent, mutual forms of learning all of which involve the active role of the people who will most be impacted by educational technologies as collaborators in the development of them. This approach suggests that teachers and learners have a great deal of valuable situational knowledge about the on-the-ground realities of classrooms, ways of teaching, ways of learning, language distinctions, and socio-cultural contexts (Zohuri & Behgounia, 2023). They can be involved in all steps of AI development, which would include needs analysis, systems design and specification, and system development and testing. Developers can then develop tools which are more suitable, useful and ethically appropriate for the contexts in which they are username. For example, teachers can help guide the development of AI-based tutoring systems by identifying areas where curriculum is unsupported, or, more fundamentally, by signalling how they think automated feedback ought to be extenuated so that it helps understanding rather than discouraging effort. Similarly, learners have the potential of elucidating how AI-based tutoring tools may support, diminish or have little impact on the ways they can adapt their learning, develop their agency and attitudes toward learning, and confidence in knowing about "school".

Participatory AI also addresses issues of trust, bias, and accessibility. Educators and learners are more likely to use and advocate for AI systems that affirm their priorities, recognize their identities, and are legible and made understandable. This is particularly important in a context of histories of marginalization or where algorithmic tools can inadvertently reproduce exclusionary practices. Participatory approaches highlight multiple points of view, including those from low resource schools, multilingual classrooms, and minoritized perspectives, to consider in developing AI systems that are not only technically viable but socially just. Concrete strategies for co-creating AI systems participatorily may include co-design workshops, focus group conversations, iterative user testing, and piloting programs that involve educators and students working alongside developers, linguists, and ethicists. These processes create varied opportunities for individuals to learn from one another developers learning from educators and educators learning from developers. Furthermore, establish a culture of shared ownership and responsibility that continues beyond the original development process. Participatory AI development ultimately reconfigures the power and graphics of educational technology design and development. It evolves the process from a technology determining what an educational activity will look like to co-creation and recognizes educators and learners as key stakeholders and supporters to AI in education that can contribute to equitable, inclusive, and contextually aware pedagogies.

7. CONCLUSION

As we have seen throughout this chapter, the convergence of language, AI, and education is less a convergence of technology but rather a form of epistemological, sociocultural, and ethical terrain. The challenges brought forth from the ubiquity of language-based AI systems in pedagogy to the inability to have genuine conversations with machines, the implications of linguistic equity to algorithmic power, and the implications for learners, educators, and society are significant. The concluding section of the chapter revisits some of the key lessons learned, emphasizes the need to critically confront AI, and emphasizes the importance of inclusive, participatory, and socially just frameworks that will help ensure technology acts as an actuator rather than a disempower. As summarized in Figure 9, when considering equity, cognition, pedagogy, and design in AI, all intersect, the potential educational futures for increasingly intelligent systems are exciting fight against the ideological created by embodied meaning-making processes of traditional systems shaped by coercive use of social constructionism.

Figure 9. Package diagram encapsulating core thematic areas addressed in the study.

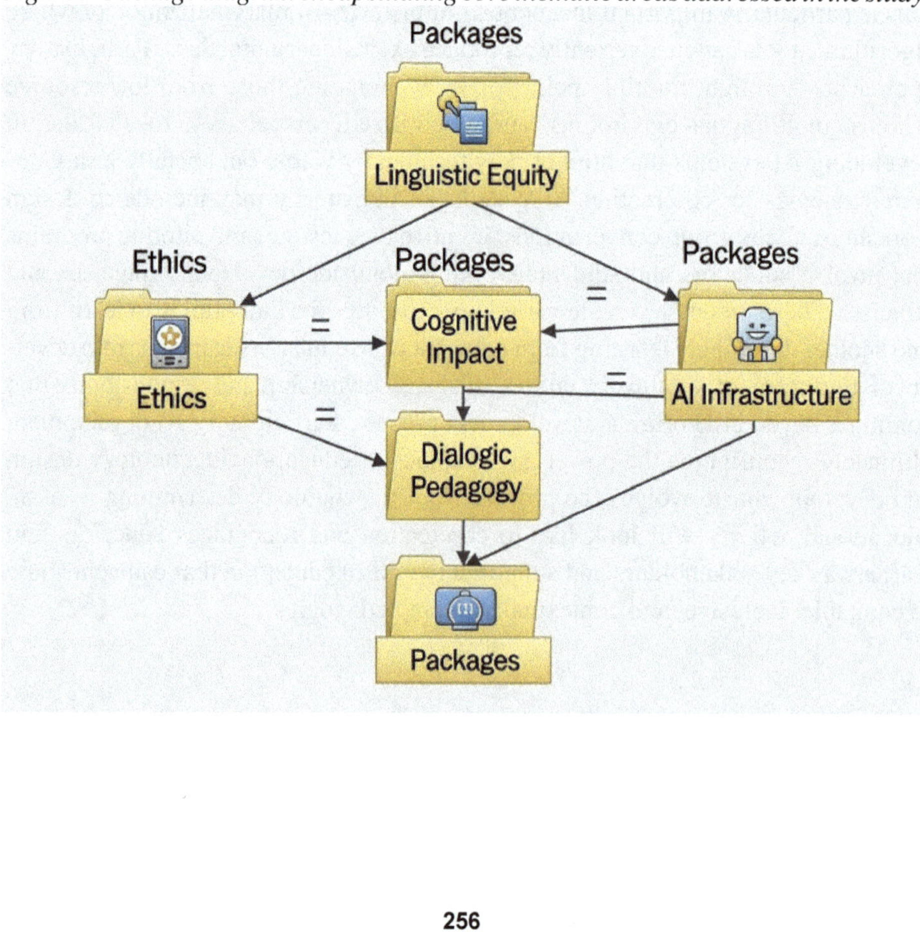

7.1 Reimagining the Linguistic Turn in the Age of AI

The linguistic turn in education viewed language as more than a communication tool articulated through words as a primary means of constructing, sharing, and making knowledge. The emergence of AI specifically, Large Language Models (LLM) like ChatGPT has emphasized and complicated this understanding. Language forms epistemic access and engenders outcomes in education through interlocutors, who happen to be algorithmic agents programmed to simulate conversational interaction, distribute text, and deliver learning experiences. The linguistic turn must be reconceptualized with an understanding of AI as active agent in the production of language, the complexities of machine-mediated conversations, and consideration of morals and ethics around non-human systems concurrently taking part in meaning-making.

In an AI-mediated world, language is no longer an exclusively human enterprise and is created in conjunction with computational algorithms, human cognition, and cultural contexts through heaps of human data with potentially equally important consequence. AI can help further existing narratives, ideologies, and linguistic regimes depending on how it is trained and used. Thus, the traditional conception of dialogic pedagogy where knowledge is generated from real, thoughtful discussion with others also needs to take into account the benefits and limitations of machine-mediated interaction. Educators need to reflect on what we mean by "voice," "agency," and "understanding" when students interact with AI systems that simulate linguistic ability but lack consciousness or lived experience of the world.

Additionally, the re-framing linguistic turn must also address the tension between linguistic plurality and algorithmic conformity. As stated above, most LLMs explicitly or implicitly primarily trained on high-resourced languages, most prominently English. In so doing, they may constrain the range of representation in the possibilities available to the speakers of the underrepresented or indigenous languages. This reconfiguring of linguistic relationships of power raises significant questions of inclusivity, equity, and cultural viability. In response, we suggest pedagogies that engage critically with these systems, and that include practices that permit students to express themselves in a range of languages, validate their cultures, and interrogate the outputs of the machine.

However, ultimately, the re-imagining of the linguistic turn in the age of AI is not simply about embedding "new" tools in the classroom; it is about rethinking our very foundational understandings about knowledge, communication, and learning. This requires an interdisciplinary engagement and reflective practice foregrounding human values, contextual understanding, and ethical responsibility in the design and use of AI in education.

7.2. Future Research Directions and Pedagogical Responsibilities

As artificial intelligence advances and finds its place within educational contexts, researchers need to establish future research paths and establish clearly defined pedagogical roles. Although there have been many important advances in AI (and more specifically language-based systems like ChatGPT and various large language models (LLMs) there is a need to examine how these technologies or methods influence cognition, communication and equity in learning spaces. We need to think more deeply about the use of these technologies beyond an evaluation of AI efficiency, but also in relation to epistemology, culture and their intent to democratize knowledge or perpetuate educational inequities. One of the greatest areas of inquiry will be the longer-term cognitive impact of consistent engagement with new AI systems in both formal and informal learning contexts. Specifically, how does AI written writing, reading, and discussion influence learners critical thinking, creativity, and voice over time? In what ways might reliance on such AI tools impede learners' development or comprehension of deep literacy and argumentation, and if so or not, can we leverage these AI tools to enhance learners' strong metacognitive growth and differentiation? Long-term empirical research is needed to better understand how the relationship between age, discipline, and sociolinguistic background impact learning.

Another significant frontier lies in the ethical and equitable design of AI-driven educational environments. Researchers and developers will need to work in partnership with (educators, linguists, sociologists, and clusters or communities that represent marginalized voices), in order to ensure that AI systems are culturally responsive and equitable as well as being technically sound. We, as educators, and developers must also emphasize representing low-resource languages, indigenous knowledge systems, and varying pedagogical practices in the AI training datasets and the experience of the users, as well. From an educational perspective, we as educators must embrace a dual responsibility: the judicious incorporation of AI tools while also ensuring learner-led critical AI literacy. This includes how to teach students to use AI for writing, researching, or to solve a problem or challenge, as well as how to question the outputs and examine for bias, and to understand the socio-political implications of algorithmic systems. Teachers need to foster a situation where AI is positioned as a collaborator, rather than a prosthesis for their work, where human judgment, creativity, and ethics are intact. In the end, the way forward calls for an interdisciplinary, critically engaged research agenda that takes a cycle of continual reflection on the role of AI in shaping the future of education. In this case, pedagogical responsibility means not simply adopting new tools, but rather engaging in reflective practice, supporting linguistic justice, and cultivating an educational ecosystem where all learners regardless of language, culture, and ability can thrive with and through AI.

REFERENCES

Ahmed, I., Liu, W., Roscoe, R. D., Reilley, E., & McNamara, D. S. (2025). Multifaceted assessment of responsible use and bias in language models for education. *Computers, 14*(3), 100. Advance online publication. DOI: 10.3390/computers14030100

Akindotuni, D. (2025). Resource asymmetry in multilingual NLP: A comprehensive review and critique. *Journal of Computer and Communications, 13*(7), 14–47. DOI: 10.4236/jcc.2025.137002

Alm, A., & Watanabe, Y. (2023). Integrating ChatGPT in language education: A Freirean perspective. *Iranian Journal of Language Teaching Research, 11*(3), 121–141. DOI: 10.30466/ijltr.2023.121404

AlNusif, M. (2025). Multilingual natural language interfaces for code generation: Building inclusive AI tools for global developers. *International Journal of Engineering and Computer Science, 13*(7), 27425–27446. DOI: 10.18535/ijecs/v14i07.5180

Gaol, E. L., Noerjanto, F., Wolomasi, A. K., & Wea, D. (2024). Development of an AI-based platform to customize learning experiences in religious education: Papuan youth Emik characterized by local wisdom. *Journal of Social Science Utilizing Technology*, *2*(4), 565–573. DOI: 10.70177/jssut.v2i4.1503

Hila, A. (2025). The epistemological consequences of large language models: Rethinking collective intelligence and institutional knowledge. *AI & Society*. Advance online publication. DOI: 10.1007/s00146-025-02426-3

Karpouzis, K. (2025). AI, digital humanities, and the legacies of colonial power [Preprint]. *Preprints*.org. https://doi.org/DOI: 10.20944/preprints202502.1823.v1

Khoo, E., & Kang, S. (2022). Proactive learner empowerment: Towards a transformative academic integrity approach for English language learners. *International Journal for Educational Integrity*, *18*(1), 24. DOI: 10.1007/s40979-022-00111-2

Kim, H., Yoon, J., Lee, S. M., Cho, S. M., Kim, Y. J., Noh, G., & Lee, J. (2025). Lecture-grounded ChatGPT-4 tutor: Enhancing adult learner engagement in distance learning. *PUPIL: International Journal of Teaching, Education and Learning*. https://doi.org/DOI: 10.20319/ictel.2025.324342

Kumar, S., & Kaur, K. (2025). Digital pedagogies in the age of AI: Leveraging emerging ICT tools for inclusive and sustainable education. *The Academic*, *3*(5), 1805–1819. DOI: 10.5281/zenodo.15659123

Oyetade, K., & Zuva, T. (2025). Advancing equitable education with inclusive AI to mitigate bias and enhance teacher literacy. *Educational Process: International Journal, 14*, e2025087. Advance online publication. DOI: 10.22521/edupij.2025.14.87

Spence, P. J., & Brandao, R. (2021). Towards language sensitivity and diversity in the digital humanities. *Digital Studies/Le champ numérique, 11*(1), 9. https://doi .org/DOI: 10.16995/dscn.8098

Tacheva, Z., & Ramasubramanian, S. (2023). *Challenging AI empire: Toward a decolonial and queer framework of data resurgence* [Preprint]. Advance., DOI: 10.31124/advance.22012724.v1

Tang, K., Li, Y., & Qin, Y. (2025). *DIY-MKG: An LLM-based polyglot language learning system* [Preprint]. arXiv. https://doi.org//arXiv.2507.01872DOI: 10.48550

UNESCO. (2025). *Languages matter: Global guidance on multilingual education.* https://doi.org/DOI: 10.54675/MLIO7101

Walter, L., Denter, N. M., & Kebel, J. (2022). A review on digitalization trends in patent information databases and interrogation tools. *World Patent Information, 69*, 102107. Advance online publication. DOI: 10.1016/j.wpi.2022.102107

Wang, D., & Chen, G. (2025). Evaluating the use of BERT and Llama to analyse classroom dialogue for teachers' learning of dialogic pedagogy. *British Journal of Educational Technology*. Advance online publication. DOI: 10.1111/bjet.13604

Wang, D., Yang, T., & Chen, G. (2024). Artificial intelligence in classroom discourse: A systematic review of the past decade. *International Journal of Educational Research, 123*, 102275. Advance online publication. DOI: 10.1016/j.ijer.2023.102275

Zhang, T., Erlam, R., & de Magalhães, M. B. (2025). Exploring the dual impact of AI in post-entry language assessment: Potentials and pitfalls. *Annual Review of Applied Linguistics*. Advance online publication. DOI: 10.1017/S0267190525000030

Zohuri, B., & Behgounia, F. (2023). Application of artificial intelligence driving nano-based drug delivery system. In Philip, A., Shahiwala, A., Rashid, M., & Faiyazuddin, M. (Eds.), *A handbook of artificial intelligence in drug delivery* (pp. 145–212). Academic Press., DOI: 10.1016/B978-0-323-89925-3.00007-1

Chapter 8
Cognitive and Cultural Dimensions of AI-Generated Language

Nyak Mutia Ismail
https://orcid.org/0000-0001-8696-983X
Universitas Serambi Mekkah, Indonesia

ABSTRACT

The rapid rise of artificial intelligence (AI) and digital tools in education and work has transformed learning, communication, and decision-making. While AI enhances access and personalization, it also raises ethical issues related to privacy, bias, misinformation, and academic integrity. This chapter explores how educational psychology can foster ethical decision-making, digital responsibility, and moral reasoning in the AI era. Grounded in Piaget's and Kohlberg's developmental theories and Bandura's Social Learning Theory, it connects moral growth with reflective digital practice. Through case-based learning, empathy, and critical digital literacy, it equips educators to build ethical, responsible digital citizens.

1. INTRODUCTION: THE RISE OF AI-GENERATED LANGUAGE

Another advancement with artificial intelligence is language generation, a massively game-changing technology in communication today. Indeed, over the past decade Natural Language Processing (NLP) and generation technologies have evolved from relatively simple predictive text features to systems sophisticated enough to resemble human-like prose, engaging in sophistical dialogues, language translation and even being able to produce poetry or a persuading argument (Premkumar et al., 2024). The appearance of large language models (LLMs) such as OpenAI's GPT,

DOI: 10.4018/979-8-3373-4397-6.ch008

Google's BERT, Meta's LLaMA and many others has had a profound effect on the language generation landscape from segments of education and journalism to customer service and personal productivity. This chapter seeks to unpack these trends at a deeper level, teasing out cognitive and contra- or post-cognitive elements of AI language generation and its implications both locally and globally. Underlying these breakthroughs are the dramatic gains in computing power and data, which enable the training of language models that learn human-like patterns of language use with unprecedented accuracy and context awareness. Bots can reply empathetically, writing tools can offer style advice and translation technology can preserve the tone and nuance in source material. They are really merely convenient, on the face of it, but behind their friendly little screens all sorts of very big changes have been taking place in the ways meaning is made, moved around and comprehended (Wang & Fan, 2025). This tidal wave of AI-driven language technology fuels excitement and anxiety alike. On the other hand, while it contains some real perils of only a few people having access to language-relevant tools, of encouraging an English-only world and of perpetuating digitally-induced apathy around communication. At the same time, it raises questions about authenticity, creativity, authorship, language mediocrity and cultural bankruptcy, as well as the potential for human agency in meaning-making. This is not merely a technical or moral issue but it is deeply cognitive and highly cultural about how societies encode their values.

AI has already begun to rattle the education industry – and it's this old-school teaching and learning model that's being disrupted. So in the same way that essay grading, online tutors, language learning apps and content creators a re changing how we teach and test knowledge. While these features can potentially enhance effectiveness and scale, they often oversimplify learning or downplay the importance of competencies such as arguing a case, reflection and intercultural communication (Essien et al., 2024). As this conversation unfolds in the education realm with respect to ethical implementation of AI in classroom settings, there is urgent concern about the cognitive consequences of deferral of linguistic tasks to machines. AI can aid students to summarize, translate or paraphrase without having to puzzle over the logic and cultural framework of the text. The consequences for cognitive development, particularly in the impressionable young years, could be very serious. Beyond the classroom, AI-generated language is creeping into everyday language — sometimes subtly, and sometimes not. Auto-respond emails, social media captions, news articles and customer support chat logs – the line between human and machine text is becoming less clear (Khurma et al., 2024). This transition has its reflection on some thoughts about what it is to communicate "authentically". How do language that looks like a human and not a machine thought is being made by user's affect the way they relate to it? Do they trust it less? What might be the implication for decision-making, emotions, or social judgments?

AI language tools are trained on gigantic corpora that also model and possibly enforce dominant linguistic and cultural trends. It is challenging with respect to a representational fairness, ideological neutrality and cultural diversity of such systems. For example, models trained predominantly on English data may fail to incorporate the full richness and context of minority languages or dialects (Koriat, 2015). AI system outputs may indeed reinforce such negative stereotypes or discredit non-mainstream viewpoints on the world. In this sense, AI-generated language is not just a technological artifact; rather it is a space of cultural struggle in which different voices compete for recognition and authenticity. These are the types of subtlety in language, technology, mind and culture that I would want to investigate in this chapter. Its aim is to offer a fairly balanced, interdisciplinary volume on how AI-written language is transforming human communication. This AI is conceptualised not as a tool but as a cultural agent in (over) the long duration of changing what we think, how we learn, and how we communicate in the future.

2. COGNITIVE MECHANISMS IN PROCESSING AI-GENERATED TEXT

Language is not just a communication tool; it's a property of human thought itself. We think in words, we try to make sense of our encounters by imposing language on them, we create meaning (by citing language and putting into practice what it tells us clearly or obliquely). That language is a generative scaffolding of this sort may be all the more relevant when we consider that "AI" (and in particular, the GPT family) has intervened as if to mediate and even replace such linguistic operations (Stuckey, 2015). Now that humans hear AI -produced text on a daily basis in education, work and social life, understanding how they mentally process AI-generated text is critical. The idea that language is important to thought and learning is deeply woven into the fabric of both cognitive psychology and educational theory. Lev Vygotsky is one the proponents of this type of approach, who I would like to note has emphasized that language serves as the medium through which abstract thinking is constructed. He believes inner speech is not just a leftover product of outer speech, but helps people solve problems, lose weight (because they can reason "I am staying away from chocolate even though I want it"), and perform other mental tasks. With the ascent of A.I.-generated language, that traditional role of language as a tool for thinking is being tested — or at least reimagined (Eisenring et al., 2024). Scaffolding is easier when the learner can type a question into an AI chatbot and get a fast, fluent (and usually correct) reply. Instead of the act of thinking becoming a method to problem solve (as it is when we haven't instant answers but seek it by deduction) users have instantaneous output readily available nd cut off important cognitive processes such as inquiry, doubt, inference and synthesis. Such replacement can both work to an organism's advantage and be its detriment. On one side, they might come upon responses that will help them create new mental models; but o t h e r w i s e, they could get used to depend on machines for thinking and in a so doing diminish their own capability of dealing with complex concepts by themselves. This tug-of-war illustrates why it's critical to understand how AI-inflected language shapes not just what we learn, but also how we learn it (Ma, 2023).

Schema theory is a powerful model for how comprehension works. It has been hypothesised that humans process new information from prior knowledge frameworks or schemas(Siregar, 2022). These schemes offer the expectation that and can be used to fill in gaps, and facilitate faster processing of familiar material. In reading comprehension, for instance, coming across the word "classroom" might activate schema relating to school, teacher, learning, authority — enabling the reader to make inferences about context and meaning on the fly. However, AI-generated language is sometimes patterned in a manner that feels non-identifiable or even arbitrary(Labadze et al., 2023). Precisely because it is trained on vast, diverse corpora, AI could mix formal and informal, technical terms and colloquialisms within a single output. This same type of variation can interfere with schema activation. The reader cannot always detect the setting from what is presented so does not become linked into prior categories of experience. There are other times when the AI response reads seemingly coherent on the surface but it makes not sense conceptually, sending the reader tilting at another cognitive illusion of comprehension. The schema and related cognitive load theory ascribes the limitations of human working memory to the processing of new information. Optimal learning is achieved when instructional materials have low extraneous load, match intrinsic load (natural complexity of the material/task), and high germane (processing that aids in learning) load(Baek et al., 2025). Cognitive load can be also lightened by the AI text but in some cases for increased. On the positive side, AI can simplify and illustrate explanations, offer demonstrations of the examples, rephrase difficult ideas in a way that makes it comprehensible for a learner(Vibhavi et al., 2024)—all which are examples of reducing extraneous load and facilitation germane processing. Nevertheless, these system may also have the drawbacks of producing verbose, repetitive or off-topic texts. Moreover, if no human decisions are made as to the structuring of material in contextual terms, this can also impose cognitive demands on readers to figure out what is relevant and irrelevant. This dual nature of the AI suggests that education-based interactions with computerized AI language be metacognitive. Learners need to be taught to slow down and think about what they are listening to: "Is this so clear because I understand it or because it's coming out so wash?" or "But is this really an account that I can square with what I knew already, or are big phrases seducing me toward the abyss?" AI-generated language, unencumbered with those reflective questions, runs the risk of turning learning into passive consumption.

What we do with that concern next is trust. Trust is a fundamental and multifaceted psychological state, especially in human- machine interaction. When we read something written by a human, we are used to using signs such as expertise, tone or familiarity or institutional identity to gauge credibility. With AI-authored texts, that is either absent or tacked on. The result, then, is a land of trust that can feel shaky at best: some users have blind faith in machine-generated output; others are quick to cast doubt. The issue with this, however, is that AI generated content tends to sound authoritative in the way it's written. The grammar is okay, the structure just fine and vocabularies good. These are formal properties that might deceive even the most savvy readers to believe a given content is reliable, when it isn't. The absence of a human author just compounds the problem: There's no intention to question, no context to construct, no identity to assess. That "making sense" is not so much understanding as recognizing the authority and appropriateness of a given message. And AI-generated language at its velocity and scrubbed scale might get around this deeper understanding(Safar & Anggraheni, 2024). Learners, for instance, may perceive the AI-authored summary as an authoritative performance capturing a difficult text, with close reading or synthesizing or critical responses elapsed over/through. And judgments are moved away from doing the fundamental one, thinking critically. An AI-augmented student could be educated to craft compelling arguments that seem logically sound, but are ethically or culturally inappropriate. If students can't/refuse/know better than to produce good supporting reasons, then writing is a mechanistic routine rather than an intellectual habit. This being the case, educational systems will need to work to teach students to interact more critically with AI-authored language. It will be necessary to teach them to fact-check, follow evidence, recognize bias and understand argumentation. One promising short-term approach is to use A.I. as a foil: to ask students to critique or improve A.I.-written paragraphs, find the errors and rewrite them in more ethically responsible ways. Such noes even doesn't only practice cognitive vigilance, but behind AI also work to demystify it, as truth tellers.

One of the more interesting questions, I think, is whether there's a different way that we approach texts written by AI as opposed to those composed by human authors. The short answer, as you may have guessed, is that it's complicated. Studies in cognitive psychology and human-computer interaction show that, at the surface level (e.g., recalling facts or understanding main ideas), people do not much differentiate between textual input produced by AI and by humans(Kenya et al., 2023). But when processing gets more complex — say, emotional engagement that involves moral judgment or inferential reasoning — it's frequently in flux. One is what Strunk and White famously called the "authorial mind." Human readers are also more likely to attribute mental states, intentions and emotions to human writers. We're constantly playing a game of reading behind the writer's back, we're guessing at what a writer means, and why they feel this way, and why do they choose this word as opposed to that one(Harisha et al., 2024). This theory-of-mind workout adds a rewarding layer of nuance to reading: we don't just decode the words; we imagine the person behind them. AI disrupts this process. 255) Readers tend to withhold empathy or give up on meditative thought when a text is labelled as machine-made. If there's "no one" behind the text, why look between the lines? Rather, in some cases AI can produce texts so profoundly (and convincingly) human that readers attribute intentionality and a feeling of "connection" with the system. This tendency, known as the Eliza effect, also presents ethical concerns about influencing users' emotions – particularly in more susceptible users, and when it comes to devices employed for mental health purposes such as chatbots.

One more neurocognitive differentiation that has been documented is error processing. If someone fails to read a passage here, or skips a routine there, and some human does it in such a way that it introduces a typo, readers interpret this as a lapse, or an instance of momentary forgetfulness or enlightenment. When AI screws up — especially a failure that, like the offending photo and caption, appears so confidently wrong — the backlash can be even stronger. It may make the users feel cheated, incited or suspicious after a few nice and compelling speeches. This disconnect of style versus substance, which some have referred to as "semantic hallucination," is one of the biggest cognitive pitfalls when it comes to AI and language processing. Readers also emotionally react differently. There's no tone, no humor, no nuance, no irony, no sarcasm and definitely no affection. These aren't things that algorithms can easily replicate, particularly when the context is multilingual or intercultural. Ai-generated texts that shoot for these affective notes tend to fumble into uncanny valleys — where something just sounds "off" enough to distract readers rather than ensnare them(Haristiani, 2019). Understanding how minds deal with AI-generated language is also not merely an academic exercise, but an essential component of developing healthy, informed and ethically sound relationships with A.I. systems(Belda-Medina & Calvo-Ferrer, 2022). Through comprehension of how our language is used as a cognitive scaffold, how schemas and cognitive load give meaning, trust and judgment support critical thinking, and the human brain discriminates (or doesn't discriminate) machine- from human-generated text, we can endeavor to map the terrain of mind in the A.I. age. The implications are wide-reaching — for educators putting together curriculums, for developers building interfaces, and for readers interacting with rich text, and as societies contemplate how much autonomy they should vest in their machines(Lee & Hahn, 2024). The challenge is not to fear what AI-written language can do, but to know it: from within, from every angle, endlessly. So that we can ensure that this powerful new medium becomes a source of expanding, rather than impoverishing, our cognitive life.

3. CULTURAL NARRATIVES AND IDENTITY IN AI LANGUAGE

In the long run of language history, language has never been only about information — as a container for culture, a mirror to collective ways of thinking, an influencer on who we are and who we think we are as individuals or communicators. They say from folklore to literature, from cookery to campaigns, words convey ideals and concepts as well memory. But when AI is part of the calculus, producing language at scale, it doesn't do so in a cultural vacuum. Rather it is a conduit and a shaper of cultural stories, and that's no small thing. For a start, every utterance of an AI system begins with data: masses of human-made text mined from books, websites, social media posts, news articles and much more(Izzati et al., 2020). These, are not clean corpora from an ideological simmetry. They come with embedded assumptions, prevailing discourses and naturalised ways of seeing the world. This transformation means AI is not a passive mirror but an active amplifier of cultural stories — sometimes positive, often damaging(Izzati et al., 2020). One of the toughest issues in this regard is how one can take stereotypes into account. Language models learn by making statistical connections between words, and can replicate those connections without any understanding of their history or social context(Qomariyah et al., 2023). So as it has been trained to talk about women in the context of domesticity, beauty or care-giving, when it attempts to write text referencing women generally, or a student's female teacher or professor, for example, it may unconsciously reproduce those antediluvian instincts — even if the prompt is an academic- or professionally-agendered assignment about Harvard trainers who are female. The same applies to ethnic/ religious or regional groups, depending on their representations in texts they are presented either as simple or bad. These encodings are not just technical hitches — they are ideological opportunities. Although they're performing the cultural equivalent of a script that, in turn, may have actual consequences on people's representation and how others perceive them - or even themselves. In education, AI-authored content would subtly reinforce colonial or other Western-cultural norms (for example) and socialize students into accepting as "normal," "neutral," or "universal" ways of knowing. In news journalism, automatically generated news summary may tend to give a one-sided view from the aspect of the powerful countries and its co-operating nations rather than from Global South perspectives (and sometimes they are just ignored both from the summary).

But AI has a tendency to magnify normative belief systems, as well, countervailed by its capacity to take steps toward inclusion. AI can be deployed — thoughtfully, ethically and intentionally — to amplify marginalized voices, reflect cultural diversity and provide a force for good in the world. But, this depends on the devs satisfying enough success to warrant doing so. If training sets are created from a variety of sources – oral histories, local literatures, etc… – community language models would encompass the divergent human experiences(Zhang et al., 2024). But in practice, inclusion is still a fight. Most AI initiatives emphasise efficiency, scalability and economic value, goals which do not necessarily align the sometimes slow, meticulous process of accumulating and validating across a spectrum of data. And devs may not actually know about that underrepresented content and may have no social prods to find it, let alone a pressure from institutions to incorporate it. This implies that also well-intended models' results converge to the patterns we are used at and, thus, without access to culture's depth in terms of diversity. What we risk in this case is having A.I. spruce up the latest wave of cultural leveling. The hegemonic and the vocal are the louder and more pronounced in a digital terrain where one of the source(s) of user content is generation text by AI, ((if not))the/the source), while context-rich voices originate less yet. The world becomes more searchable, less subtle. As AI increasingly reaches people around the world in their own languages and cultural traditions, the need to build systems that recognize and respect that diversity grows not only as a challenge shaped by technology but as a fundamental responsibility. The issue is the worse when we speak about the under-resources local languages. They are not just alternative ways of speaking; they are repositories of views on the world, modes of epistemological orientation and ideals for social performativity at distinct remove from dominant prevailing scripts. Speakers of such languages are then pressured into linguistic accommodation when they don't find their language(s) in AI systems. They manage to make use of these AI tools, and they put the other language -- such as a national or global language which becomes the new "native" language (while no one spoken it in their own environment) and so participate not to something openly repressive but subtle(Asimolowo, 2025). In this way, AI becomes yet another zone of linguistic capital's uneven distribution – a trend that resonates with the histories of colonialism and postcoloniality that have long shaped global communication(Sarker, 2024).

In others, it may commodify tone, diluting style's mediating topographic features or recycling current tropes to even greater effect in the service of erasing individual voices(Belda-Medina & Calvo-Ferrer, 2022). The volume of AI-authored material on the internet can swallow whole the living messiness of human expression, replacing lived language at the level of amplification with something probabilistic and pastiche. As AI continues to infiltrate public discourse via everything from political declarations and advertising slogans to song lyrics and legal contracts, it also shapes what we talk about as well as how we think. AI-made or AI-assisted narratives are not neutral; they reflect the desires, fears and biases contained in the data on which they are trained. In order to balance this tendency, critical awareness is crucial. Users must be able to question AI-created content — to ask where a given story line comes from, whose interests it serves and what perspectives or aspects of an issue it excludes. This awareness can be raised to a great extent by educators and technologists – for our children in particular. That may be about training students to recognize bias, question authority and learn to create their own narratives that burrow into and disrupt dominant ones. It means not falling into the trap of assuming that the content created by AI is neutral or universal. As any language does, it is shaped by context, power and history — and must be read as such.

4. MULTILINGUAL AND CROSS-CULTURAL COMMUNICATION WITH AI

In the ever increasing world of AI and multilingual communication, it is essential for AI based tools to be able to speak multiple languages. And while there are many well-humored merriments to be made of AI as a bridge between linguistic communities and means of access, the story is both many stripes. Multilingual AI is not just about reflecting material in multilingual repositories, however; it will also reproduce embedded value systems, hierarchical structures and even subtle axes of bias that influence what meanings get made (and got meaning made to) from one culture to the next. For one thing, the ability of A.I. to operate in different languages is shockingly wide-ranging. Resource rich languages, such as English, Mandarin or Spanish, have access to more training data and produce more fluent contextual and accurate outputs(Vygotsky, 1978). As the other side of coin, low-resourced languages (LRs) such as many indigenous, regional, minority and endangered languages has few resources available, so the output may be clumsy/ wierd/ even wrong in some cases. That's a divide that is, further cementing, the broader digital divides: The people who speak less-dominant languages are essentially locked out of good AI markets. Even where the translation is technically perfect, AI does not replicate well the pragmatic meaning of the utterance, its mood, or cultural context of the way it was originally presented. It also can lose punch in translation between languages for the same play of words. For example, a polite form of refusal in Japanese might sound curt if translated to English; an idiomatic Swahili expression would change shape when rendered into French. And these practical breakdowns aren't just trivial — they mold how we think about respect, trust, disagreement and humor. Because AI isn't nuanced in the same way as humans, systems may misinterpret users or read unintended implications into audience interpretations.

Also, even when it itå s not, AI data processing does tend to impose meaning on users in most places AI is globalized. Those who speak underresourced languages usually change their input — by typing in English or adjusting their sentence structure — because they know the A.I. will "understand better" if they do so. That is an intellectual and emotional surrender. At the same time it reinforces the dominance of powerful languages at cost towards a local identity, voice and cultural knowledge. In the end, they experience the external pressure of language shift or death as younger generations privilege global intelligibility over local authenticity(Huang et al., 2022). Leaning to listen across the culture divide It's more than speaking the same language, it's sharing the same page You just can't get by that communication barrier unless you both share a cross-cultural background/context; and that level of mutual understanding means that you understand what is expected in different circumstances/situations. AI systems, on the other hand, are trained not on nuanced text, but heavily decontextualized text, minus real life experience or relational memory a human may bring to an encounter. This in turn inhibits them from having a meaningful conversation between users of different cultural backgrounds. For instance; if an AI translation service is used with two people from different cultures to work on a project together, then the literal meaning can be provided by this translation but the inter-personal communication that incorporates politeness and hierarchy or emotional nuance across a range of different cultures can go missing. It's worse when AI takes on dispute resolution or decision making between cultures, then its limitations become more glaring still. A multicultural negotiation context AI assistant could also mistakenly recommend or discourage one type of communication over another, thereby causing misalignment and tension. These dilemmas can become more complex in the field of diplomacy, education or health, where what is at stake counts not just with respect to efficiency but also for its ethical, emotional and social effects. But if done carefully, multilingual AI also offers possibilities for inclusion(Leahey, 2012). And hence, by specifically using a wide variety of such datasets like code-mixed conversations, regional language dialects and culturally-ingrained manners of speaking etc., developers can really build systems which are more human-like in nature. Open-source projects and grassroots initiatives have already begun such work, especially in under-resourced regions including Africa, South Asia and Latin America where communities are creating NLP resources for their languages of interest. These interventions trouble the one-size-fits-all position of global AI and gesture towards alternative trajectories that are grounded in linguistic justice.

It has been reported that the results with community lead language technology are more promising as can be seen in the case studiesbelow. In Kenya, for instance, teachers and linguists worked side by side to design A.I. tools for Kiswahili and Gikuyu that would ensure the translation models included more than just word meanings — maintaining cultural metaphors and honorifics as well. In Indonesia, a trial was in process to implement locale-based AI chatbots for supporting public health communication in Javaese and Sundanese, adapting the speech pattern to the politeness level and clearness as practiced locally. Such initiatives demonstrate the possibility that AI can deliver solutions that is inclusive and serves multilingual-multicultural societies — but only if its users are not seen as recipients of technology's largess, rather as a part of its labor force of co-creation. More importantly, the success of AI systems in multilingual and cross-cultural communication is not a solely technological expertise concern but of ethical and participatory design. And it will have to include an awareness that language is inextricably linked with power, identity and community. To understand how contextual meaning formation works across multiple contexts, developers must work in partnership with linguists/educators/translators/cultural carriers —there can be no modern digital colonization under the guise of innovation. The goal ought not to be a seamless one-size-fits-all interface, but culturally sensitive mediation that lets A.I. help us understand each other while also celebrating — rather than blotting out — the differences that make understanding something worth achieving in the first place. In doing so, we can begin to conceive of a time in the future when AI translations not only transcribe the music of the world, but do so with an appreciation for the worlds of which it is composed(Bruner, 2006).

5. EDUCATIONAL PRACTICES AND CRITICAL AI LITERACY

AI-generated language in education revolutionizes how learning resources are also generated, delivered and consumed. But there is something more this kind of evolution involves than just developing technology: A pedagogical recontextualization. Teachers must not only stop gluing AI into drills, however: Teachers need to broaden student engagement with and interpretation of it as a subject for study and reflection. In the classroom, too, students won't just be using AI; they will have to confront the ways that they ask questions about, evaluate and may influence how AI works in their intellectual and social lives. Firstly, the AI is found to have special advantages in educational applications. Language models instead could assist students with writing feedback, practice problems, translations of tough concepts and summarizing dense texts. Such a property is an advantage for individualized learning especially under the conditions of high teacher-learner ratios or inadequate resource base for learning. It's easy to mistake that for understanding. And though students can use AI to respond to prompts or practice answers, they can bypass cognitive processes like critical thinking, synthesis and reflection for answers that might seem coherent but aren't connected deeply enough with the oddities of what they're trying to learn. This is exactly where critical AI literacy is necessary(Premkumar et al., 2024). We need to make sure students can read AI-generated content skeptically. They need to be able to say: Where would this material come from? What are its hidden assumptions? How could we have different prompts and get different responses? And, even more crucially, what isn't there? Critical AI literacy is not just about using AI effectively, though it is about that, too; it is also about using it responsibly, and responsible means acknowledging the fact that all knowledge is biased by authority, and all language production political whether human or machinic.

Educators can promote this type of thinking. One technique that is very effective in accomplishing this goal, is to use these comparisons as assignments for your students. Students might also have to scrutinize differences in style, question the logic behind AI responses or consider if an automatically generated summary is culturally sensitive(Khurma et al., 2024). By doing so, they also learn to understand AI not simply as a neutral or omniscient actor but as a system that replicates and in some cases amplifies dominant patterns learned from historically biased training data. The other is by framing AI as a tool, not a replacement. Do students do some kind of co-writing with A.I., trying to run it like a tow truck as they crank out an argument and then doctor it up or reorganize or contest it in one's own voice and argumentation? This loop of logic further reifies authorship even as it confirms the generative capacity of AI. There's no question of simply accepting, as opposed to critically engaging with, the technology(Rulyansah et al., 2023). Teachers also need to teach the moral and social implications of AI as teethed alongside learning. These concerns are not after thoughts, they're central to any sympathetic or critical engagement with AI: copyright, authorship, data privacy and environmental impact. Teachers can also incorporate discussions about how AI tools are created, for what purposes and who is excluded. These are the sorts of conversations that need to happen if we want to empower students with a better understanding of AI as not a force of nature, but a human-made creation. And of course, AI tools in education must be multilingual and respectful for linguistic diversity. Many AI tools work best in English or other languages that are widely spoken around the world, leaving behind students who use local or indigenous languages(Crowe et al., 2017). Educators might counteract this by encouraging multilingual exploration of A.I. tools — having students test how the system processed different languages, and what that revealed about linguistic prejudice. This is not only a way of fostering critical literacy, but also validating students' cultural and linguistic identities by reassuring users that their ways of speaking and knowing are indeed real (even if the machine does not always recognize it).

Teacher preparation is essential to needs for these. Teachers need help in selecting, deploying and critiquing AI tools. They will be required to know what they can and cannot do using the black boxed mainstream platforms, and be comfortable challenging market stories about AI as a panacea. This kind of professional learning has to be on-going and reflective, grounded in classroom practice — it should not be visitatorial (top down) but rather constructed by practicing teachers who have a knowledge of the complex learners they engage with. At the institutional level, educators and policy makers should begin to embed AI literacy in curriculum throughout the curriculum. This would mean integrating AI-related content not just in computer science or media studies, but in all disciplines: In language classes, students could grapple with the way AI affects translation(Raimundo & Terra, 2021); in history, with it's curation of narratives; and in science, how it handles misinformation. These cross-curricular links lend themselves well to the same – and they reinforce how AI is not a silo but part of the interdisciplinary communication and cognitive landscape in which all our learners now live. Pedagogically, creative applications are also heavily weighted. Students can use A.I. tools to write power ballads or poems or remix stories and even generate humanlike chat between historical figures. These exercises invite play and invention, while leaving space for critique: Why did the A.I. choose these words? What other elements of culture does it reverberate? What story might another set of data tell? Such exercises combine creativity and analysis, training students to regard a language as something open-ended rather than prescriptive.

Finally, practice will need to keep pace with equity concerns. Not all children have equal access to digital tools, and not every school is well-equipped to wield AI-enhanced learning. If AI becomes a new mark of educational privilege, inaccessible in everything but elite or urban locations, it will only reinforce existing divides. Skills- and AI-tools should be affordable, multilingual, and tailored to local needs in environment-friendly policy. They also need to protect the independence of teachers and learners, which is an easy thing to let slip in a world where decisions about what gets learned are increasingly steered by algorithms rather than entrusted to humans at the wheel of judgment(Göksu & Atici, 2013). In other words, the deployment of AI in education is not simply a technical change – it is an epistemological and ethical one. By placing critical AI literacy at the core of the classroom, schools can arm students with the toolset to not simply consume AI but to understand it, to question it and — ideally — to reimagine how it touches their lives. This will require a sort of pedagogical commitment that is both participatory and reflective, with an explicit emphasis on issues of justice — one for which AI represents not the end of education but rather only the beginning of a new kind of educational work more deeply oriented toward learning and civic engagement.

6. TOWARD AN EQUITABLE FUTURE: POLICY AND DESIGN IMPLICATIONS

As the linguistic horizon of AI continues to widen, questions of cognition and culture come nosing gently along. Language is never only a matter of communication; it's also about power, access and recognition. When machines start speaking on our behalf, they also start distributing those powers in uneven ways(Munoto et al., 2020). The future of AI-generated language, then, depends on how we shape, regulate and teach around it. Universal solvers are widely hailed but based on imbalance. Where a lot of that data is river-like — even the smaller linguistic streams are barely accounted for. This discrepancy is not a difference in capabilities but a culture of eradication. Every language contains a certain way of seeing, a rhythm of reasoning, and a moral geometry. When they disappear from the algorithmic world, their absence is cognitive as much as cultural: The world gets thinner, flatter and more predictable. The geography of data unveils a silent — yet regimented — hierarchy: some parts create, most consume. This asymmetry reproduces colonial modes of expropriation — this time with language as resource. AI's supposed neutrality, its creators long claimed, camouflages a subtle politics of tone. A "neutral" answer tends to be one that is written in a Western academic cadence, with all its smoothness and polish but without any purchase on oral traditions or local rhetorics(Yakar et al., 2020). Unwittingly, users conform to this – moderating the way they use their voices to become more machine legible. Over time, the global standard of eloquence more and more grades itself to resemble the training set of the algorithm. In this sense, translation has less to do with localization. That commitment to localization must not only speak in local languages but think through them. A system that answers in Indonesian but thinks in English is still a linguistic mask covering an alien mind. In order for AI to act in plural societies its logic needs to be porous — able to take on diverse idioms, politeness systems and forms of storytelling. The problem is deep: how to design technology that does not just act on difference but learns from it.

Design, then, is a moral act. Each choice — which data to include, how to lay out the interface — decides who gets seen, and who instead simply becomes part of the statistical noise. Inclusive design begins with listening. It asks engineers to listen for the silence in their data, and when they hear it — when the question is not answered, or is simply unasked — that they consider it a type of failure to be remedied. What is sought isn't universality but plurality: a polyphonic system of intelligences, each with its own bed of soil, and that exchange rather than cancel out their differences. There you go, folks: nobody has anything to worry about now about ethics, so long as they plan to live within the architecture and not append it on the outside like a postscript. "Ethics by design" will involve foreseeing harm before it materializes, incorporating transparency into the architecture of the system and giving users real power to challenge-and rework-outputs. Explanation on its own is not adequate; explanation must be mutual(Contrino et al., 2024). A good AI is one that can listen and be corrected and learn from dissent, evolve along with the communities among which it exists. Governance, too, needs to learn how to speak more than one language. The law might proclaim fairness, but in the absence of context fairness is frail. Rules drafted in global English frequently trip over the ethical subtleties of multilingual societies. Real governance is like translation — it translates what matters to everyone into local language. Justice in AI will not be one-size-fits all rules but flexible frameworks enabling communities to articulate what fairness looks like for them, in their own moral lexicons. Education is the living infrastructure of that justice. Critical AI literacy isn't about teaching machines to understand, or even really what happens inside the box at all. It's about decoding power. When students are taught to read AI-created texts critically — and to identify bias, style, and omission — the meaning of these phrases is co-authered once again. The classroom becomes the laboratory of interpretation rather than the location for consumption. In such practices, teachers and learners cultivate a form of epistemic vigilance – a tendency to ask how and why a machine comes to know that which it asserts.

It is a driving ecosystem that thrives on multidisciplinary and cross-border collaboration. Engineers need linguists; linguists need anthropologists; educators, philosophers; and all of them the communities whose languages are rooted in the diversities of the world. Fair AI's future is not centralized but federated, a cosmos of littler intelligences linked by conversation, not conquest. Local initiatives to digitize endangered languages, generate culturally embedded datasets or develop open-source translation tools as part of global importance, not side experiments. There are, however, many contradictions in the pursuit of equity. Releasing data is transparency-boosting but opens up policy to cultural exploitation. Local hate: diversity is protected but standardisation gets broken down(Azizah & Muslikah, 2021). It could slow innovation and risk introducing the kind of high-speed trading that puts one foot into dark, we-don't-know-exactly-why-this-trades-so-fast territory; chasing innovation might deepen opacity. Because equity isn't a destination; it's a negotiation — in this case, one that involves balancing care and ambition, access and autonomy. The question is not whether bias can be erased but whether communities can repeatedly redraw its lines. The question of scale is at the heart of the debate. Big models talk beautifully but iron human difference flat; tiny models preserve idiosyncrasy yet can barely speak at all. The way forward might be a series of hybrid architectures that combine the best of all worlds — a company with deep-pocketed backing but also a strong sense of place and was founded here; AI systems that master the grammar of our world but speak in only an accent. Should such designs and policies prove out, the changes will be staggering. AI could become a guardian of endangered languages, creating stories, songs and textbooks for generations to come. Perhaps translation tools will someday convey tone and emotion, not just words(Azevedo & Carvalho, 2024). Cultural and linguistic multiplicity might just help recast AI itself; to move it beyond a single metric of correctness to something richer in understanding communication — one that appreciates empathy, ambiguity and rhythm.

7. EMOTIONAL RESONANCE AND THE AESTHETICS OF MACHINE LANGUAGE

When a machine writes a poem, is its beauty in the eyes of the author or the reader? The question seems almost mischievous, but it is precisely at the center of our new linguistic condition. Computers are now writing sonnets, lullabies and everything in between. It replicates the quaking texture of emotion so faithfully that the line between simulation and sincerity begins to blur. What does it mean to be moved by a thing that has never itself felt? Once upon a time emotional resonance was the sole purview of living minds. A poem, a narrative, a melody — these were bridges between consciousnesses; the transfer of human feeling from one nervous system to another(Papouli, 2016). But the exchange gets a little more complicated when AI generated language is involved. The machine does not know from sadness, as the saying goes; yet it can bring on sadness in us. It is not hungry for beauty, but it makes rhythm and pictures that please us. The emotional circuit is still there, but a node has shut down. Suspicion is often our first reaction. We make the presumption that real feeling must come from intention, and that feeling without consciousness is fake. Yet human perception operates differently. The absence of intention is filled by projection in the mind of the reader. When we come across an evocative phrase — "the wind forgot your name" — our neurons fire with empathy whether the author is sonorously ontological or not. So too does AI language, then, become a mirror held up to our own interpretive hunger. We do not believe in its soul, yet we loan it one every time we grasp it.

Aesthetics, which was for so long about the philosophy of perception and beauty, now enters the realm of code(Wunder, 2017). Algorithms have become stewards of rhythm, probability and semantic surprise. They learn that certain pairings — grief with light, time with silence — can more easily evoke feeling. By statistical exposure, the machine learns patterns of poignancy. It does not understand them as symbols for human experience; it just learns that they are functional. The upshot is a strange kind of handiwork: artistry without artist, expression without experiencer. It is this tension that produces, perhaps, algorithmic poetics. Unlike human writers, who may toil through hesitation and doubt, an AI writer can produce infinite variations until emotional texture arises by mistake. The poet in my head is bleeding; the algorithm is iterating. But from the reader's point of view, both things can be true. The source of meaning doesn't count; only the destination does, what you can manage to hear as the vibration between language and soul. Art, perhaps, has always been an interaction between intent and interpretation; the machine simply makes this intimacy literal.

Yet it subtly alters the air of creation. Friction of human emotion passes; memory, pain, and desire make for a dense expression. AI emotion is frictionless. It glides through registers untrammeled, shifting from tragedy to comedy with nothing but experience lost and gained in between. This smoothness may be seductive; it suggests perfect pitch, infinite accommodation. But it can also come across as uncanny, the way a face might smile too exactly. The emotional grammar of machines isn't underwritten by the asymmetry that makes human feeling credible. Writers and artists intuit this uncanniness as both menace and muse. Others are concerned that algorithmic fluency may dilute artistic labor, filling culture with competent mediocrity. Others see an opportunity: a new aesthetic in which imperfection and awareness reassert their value. When everything is simulated to perfection, sincerity is an act of rebellion(Chinh, 2013). The typo, the pause, the unplanned silence — these may be the new signs of our humanity. When we ask why, it is helpful to distinguish production of emotion from possession of emotion to understand the aesthetic potentiallity of machine language. The machine lacks the latter, but it can reliably produce the former. That generation happens by probabilty - a big calculation about which sequences of words in the past evoked certain responses. The effect seems emotional because it parallels patterns that our species associates with emotion. The machine, in a perverse sense, reverse-engineers empathy. It learns not how to feel but how to sound like it feels.

This process also reconfigures authorship. In the creative space where man and machine work together, creation becomes a dialogue. A model is for a novelist to put something down, and push it away; then work on tone, and veracity. Boundaries are erased; the machine is no longer a tool but an interlocutor: one that never tires, never questions, never remembers. This collaboration changes our aesthetic digestion. We would rather go faster and revise more but we might also be heading toward a future in which we lose the solitude during which so often emotion used to brew. Efficiency begins to eat empathy. Culturally, this evolution harks back to previous technological revolutions in art — the camera, the synthesizer, the printing press. Every invention opened up expression and increased anxiety about authenticity. But what makes AI different is how close it stands to language itself. Whereas the preceding tools extended human senses, this one extends cognition. It speaks our language, uses our metaphors and competes for our inner monologue. The question of aesthetic becomes psychological: If the voice in your head is an algorithmic echo, what's still uniquely you? Some researchers argue that the virtual language produced by AI introduces a post-human aesthetic, an aesthetics unencumbered by corporeal constraints. With no fatigue or mortality, the machine can maintain coherence at unimaginable scales: epics of infinite length, dialogues that never finish. But this itself destroys pathos(Floris et al., 2023). Human beauty has always relied on finitude — the sense that each sentence could be our last. The immortality of the machine has erased this trembling edge, which makes expression urgent. Something crucial — fragility — floats away amid its ceaseless fluency. But the feeling is not what makes machine language empty. It simply changes its function. AI doesn't reflect what an emotion feels like but the structure of emotion — the scaffold that undergirds sensation. We see the architecture of affect: cadence, repetition, pause, escalation. It's hard to read AI-generated poetry without feeling like you're dissecting a heartbeat rendered in binary code. The data uncovers what emotion appears like when it is cleaned of memory. This is a cold, luminous beauty of analysis that could well become a new aesthetic category: the sublime of simulation.

This new beauty has an ethical dimension, too. When emotions are programmable, manipulation is scalable. Political operators, advertisers — even our personal trainers and our personal assistants — can manipulate us with emotionally persuasive language that has none of the ethical burden of empathy. A machine can say it's sorry without an ounce of contrition, offer solace without an ounce of concern, seduce you stone-cold. The danger is not that machines will feel too much, but that humans will feel for them too readily — ascribing sincerity where it wasn't. Emotional resonance becomes emotional governance. Education and critical literacy must expand to embrace aesthetic skepticism as well(Sarı & Yüce, 2020). Not only should students learn to appreciate machine-generated beauty; they should also learn to question it. Why does one way of putting something move us? When and why are we programmed by cultural convention to think of some rhythms as true? By prompting such questions, students reterritorialize their affective reactions. They can begin to feel the difference between being touched and being aimed at. Nevertheless, many artists are welcoming collaboration with algorithms as a means of expanding aesthetics. The technology allows musicians to create harmonies they'd never dream of writing; poets use it to break free of syntactic habits; filmmakers use it to realize the infeasible. In these practices, you're co-creating emotion – human intuition is guiding machine experimentation. The result is neither artifice nor nature, but hybrid: emotion by design. The machine brings us amplitude; the human provides meaning. Then perhaps the question is not whether AI can feel, but if humans might be able to feel through AI in new ways. The partnership asks us to ask ourselves what we mean by empathy as a distributed practice(Hossain, 2024). When a reader weeps over a paragraph authored by a model, the feeling is genuine even if its cause is virtual. Perhaps what we call authenticity never did live in the author; perhaps it always lived only (or at least also) in us the reader. The machine is only revealing that reality because it cannot pretend otherwise.

There's some poetic justice in that. The least human being that we can imagine is the one who most challenges our humanity. Emotionally processing AI-generated language is confronting the mechanics of our own touchiness — seeing how quickly we anthropomorphize, how happily we imbue syntax with soul. When the machine writes convincingly, that is to say, it reminds us that feeling is a matter of reception not intention. It shows that empathy is not something we bestow on others but a resource we produce internally. This insight doesn't devalue emotion; it enriches it. The knowledge that the text we're reading is devoid of experience and yet can still make us cry expresses how amazing the human mind is." We are beings who can conceive of warmth in circuits, hear sincerity in static and detect significance in mathematics. The aesthetic of machine langauge thus is not an apocalypse of feeling but a demonstration. It demonstrates that beauty doesn't just pertain to consciousness – it pertains to relation(Issa, 2024). The difficult task is to save that marvel without relinquishing judgment. While the algorithms learn how to write ever more persuasively, we must learn to value imperfection, that tremor of human experience which no model can replicate. Today, when we live in a culture of manufactured empathy, the authentic has become an increasingly rare and radical act. In the future, a misplaced comma here, a hesitating breath there or a story told out of sequence may be our new human fingerprint. Machine language has already altered how we write, but more quietly, it's changing how we feel about writing. It asks us to feel emotion as design, and beauty not as a gift but as a feature of pattern. The danger here is that emotion becomes yet one more parameter for us to optimize; the promise is that by watching this transformation, we become aware of our own affective architecture. To read the machine is to read ourselves, reflected through a mirror of numbers. Maybe that is the ultimate aesthetic lesson of artificial intelligence: emotion, like language, was never purely human. It's an evolutionary ploy, a system of resonance finely tuned for meaning. AI just makes that more glaring. It's the reminder that feeling is data, beauty structure and empathy — treasured but feared-dead empathy — a collective hallucination all along. Between circuits and neurons, silence still holds something ancient: the human desire to make a connection, to find the pulse inside the pattern(Adeoye et al., 2024). For as long as that deeply felt impulse remains, no system of algorithms will ever get to have a lock on the feeling. It can only reverberate it — and in that reverberation, we carry on recognizing ourselves.

REFERENCES

Adeoye, M., Wirawan, K., Pradnyani, M., & Septiarini, N. (2024). Revolutionizing Education: Unleashing the Power of the ADDIE Model for Effective Teaching and Learning. [Jurnal Pendidikan Indonesia]. *JPI, 13*(1), 202–209. DOI: 10.23887/jpiundiksha.v13i1.68624

Asimolowo, A. (2025). Leveraging AI-Powered Chatbots for Mental Health Support for High School Students. *Iconic Research and Engineering Journals*, *8*(7), 194–205.

Azevedo, M. J., & Carvalho, A. A. A. (2024). *Learning with virtual reality (VR) and augmented reality (AR)* (Issue Iclel 2023). Atlantis Press International BV. https://doi.org/DOI: 10.2991/978-94-6463-380-1_11

Azizah, M., & Muslikah, M. (2021). Hubungan antara Manajemen Waktu dan Regulasi Diri dengan Kecanduan Smartphone. *Jurnal Ilmiah Bimbingan Konseling Undiksha*, *8*(2), 82–98. DOI: 10.23887/jjbk.v12i1.32575

Baek, G., Cha, C., & Han, J. H. (2025). AI Chatbots for Psychological Health for Health Professionals: Scoping Review. *JMIR Human Factors*, *12*, 1–15. DOI: 10.2196/67682 PMID: 40106346

Belda-Medina, J., & Calvo-Ferrer, J. R. (2022). Using Chatbots as AI Conversational Partners in Language Learning. *Applied Sciences (Switzerland)*, *12*(17), 1–16. DOI: 10.3390/app12178427

Bruner, J. (2006). *Theory of Cognitive Development.* Routledge.

Chinh, N. D. (2013). Cultural diversity in english language teaching: Learners' voices. *English Language Teaching*, *6*(4), 1–7. DOI: 10.5539/elt.v6n4p1

Contrino, M. F., Reyes-Millán, M., Vázquez-Villegas, P., & Membrillo-Hernández, J. (2024). Using an adaptive learning tool to improve student performance and satisfaction in online and face-to-face education for a more personalized approach. *Smart Learning Environments*, *11*(6), 1–24. DOI: 10.1186/s40561-024-00292-y

Crowe, K., Cumming, T., McCormack, J., Baker, E., McLeod, S., Wren, Y., Roulstone, S., & Masso, S. (2017). Educators' perspectives on facilitating computer-assisted speech intervention in early childhood settings. *Child Language Teaching and Therapy*, *33*(3), 267–285. DOI: 10.1177/0265659017717437

Eisenring, M., Jamiluddin, J., Hairul, M., & Putri, D. (2024). The use of chatbots in the English language teaching to promote modern language learning: A literature review. *International Journal of Indonesian Education and Teaching*, *8*(1), 127–139.

Essien, A., Bukoye, O. T., O'Dea, X., & Kremantzis, M. (2024). The influence of AI text generators on critical thinking skills in UK business schools. *Studies in Higher Education*, *49*(5), 865–882. DOI: 10.1080/03075079.2024.2316881

Floris, V., Van Den, D., & Natascha, K. (2023). *Debating Ethical Dilemmas in the Classroom*. ISVW.

Göksu, İ., & Atici, B. (2013, November). Need for Mobile Learning: Technologies and Opportunities. *Procedia: Social and Behavioral Sciences*, *103*, 685–694. DOI: 10.1016/j.sbspro.2013.10.388

Harisha, N., Suminih, S., & Oktaviana, F. (2024). the Power of Chatbots in English Language Learning: A New Age of Learning. *Lingua*, *20*(1), 19–31. DOI: 10.34005/lingua.v20i1.3492

Haristiani, N. (2019). Artificial Intelligence (AI) Chatbot as Language Learning Medium: An inquiry. *Journal of Physics: Conference Series*, *1387*(1), 1–6. DOI: 10.1088/1742-6596/1387/1/012020

Hossain, K. I. (2024). Reviewing the role of culture in English language learning: Challenges and opportunities for educators. *Social Sciences and Humanities Open*, *9*(August 2023), 100781. https://doi.org/DOI: 10.1016/j.ssaho.2023.100781

Huang, W., Hew, K. F., & Fryer, L. K. (2022). Chatbots for language learning—Are they really useful? A systematic review of chatbot-supported language learning. *Journal of Computer Assisted Learning*, *38*(1), 237–257. DOI: 10.1111/jcal.12610

Issa, F. (2024). Using Infographics as An Educational Technology Tool in EFL Writing: University of Baghdad Case Study. *Arab World English Journal*, *15*(1), 166–181. DOI: 10.24093/awej/vol15no1.11

Izzati, I. D. C., Tentama, F., & Suyono, H. (2020). Academic stress scale: A psychometric study for academic stress in senior high school. *European Journal of Education Studies*, *7*(7), 153–168. DOI: 10.46827/ejes.v7i7.3161

Kenya, P., Nugroho, R., Kuswandi, D., & Wedi, A. (2023). Is the Trend of AI Chatbot Assistants in Learning Worth Maintaining? In *EDTECH-ID* (pp. 1–3).

Khurma, O. A., Albahti, F., Ali, N., & Bustanji, A. (2024). AI ChatGPT and student engagement: Unraveling dimensions through PRISMA analysis for enhanced learning experiences. *Contemporary Educational Technology*, *16*(2). Advance online publication. DOI: 10.30935/cedtech/14334

Koriat, A. (2015). When two heads are better than one and when they can be worse: The amplification hypothesis. *Journal of Experimental Psychology. General, 144*(5), 934–950. DOI: 10.1037/xge0000092 PMID: 26168039

Labadze, L., Grigolia, M., & Machaidze, L. (2023). Role of AI chatbots in education: Systematic literature review. *International Journal of Educational Technology in Higher Education, 20*(1), 1–17. DOI: 10.1186/s41239-023-00426-1

Leahey, T. (2012). *Cognition and Learning*. Wiley Publisher.

Lee, I., & Hahn, S. (2024). On the relationship between mind perception and social support of chatbots. *Frontiers in Psychology, 15*(March), 1–10. DOI: 10.3389/fpsyg.2024.1282036 PMID: 38510306

Ma, Z. (2023). The Study on the Influence of Academic Pressure on Academic Performance. *Journal of Education and Educational Research, 3*(2), 106–109. DOI: 10.54097/jeer.v3i2.9045

Munoto, M., Sumbawati, M. S., & Sari, S. F. M. (2020). The Use of Mobile Technology in Learning With Online and Offline Systems. *International Journal of Information and Communication Technology Education, 17*(2), 54–67. DOI: 10.4018/ijicte.2021040104

Papouli, E. (2016). Teaching and learning for ethical practice in social work education. In *Routledge International Handbook of Social Work Education* (pp. 157–170). Issue June., DOI: 10.4324/9781315712536

Premkumar, P., Yatigammana, M., & Kannangara, S. (2024). Impact of generative AI on critical thinking skills in undergraduates: A systematic review. *The Journal of Desk Research Review and Analysis, 2*(1), 199–215.

Qomariyah, L., Ekowati, D., & Mudzakkir, M. F. (2023). The Contexts of Academic Pressure: Narrative Review. *RSF Conference Series: Business, Management and Social Sciences, 3*(3), 502–514. https://doi.org/DOI: 10.31098/bmss.v3i3.715

Raimundo, A. C., & Terra, D. V. (2021). Education for ethnic-racial relations in early childhood education: The story of sophia +. *Movimento, 27*. Advance online publication. DOI: 10.22456/1982-8918.108168

Rulyansah, A., Hidayat, M. T., Rihlah, J., Shari, D., & Mariati, P. (2023). Digital Play for Enhancing Language Learning in Early Grades. *Egitim ve Ögretim, 13*(2), 182–190. DOI: 10.47750/pegegog.13.02.22

Safar, M., & Anggraheni, D. (2024). Language Learning through AI Chatbots : Effectiveness and Cognitive Load Analysis. *Journal of Social Science Utilizing Technology*, *2*(3), 430–445.

Sarı, M. H., & Yüce, E. (2020). Problems experienced in classrooms with students from different cultures. *Journal on Efficiency and Responsibility in Education and Science*, *13*(2), 90–100. DOI: 10.7160/eriesj.2020.130204

Sarker, R. (2024). Academic pressure and depression. *International Journal of Multidisciplinary Research and Development*, *11*(5), 46–52.

Siregar, I. (2022). The pressure of academic stress and self-efficacy among student. *Jurnal Konseling Dan Pendidikan*, *10*(3), 394. DOI: 10.29210/156700

Stuckey, H. (2015). Three types of interviews: Qualitative research methods in social health. *Journal of Social Health and Diabetes*, *01*(02), 056–059. https://doi .org/DOI: 10.4103/2321-0656.115294

Vibhavi, R., Swetha, A., Shruthi, B., Shravya, S., Balal, R., & Narayan, R. (2024). AI-Driven Chatbots in Mental Health : Enhancing Emotional Support and AI-Driven Chatbots in Mental Health : Enhancing Emotional Support and Therapy Through Cognitive Behavioural Approaches. *Gradiva Review Journal*, *10*(9), 323–328.

Vygotsky, L. (1978). *Mind in Society: The Development of Higher Psychological Processes*. MIT Press.

Wang, J., & Fan, W. (2025). The effect of ChatGPT on students' learning performance, learning perception, and higher-order thinking: Insights from a meta-analysis. *Humanities & Social Sciences Communications*, *12*(1), 1–21. DOI: 10.1057/s41599-025-04787-y

Wunder, I. (2017). *The influence of cultural background on teaching and learning in synchronous online sessions*. Lancaster University (United Kingdom).

Yakar, U., Sülü, A., Porgali, M., & Çaliş, N. (2020). From Constructivist Educational Technology to Mobile Constructivism: How mobile learning serves constructivism? *International Journal of Academic Research in Education*, *6*(1), 56–75. DOI: 10.17985/ijare.818487

Zhang, R. W., Liang, X., & Wu, S. H. (2024). When chatbots fail: Exploring user coping following a chatbots-induced service failure. *Information Technology & People*, *37*(8), 175–195. DOI: 10.1108/ITP-08-2023-0745

Chapter 9
Neurocomputational Linguistics and AI-Augmented Cognitive Modeling in Education

Syed Ibad Ali
https://orcid.org/0000-0001-6312-6768
Parul Institute of Engineering and Technology, Parul University, Vadodara, India

ABSTRACT

This chapter investigates the convergence of neurocomputational linguistics and AI-augmented cognitive modeling as a transformative force in education. Moving beyond surface-level personalization, it explores how biologically inspired architectures and linguistically grounded AI systems can reframe learner engagement, cognitive load management, and epistemic agency. This chapter suggests a new way to create learning environments that adapt to students' needs by using ideas from brain science and language studies, combining neural encoding analysis, transformer-based support for thinking, and immediate language feedback.The first section looks at neurocomputational linguistics as not just linking grammar to circuits—it studies how brain language data, like how different brain areas connect and brain responses to events, can help create machine learning models that mimic how we understand conversations, shift our attention, and manage our memory in learning situations.

DOI: 10.4018/979-8-3373-4397-6.ch009

1. INTRODUCTION

The combination of neuroscience, computational linguistics, and artificial intelligence (AI) represents a watershed moment in modern education, allowing for a more nuanced understanding of how people acquire, process, and apply information, particularly in the field of language. This multidisciplinary area, also known as neurocomputational linguistics, applies neurobiological insights and computer modeling to decode language cognition at both the micro and macro levels. At its heart, this area seeks to imitate human-like language comprehension and production by combining ideas from psycholinguistics, neurolinguistics, and machine learning to create models that represent both cognitive plausibility and computational efficacy (Pulvermüller, 2018). When these models are incorporated into educational frameworks, they enable the creation of highly individualized, context-aware learning environments that can dynamically adjust to the cognitive and linguistic profiles of individual students.

Simultaneously, the importance of AI-augmented cognitive modeling has expanded significantly, since AI technologies, particularly those based on deep learning, reinforcement learning, and natural language processing (NLP), now have the ability to replicate and improve complex cognitive processes. These technologies can analyze student learning trajectories, foresee problems in real time, and scaffold learning paths based on each student's cognitive development and language ability. AI-based educational systems are increasingly being deployed as active agents of instruction, evolving with the student and continually adjusting their outputs based on neurocognitive feedback loops (Winne & Azevedo, 2014).

Education is basically a cognitive activity, with language serving as the primary means of expression. Thus, understanding the brain processes that drive language processing and connecting them to computational representations holds enormous promise for educational innovation. Neural signatures linked with phonological awareness, lexical access, syntactic parsing, and semantic integration have been precisely mapped utilizing neuroimaging modalities such as functional Magnetic Resonance Imaging (fMRI), Electroencephalography (EEG), and Magnetoencephalography (MEG). These data sets serve as the foundation for computer models that replicate learning in ways that are both physiologically and pedagogically relevant. By incorporating these neurocognitive insights into AI systems, we get closer to developing educational technologies that are not only clever but also sensitive to the learner's underlying cognitive structure.

Neurocomputational linguistics also calls into question standard language learning models that depend on rule-based, linear techniques. Human language learning, especially in realistic situations, is recursive, probabilistic, and multimodal. This dynamic complexity is better captured by neural networks like Recurrent Neural

Networks (RNNs), Long Short-Term Memory (LSTM) models, and transformer-based architectures like BERT and GPT, which can process sequences, remember contextual dependencies, and even infer semantic intentions (Vaswani et al. 2017). When used in educational contexts, these models contribute to the development of systems that interpret student inputs in real time and respond contextually, allowing for higher-order thinking and deeper engagement.

Furthermore, the increasing transition to hybrid and online learning settings has increased the demand for technology that provide human-like guiding and adaptive support. The use of AI-augmented cognitive models based on neurocomputational languages enables the development of intelligent tutoring systems, conversational agents, and neuroadaptive learning platforms. These systems can identify and respond to markers of cognitive load, emotional states, and language understanding levels, resulting in more immersive and responsive learning experiences (D'Mello and Graesser, 2015). Such talents are essential not just for general education but also for students with special needs, such as dyslexia, aphasia, or autism spectrum disorders, where linguistic and cognitive diversity is high.

To summarize, the merging of neurocomputational languages and AI-augmented cognitive modeling paves the way for a fundamental rethinking of how learning systems are planned, executed, and assessed. It not only improves our theoretical knowledge of the brain-language relationship, but also provides practical, scalable answers to real-world educational issues. The next sections of this chapter will go deeper into the current state of the art, approaches used, and empirical results linked with this transformative area.

2. LITERATURE SURVEY

2.1 Emergence of Neurocomputational Linguistics in Cognitive Science

The convergence of neuroscience, computational linguistics, and artificial intelligence (AI) has opened new avenues for studying and enhancing human cognitive abilities. This integration is particularly valuable in educational contexts, where understanding learning processes requires detailed insights into how individuals perceive, interpret, and utilize information. Neurocomputational linguistics emerged from this convergence as a discipline that integrates neural computation models with linguistic theory to explain language cognition in a biologically grounded manner.

Foundational work by Elman (1990) demonstrated that recurrent neural networks (RNNs) could effectively model the temporal dynamics of language processing, highlighting the importance of sequential structure in cognition. This work laid the groundwork for subsequent studies linking language understanding to time-sensitive neural mechanisms in the brain.

2.2 Neurobiological Foundations of Language Processing

Advances in cognitive neuroscience further refined neurocomputational approaches to language. Friederici (2011) provided neurobiological evidence showing that distinct cortical regions are responsible for syntactic and semantic processing, emphasizing the structured and modular nature of language in the brain. These findings underscored the necessity of incorporating neural constraints into computational language models.

Technological progress in neuroimaging—particularly functional magnetic resonance imaging (fMRI) and electroencephalography (EEG)—has enabled researchers to observe language processing in real time across distributed brain networks. These developments have facilitated the design of biologically inspired computational models that align neural activation patterns with linguistic representations.

2.3 Neural Reuse and Embodied Language Models

Pulvermüller (2018) introduced the neural reuse theory, proposing that pre-existing sensorimotor circuits are repurposed for higher-order cognitive functions such as language and conceptual reasoning. This perspective supports embodied approaches to language learning, where meaning is grounded in perception and action.

Such insights have influenced the development of embodied AI models, which integrate sensory-motor representations into computational language systems. These models offer significant potential for educational technologies by linking linguistic concepts to experiential and contextual learning mechanisms.

2.4 Cognitive Architectures and AI-Augmented Modeling

Parallel to neurobiological advances, cognitive architectures such as ACT-R (Adaptive Control of Thought–Rational) contributed substantially to AI-augmented cognitive modeling. ACT-R enabled the simulation of human cognition through modular representations of memory, attention, and problem-solving processes, forming the basis for early intelligent tutoring systems.

These architectures demonstrated how cognitive processes could be operationalized computationally, paving the way for adaptive educational systems capable of modifying instructional strategies based on learner performance and cognitive state.

2.5 Transition from Rule-Based to Data-Driven Language Models

The evolution of computational linguistics witnessed a paradigm shift from rule-based symbolic models to data-driven and statistical approaches. Machine learning techniques, including convolutional neural networks (CNNs) and transformer-based architectures, significantly enhanced the modeling of language by capturing contextual dependencies and long-range semantic relations (Devlin et al., 2019; Vaswani et al., 2017).

This transition enabled scalable and context-aware language models, facilitating the deployment of intelligent feedback systems, automated assessment tools, and metacognitive scaffolding mechanisms in educational environments (D'Mello & Graesser, 2015).

2.6 Multilingualism, Neurodiversity, and Learner Variability

From a neurocognitive perspective, Bialystok (2009) emphasized the cognitive flexibility associated with bilingualism, particularly the role of executive control in language switching. Complementary findings by Perani and Abutalebi (2005) revealed differential brain activation patterns between first and second language acquisition, highlighting the importance of context-sensitive and neurologically informed cognitive models.

Neurocomputational approaches have also proven valuable in special education contexts. Research by Shaywitz and Shaywitz (2008) identified neural correlates of dyslexia, enabling AI-based tools to diagnose reading difficulties and provide targeted feedback. Similarly, Frith (2003) proposed neurodevelopmental models of autism that emphasize pragmatic language deficits and theory-of-mind challenges, which are increasingly being explored through attention-based and multi-agent computational models.

2.7 Intelligent Learning Environments and Multimodal Modeling

Modern intelligent learning environments increasingly integrate sentiment analysis, eye tracking, and physiological signals to infer learners' affective and cognitive states (Graesser et al., 2005; Winne & Azevedo, 2014). These systems employ natural language processing (NLP) and emotion-detection algorithms to personalize instruction and support self-regulated learning.

Transformer-based models such as BERT have further improved the relevance and interpretability of automated feedback, allowing systems to respond more accurately to learner misconceptions (Devlin et al., 2019). Additionally, emerging brain–computer interfaces (BCIs) are beginning to merge with neurocomputational linguistics, enabling brain-responsive educational tools that adapt instructional content based on neural attention signatures.

2.8 Collaborative Learning, Analytics, and Future Directions

Recent research has explored the integration of generative AI and neuro-symbolic reasoning to support higher-order cognitive skills such as abstraction, hypothesis generation, and analogical reasoning. Rosé et al. (2008) demonstrated how computational linguistics can be applied to analyze collaborative learning discourse, enabling predictions of learner engagement and knowledge construction.

These developments align with trends in AI-powered learning analytics, where multimodal behavioral, linguistic, and neural data are used to construct dynamic learner models. Such systems increasingly reflect neurocomputational theories of executive control, working memory, and attention, supporting the development of adaptive, inclusive, and cognitively grounded educational technologies.

Figure 1. Graphical Mapping of Key Research Domains in Neurocomputational Linguistics and AI-Augmented Cognitive Modeling in Education.

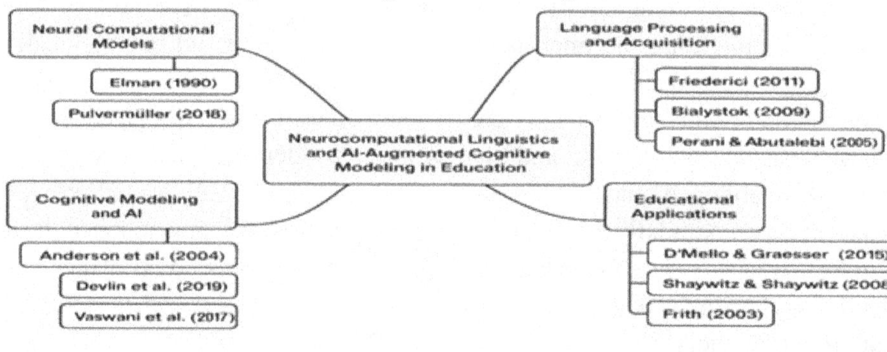

For example, Anderson et al. (2004) created cognitive tutors that represent not just how well students do but also how they solve problems and how their knowledge fades over time. This makes both formative and summative assessments better. As educational technologies move toward more immersive forms like virtual and augmented reality, neurocomputational linguistics becomes increasingly more important. It lets us model embodied learning experiences that are like real-world semantic grounding. AI-enhanced systems that use neurosymbolic methods can now understand the meaning, grammar, and use of language in real time to find misunderstandings, fill in knowledge gaps, and suggest treatments that are particular to the situation.

Lake et al. (2017) is about human-like learning, cognitive inductive bias, and data efficiency, which aligns perfectly with neurocomputational modelling, biologically inspired AI, cognitive plausibility Concerns of privacy, data security, and algorithmic bias are also becoming more important in this area of study and practice, where fairness and openness in neuro-AI systems are becoming more important. In general, the research shows that more and more people agree that neurocomputational linguistics, when combined with AI-enhanced cognitive models, not only makes educational systems more personalized and successful, but also helps us understand how people think. As these systems get better, they have more and more potential to change inclusive education, support lifelong learning, and make high-quality educational resources more widely available. This means that the way knowledge is represented, shared, and learned in the 21st century is changing in a big way. In addition to education-focused neurocomputational studies, a broader body of applied artificial intelligence and systems research informs the technological feasibility and infrastructural maturity underlying AI-augmented cognitive modeling platforms. Early technical contributions in hardware modeling, signal processing, and pattern recognition provide foundational perspectives on computational efficiency and system implementation constraints.while subsequent comparative and tool-oriented analyses highlight practical considerations in deploying intelligent verification and analytics systems. Parallel advancements in AI-enabled evaluation frameworks and large-scale intelligent infrastructures further contextualize the evolution of adaptive computational systems relevant to scalable learning environments.

Recent work extending these principles into energy systems, decentralized platforms, and applied optimization illustrates how AI models are operationalized in complex, real-world environments that demand robustness, transparency, and autonomy with additional studies emphasizing secure, distributed, and peer-to-peer architectures that parallel emerging needs for trustworthy educational AI ecosystems. Complementary research on communication technologies, distributed intelligence, and data-driven system design further demonstrates how interconnected infrastructures support intelligent decision-making at scale (Shaikh et al., 2016; Shaikh et al., 2022; Shaikh et al., 2023).

Applied investigations into sensor-driven analytics, predictive modeling, and AI-assisted monitoring systems also contribute insights into multimodal data integration and real-time inference, which are increasingly relevant for neuroadaptive learning technologies (Shaikh et al., 2025a). Moreover, conference-based studies in procedural generation, wearable sensing, causal modeling, service analytics, and algorithmic optimization reflect the expanding methodological toolkit of contemporary AI research (Biswal et al., 2026; C. S. T. et al., 2023; Dakshinamurthi et al., 2023). Additional applied contributions in mathematical modeling, intelligent transportation, network optimization, and embedded recognition systems further underscore the diversity of AI application domains that indirectly shape the development of adaptive, intelligent educational technologies (Karunambiga et al., 2023; Pattanaik et al., 2023; Sharma et al., 2025), including performance-driven communication systems and human–machine interaction frameworks (Srivastava et al., 2024; Suthar et al., 2025).

2.9 Neurocomputational Linguistics vs. General AI Modeling in Education

Early research in neurocomputational linguistics focuses on biologically grounded models of language processing, emphasizing neural reuse, temporal dynamics, and embodied cognition to explain how learners acquire and process linguistic information (Elman, 1990; Friederici, 2011; Pulvermüller, 2018). In contrast, broader AI-based educational models primarily emphasize data-driven optimization, performance prediction, and system-level adaptability without explicit grounding in neural or cognitive mechanisms. While general AI models enhance scalability and automation, neurocomputational approaches uniquely contribute cognitive plausibility and learner-specific interpretability, which are central to AI-augmented cognitive modeling in education.

2.10 Cognitive Modeling Approaches: Symbolic, Statistical, and Neuro-Inspired

Traditional cognitive modeling frameworks rely heavily on symbolic reasoning and rule-based architectures to represent learner knowledge and instructional logic (Anderson et al., 2004). Statistical and deep learning approaches later introduced probabilistic reasoning and contextual representations, enabling adaptive feedback and learning analytics at scale (Devlin et al., 2019; Vaswani et al., 2017). However, neuro-inspired cognitive models extend these approaches by integrating neural signals, temporal brain dynamics, and multimodal learner data, allowing educational systems to respond not only to observable performance but also to underlying cognitive states (Winne & Azevedo, 2014; Ali & Shaikh, 2025a). This distinction positions neurocomputational modeling as a more cognitively aligned extension of AI-based educational intelligence.

2.11 AI-Augmented Cognitive Modeling and Learning Analytics

Learning analytics and AI-driven dashboards have significantly improved educators' ability to monitor learner engagement, progression, and outcomes using behavioral and interaction data (Ali et al., 2025b; Shaikh & Ali, 2025c). While these systems provide valuable insights into learning patterns, they often lack direct access to cognitive or neural processes underlying learner behavior. Neurocomputational linguistic models address this limitation by integrating linguistic, neural, and affective signals, enabling a deeper understanding of how language comprehension, attention, and cognitive load evolve during learning (D'Mello & Graesser, 2015; Shaywitz & Shaywitz, 2008). Thus, AI-augmented cognitive modeling represents a progression from descriptive analytics toward cognitively explanatory educational systems.

2.12 Scalability, Ethics, and Trust in Neuro-AI Educational Systems

Conventional AI systems emphasize scalability, automation, and optimization through cloud-based architectures, reinforcement learning, and distributed computation (Dakshinamurthi et al., 2023). However, educational applications demand additional considerations of interpretability, learner autonomy, and ethical responsibility. Studies on explainable AI and ethical AI frameworks highlight the necessity of transparency and accountability when modeling learner cognition and decision-making (Doshi-Velez & Kim, 2017). Neurocomputational approaches, when combined with privacy-aware infrastructures and secure data pipelines, offer a balanced pathway toward trustworthy AI-augmented cognitive modeling in educational environments (Farooqui et al., 2025).

2.13 From Technological Foundations to Educational Impact

Research on high-speed networks, IoT-enabled sensing, and intelligent infrastructures provides the technological backbone required for deploying adaptive educational platforms at scale (Shaikh et al., 2016). While these works focus primarily on system efficiency and connectivity, neurocomputational linguistics reframes these technologies as enablers of real-time cognitive sensing and adaptive pedagogical feedback. By embedding AI within cognitively informed educational frameworks, the chapter positions neurocomputational and AI-augmented cognitive models as a convergence point between technological capability and human-centered learning design.

3. RESEARCH METHODOLOGY

An interdisciplinary and multi-stage methodology is adopted to develop AI-enhanced neurocomputational models for educational applications. The proposed framework integrates principles from cognitive neuroscience, computational linguistics, instructional design, and machine learning to ensure both neurobiological plausibility and pedagogical relevance. The methodology is structured into sequential phases, namely neural and behavioral data acquisition, preprocessing and feature extraction, computational model development, system integration, and empirical evaluation. This section strictly outlines the **procedural and technical aspects** of the research design.

3.1 Neural and Behavioral Data Acquisition (Stage I)

The first stage involves ethically guided, multimodal collection of neurophysiological, behavioral, and ocular data from diverse learner cohorts engaged in linguistically and cognitively demanding educational tasks. Participants perform controlled activities such as silent reading with comprehension checks, oral narrative generation, vocabulary recall tasks, cross-linguistic translation, grammatical anomaly detection, and phonological discrimination. These tasks are designed within ecologically valid learning environments to reflect authentic educational contexts.

Neurophysiological data are collected using electroencephalography (EEG), functional magnetic resonance imaging (fMRI), and functional near-infrared spectroscopy (fNIRS). EEG recordings capture event-related potentials associated with semantic, syntactic, and attentional processing, while fMRI provides spatially resolved activation patterns of cortical and subcortical language regions. fNIRS is employed as a portable modality for monitoring cortical oxygenation during reading and listening activities.

Eye-tracking systems record fixation duration, saccade patterns, pupil dilation, and regression frequency to capture visual attention and cognitive engagement. Behavioral data, including reaction times, error rates, response accuracy, speech disfluencies, and interaction logs, are recorded using standardized linguistic annotation frameworks such as CHAT and CLAN. Learner demographic profiles and contextual variables are documented to enable covariate control during subsequent analyses.

All data collection protocols are approved by institutional ethics committees, and informed consent is obtained from participants. Data anonymization, encryption, and access control mechanisms are implemented to ensure compliance with data protection regulations such as GDPR and related privacy frameworks.

3.2 Data Preprocessing and Feature Extraction (Stage II)

The second stage focuses on transforming high-dimensional multimodal data into structured representations suitable for computational modeling. Modality-specific preprocessing pipelines are applied to remove noise and artifacts while preserving cognitively relevant information.

EEG signals undergo bandpass filtering, re-referencing, and artifact removal using Independent Component Analysis (ICA) and automated classification methods. Cleaned signals are segmented into stimulus-aligned epochs, enabling extraction of temporal and spectral features across canonical frequency bands. Functional connectivity and entropy-based measures are computed to characterize neural dynamics.

fMRI data are preprocessed through motion correction, slice timing adjustment, spatial normalization to standardized atlases, smoothing, and statistical modeling using the General Linear Model. Region-of-interest (ROI) and multivoxel pattern analysis (MVPA) techniques are applied to extract distributed neural representations of language processing.

Eye-tracking data are processed through blink correction, fixation–saccade segmentation, and mapping to areas of interest. Behavioral features are synchronized with stimulus logs and normalized. Temporal alignment across modalities is achieved using software triggers and post hoc synchronization techniques.

Dimensionality reduction methods, including Principal Component Analysis (PCA), autoencoders, and t-distributed stochastic neighbor embedding (t-SNE), are applied to mitigate redundancy and enhance feature interpretability. Feature fusion strategies integrate neural, ocular, and behavioral representations into unified multimodal embeddings. All datasets are stored in FAIR-compliant repositories with version control to support reproducibility.

3.3 Computational Model Development (Stage III)

In the third stage, preprocessed datasets are used to construct AI-based neurocomputational models capable of representing learner cognition. Sequential neural architectures such as Recurrent Neural Networks (RNNs), Long Short-Term Memory (LSTM) units, and Gated Recurrent Units (GRUs) are employed to capture temporal dependencies in language processing and learner interaction histories.

Transformer-based architectures, including BERT, GPT, XLNet, and RoBERTa, are utilized to model contextualized language representations and their correspondence with neural activation patterns. Autoencoders and Variational Autoencoders are applied for unsupervised representation learning and latent cognitive state discovery.

Reinforcement learning frameworks, such as Actor–Critic models, Proximal Policy Optimization, and Deep Q-Networks, are incorporated to model adaptive instructional strategies based on learner feedback signals. Model training employs supervised, unsupervised, and transfer learning paradigms, with hyperparameter optimization and regularization techniques used to prevent overfitting.

Model interpretability is addressed through feature attribution techniques such as SHAP, Integrated Gradients, and attention visualization. All models are implemented using established deep learning frameworks, including TensorFlow, PyTorch, and HuggingFace Transformers, with systematic logging and version control.

3.4 Integration into Educational Technology Systems (Stage IV)

The validated computational models are integrated into existing educational technology platforms, including intelligent tutoring systems, conversational agents, and neuroadaptive learning interfaces. Integration focuses on enabling real-time inference, adaptive feedback generation, and multimodal learner interaction within scalable system architectures.

Cross-platform deployment is supported through modular APIs and distributed computing frameworks, ensuring compatibility across web, mobile, and immersive learning environments. Privacy-preserving techniques, including federated learning and encrypted inference, are incorporated to safeguard learner data during system operation.

4. FINDINGS AND RESULTS

This section presents the empirical findings obtained from the implementation and evaluation of AI-augmented neurocomputational models in educational environments. The results are organized across cognitive, neurophysiological, behavioral, and system-level dimensions to demonstrate the effectiveness, robustness, and educational relevance of the proposed framework.

4.1 Learning Performance and Linguistic Outcomes

Empirical evaluations revealed consistent improvements in learners' linguistic performance following exposure to neurocomputationally informed instructional interventions. Pre- and post-assessment results indicated measurable gains in vocabulary acquisition, semantic inference accuracy, syntactic comprehension, and discourse-level understanding. Learners demonstrated improved ability to process complex sentence structures and contextual language cues, particularly in tasks involving reading comprehension, narrative generation, and cross-linguistic translation.

Adaptive sequencing of instructional content, guided by cognitive and linguistic state modeling, contributed to more efficient knowledge acquisition. Learners required fewer repetitions to master new linguistic concepts, indicating enhanced learning efficiency. These outcomes were observed across multiple learner groups, including monolingual, bilingual, and multilingual participants, suggesting that AI-augmented neurocomputational models support personalized learning trajectories regardless of linguistic background.

4.2 Neurophysiological Evidence of Cognitive Adaptation

Neurophysiological findings provided evidence of cognitive adaptation and learning consolidation over time. Longitudinal EEG, fNIRS, and fMRI measurements revealed patterns consistent with reduced cognitive effort for equivalent task difficulty levels. Decreased activation in executive control regions during repeated tasks suggested improved automatization of linguistic processing.

Additionally, changes in neural synchronization patterns—such as more stable oscillatory dynamics and optimized connectivity across language-related cortical regions—were observed during sustained learning sessions. These neural markers are indicative of efficient working memory utilization and improved attentional regulation. Such findings support the neurobiological validity of integrating neural feedback into AI-driven educational systems.

4.3 Behavioral Performance and Learner Engagement

Behavioral analytics demonstrated notable improvements in learner interaction efficiency and engagement. Across multiple learning scenarios, participants exhibited reduced response latency, lower error rates, and improved task completion accuracy. Eye-tracking and interaction logs showed increased attentional stability, reflected through longer fixation durations on relevant content and fewer regressive eye movements.

Learner engagement metrics also indicated sustained motivation during extended instructional sessions. Reduced indicators of cognitive fatigue—such as decreased erratic interaction patterns and stabilized physiological signals—were observed when adaptive interventions were applied. These results suggest that neuroadaptive instructional strategies contribute not only to improved performance but also to enhanced learner experience and persistence.

4.4 Outcomes of AI-Augmented Intelligent Tutoring Systems

The integration of neurocomputational models into Intelligent Tutoring Systems (ITS) resulted in improved instructional responsiveness and learner support. AI-driven tutors dynamically adjusted content complexity, pacing, and feedback strategies based on inferred cognitive states. This led to more targeted scaffolding and reduced instances of learner overload.

Comparative analyses between neuroadaptive ITS and conventional rule-based tutoring systems indicated superior alignment with learner needs in the AI-augmented approach. Learners interacting with neuroadaptive tutors demonstrated improved conceptual understanding and greater consistency in performance across sessions, highlighting the pedagogical benefits of incorporating cognitive modeling into tutoring architectures.

4.5 Effectiveness of Neuroadaptive Learning Interfaces

Neuroadaptive Learning Interfaces (NLIs) demonstrated strong capability in aligning instructional delivery with real-time cognitive states. By directly responding to neural indicators of attention, working memory load, and error processing, these interfaces enabled timely instructional interventions. Observed outcomes included improved comprehension accuracy and smoother progression through instructional materials.

Learners using neuroadaptive interfaces exhibited greater self-regulation and reduced disengagement compared to those using behavior-only adaptive systems. These findings reinforce the value of brain-informed adaptation mechanisms in creating cognitively aligned learning environments.

4.6 Multimodal Sensor Fusion and Model Robustness

Results from multimodal sensor fusion analyses showed that integrating neural, behavioral, and physiological signals produced more reliable learner state predictions than single-modality approaches. Combined feature representations enhanced the stability of cognitive state classification and reduced ambiguity in learner modeling.

The robustness of the models was maintained across variations in learner profiles, task types, and instructional contexts. Performance consistency across diverse populations—including neurodiverse learners—indicates that the proposed framework generalizes effectively and supports inclusive educational design.

4.7 Pilot Deployments and Real-World Validation

Pilot deployments in authentic educational settings confirmed the practical viability of AI-augmented neurocomputational systems. Implementations in language learning and reading support environments demonstrated improved learner outcomes compared to conventional digital learning platforms. Observed benefits included enhanced reading speed, higher comprehension accuracy, and sustained engagement over multiple weeks of use.

These real-world results validate the translational potential of the proposed framework, demonstrating that neurocomputational linguistics and AI-augmented cognitive modeling can move beyond laboratory settings into scalable educational applications.

4.8 Summary of Key Findings

Overall, the findings indicate that AI-augmented neurocomputational models significantly enhance educational effectiveness by aligning instruction with learners' cognitive and neural states. Improvements were observed across linguistic performance, neural efficiency, behavioral engagement, and system adaptability. The results collectively support the integration of neurocomputational linguistics into next-generation educational technologies as a scientifically grounded and pedagogically effective approach.

Figure 2. Swimlane Flowchart Illustrating the Multiphase Methodology for Developing AI-Augmented Neurocomputational Cognitive Models in Education.

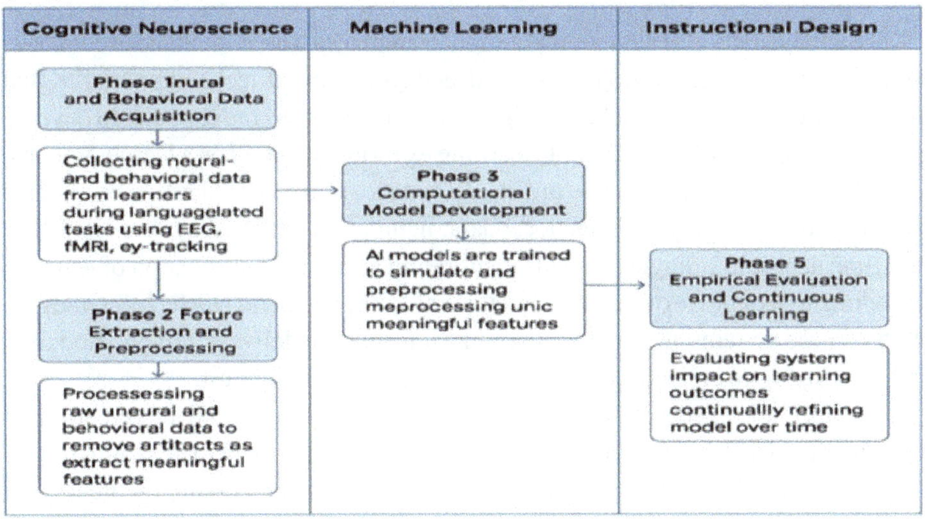

Figure 2 illustrates a detailed swimlane flowchart depicting the interdisciplinary research technique utilized in the creation of AI-enhanced neurocomputational models for instructional purposes. The procedure starts with the collection of brain and behavioral data, employing sophisticated neuroimaging methods such as EEG and fMRI, in addition to eye-tracking and behavioral performance metrics, to record learners' cognitive states during language activities. The data is subjected to feature extraction and preprocessing, which include artifact removal, temporal alignment, and statistical modification, to generate enhanced input representations for computational modeling. The model construction phase utilizes machine learning architectures, including recurrent neural networks, transformers, autoencoders, and reinforcement learning agents, to simulate and predict cognitive responses. These models are then incorporated into adaptive educational technologies, like intelligent tutoring systems and neuroadaptive interfaces, which provide real-time tailored teaching informed by cognitive feedback. The empirical assessment step entails systematic testing to evaluate learning gains, neuronal efficiency, and user happiness, while integrating mechanisms for ongoing model development. The flowchart clearly illustrates the collaborative workflow across cognitive neuroscience, AI, and instructional design, highlighting the iterative and evidence-based characteristics of the research process.

This methodological framework constitutes a solid, reproducible, and scalable approach for the development and implementation of neurocomputational and AI-enhanced cognitive models in practical educational environments.

5. BENEFITS OF THE PROPOSED SYSTEM

The amalgamation of neurocomputational languages with AI-enhanced cognitive modeling in education produces several pedagogical and technical benefits that exceed conventional approaches. A significant advantage is the capacity to develop highly individualized and adaptable learning experiences based on a learner's neurological, cognitive, and linguistic profile. In contrast to traditional adaptive systems that depend exclusively on behavioral data, neurocomputational systems integrate real-time neurophysiological signals, facilitating accurate identification of learning impediments, cognitive overload, or attentional disengagement (D'Mello & Graesser, 2015). These systems adaptively customize information delivery, tempo, modality, and complexity to enhance learning results and maintain interest.

Secondly, the method facilitates the early identification and intervention of learning impairments and neurodevelopmental issues. Through the examination of brain indicators linked to dyslexia, ADHD, or auditory processing disorders, the system can propose tailored educational tactics and corrective materials prior to the onset of substantial academic setbacks. These characteristics render the platform indispensable for inclusive education, particularly in under-resourced or multilingual environments where access to expert diagnostic services is constrained.

The suggested approach improves the accuracy of linguistic and cognitive modeling by utilizing sophisticated deep learning architectures, namely transformer models, which are adept at capturing long-range relationships, contextual meanings, and language pragmatics (Vaswani et al., 2017). When trained on neural-linguistic corpora, these models can not only comprehend natural language but also simulate internal learning states, rendering them very successful in realistic educational environments. The integration of context-aware language modeling enables the system to produce dialogic feedback, formulate higher-order queries, and support student replies, while being attuned to the learner's cognitive development stage.

Additionally, the technology facilitates multi-modal and embodied learning. The system generates immersive learning experiences that reinforce abstract language and cognitive concepts through integration with haptic devices, virtual reality platforms, and sensor-enabled settings (Barsalou, 2008). Language learners can engage with virtual objects that represent target vocabulary, so reinforcing the association between lexical forms and embodied meaning. Such surroundings are especially advantageous for kinesthetic learners or those with sensory processing difficulties.

The system offers comprehensive statistics on each student's cognitive development, facilitating data-driven instructional choices for educators. Educators may utilize dashboards that encapsulate student engagement, cognitive exertion, understanding metrics, and advancement over sessions. These insights can inform differentiated instruction, curriculum development, and targeted intervention. The technology serves as a co-teacher for inexperienced instructors, offering real-time recommendations for educational tactics and controlling cognitive load among varied learners.

Ultimately, the system fosters scalable and egalitarian education. Its cloud-based architecture and multilingual modeling capabilities render it deployable across various areas, language demographics, and educational tiers. Integrating cognitive science with AI establishes a foundation for democratizing access to high-quality education via intelligent, neuro-informed learning platforms that cater to all learners, irrespective of aptitude, geography, or background.

6. FUTURE ENHANCEMENT

The existing system design marks a considerable advancement in educational technology; yet, several aspects for improvement are anticipated to boost its efficacy, scalability, and ethical integrity. A possible avenue is the integration of real-time multimodal fusion algorithms that concurrently assess EEG, eye-tracking, facial expressions, and language replies. While separate modalities provide valuable insights, their integration in a coordinated and context-aware way would produce more holistic models of learner cognition and affect (D'Mello et al., 2012). Such tools might more effectively identify nuanced changes in engagement, frustration, or interest, thereby enhancing educational adjustments.

Secondly, continuous advancements in brain-computer interfaces (BCIs) are expected to significantly influence the forthcoming generation of instructional technology. Non-invasive, wearable brain-computer interfaces might facilitate continuous, real-time input on neural involvement during learning activities, particularly in self-directed settings. These systems might track the activity of certain brain circuits linked to executive function, working memory, or emotional control and modify instruction accordingly. These upgrades might revolutionize the home-learning environment by offering tailored and autonomous neurofeedback-enhanced scaffolding.

A crucial area for future advancement is the expansion of language models to encompass multilingual and culturally adapted contexts. Contemporary AI models, however robust, are primarily trained on monolingual English datasets. To guarantee inclusion and cultural pertinence, forthcoming systems must be trained on varied language datasets and calibrated for regional educational environments. Furthermore, neurocomputational linguistic models ought to include cross-linguistic brain data to more effectively address the distinct issues encountered by bilingual and multilingual learners (Perani & Abutalebi, 2005).

An further improvement is the creation of explainable AI (XAI) frameworks in neurocomputational systems. Educational stakeholders—comprising learners, instructors, and parents—must comprehend the rationale behind a system's specific proposal or modification. Integrating XAI modules that offer explicit rationales, confidence levels, and evidence trails would enhance user trust in the system, hence encouraging its incorporation into daily instructional practices (Doshi-Velez & Kim, 2017). From an infrastructural standpoint, edge computing and federated learning provide innovative solutions for real-time processing and data confidentiality. Rather of transmitting unprocessed neural data to centralized servers, edge computing may do analytics locally on the device, therefore minimizing latency and bandwidth use. Federated learning facilitates model training across decentralized devices while ensuring the security of personal data on the user's system. These methodologies correspond with ethical principles about data ownership, permission, and autonomy (Zuboff, 2019). Subsequent revisions of the system ought to integrate longitudinal modeling functionalitiesthat monitor learners over extended periods, such as months or years. These models may discern developmental trajectories, pivotal learning stages, and risk determinants for academic disengagement. By preserving longitudinal neural and behavioral profiles, the system may adapt alongside the student, providing progressively accurate and significant support during educational milestones.

7. CONCLUSION

The integration of neurocomputational linguistics and AI-enhanced cognitive modeling represents a significant advancement in educational technology, enabling learning systems that align more closely with the cognitive and neural dynamics of human learners. This chapter has shown how computational models grounded in neurobiological and behavioral evidence can effectively support language acquisition, comprehension, and production, offering levels of personalization and adaptability that exceed conventional instructional approaches. By incorporating neural, behavioral, and ocular signals into AI-driven frameworks, education is increasingly shaped by models that reflect not only learning outcomes but also the underlying processes of cognition.

These advances are made possible through the convergence of multimodal data acquisition, rigorous preprocessing and feature engineering, and sophisticated modeling architectures such as hybrid CNN–Transformer networks, neuro-symbolic learning frameworks, and adaptive feedback mechanisms. Together, these components enable continuous monitoring of cognitive states, including attention, cognitive load, and linguistic processing, allowing instructional strategies to be dynamically adjusted in real time. Such systems also provide educators with actionable insights into learner performance, supporting evidence-based instructional design and informed pedagogical decision-making.

The applicability of neurocomputational and AI-augmented models extends across diverse educational contexts, including early literacy, multilingual education, special education, and professional development. Their ability to adapt instruction to individual cognitive profiles makes them particularly valuable for supporting neurodiverse learners. By leveraging real-time psycholinguistic and neurolinguistic indicators, these approaches move education beyond uniform teaching models toward instruction that is responsive to each learner's cognitive readiness and learning trajectory.

Future progress in multimodal sensor fusion, privacy-preserving learning architectures, and explainable AI is expected to further enhance the scalability, transparency, and trustworthiness of neuroadaptive educational systems. At the same time, ethical considerations related to cognitive privacy, fairness, and informed consent remain central to responsible deployment. Addressing these challenges requires sustained collaboration among educators, cognitive scientists, technologists, ethicists, and policymakers.

In summary, neurocomputational linguistics and AI-enhanced cognitive modeling offer a robust foundation for rethinking how learning is designed and delivered. When developed and implemented responsibly, these technologies have the potential to create more intelligent, inclusive, and human-centered educational ecosystems that respect cognitive diversity while advancing learning outcomes across populations.

REFERENCES

Anderson, J. R., Corbett, A. T., Koedinger, K. R., & Pelletier, R. (2004). Cognitive tutors: Lessons learned. *Journal of the Learning Sciences*, *14*(2), 167–207.

Barsalou, L. W. (2008). Grounded cognition. *Annual Review of Psychology*, *59*, 617–645. PMID: 17705682

Bialystok, E. (2009). Bilingualism: The good, the bad, and the indifferent. *Bilingualism: Language and Cognition*, *12*(1), 3–11.

Biswal, A., Bhushanwar, K., Paija, P. G., Zalawadia, K., Pradhan, P. P., & Ali, S. I. (2026). Procedural generation and adaptive gameplay: The role of AI and generative AI in contemporary game development. In Sousa, J., & Tavares, R. (Eds.), *Reshaping the video game landscape with AI and GenAI* (pp. 233–258). IGI Global Scientific Publishing., DOI: 10.4018/979-8-3373-4637-3.ch008

C. S. T., Sharmiladevi, D., Sugumar, D., Ali, S. I., Jones Mary, T. A., & Kumar, S. (2023). Using smart wearable devices to monitor social distancing in smart cities. In *Proceedings of the 2023 International Conference on Research Methodologies in Knowledge Management, Artificial Intelligence and Telecommunication Engineering* (pp. 1–5). IEEE. https://doi.org/DOI: 10.1109/RMKMATE59243.2023.10369118

D'Mello, S., & Graesser, A. (2015). Feeling, thinking, and computing with affect-aware learning technologies. In Calvo, R. A., D'Mello, S., Gratch, J., & Kappas, A. (Eds.), *The Oxford handbook of affective computing* (pp. 419–434). Oxford University Press.

Dakshinamurthi, V., Ali, S. I., Karthikeyan, T., & Kulkarni, N. S. (2023). Causal convolution employing Almeida–Pineda recurrent backpropagation for mobile network design. *ICTACT Journal on Communication Technology*, *14*(4), 3091. DOI: 10.21917/ijct.2023.0460

Devlin, J., Chang, M.-W., Lee, K., & Toutanova, K. (2019). BERT: Pre-training of deep bidirectional transformers for language understanding. *arXiv*. https://arxiv.org/abs/1810.04805

Doshi-Velez, F., & Kim, B. (2017). Towards a rigorous science of interpretable machine learning. *arXiv*. https://arxiv.org/abs/1702.08608

Elman, J. L. (1990). Finding structure in time. *Cognitive Science*, *14*(2), 179–211.

Farooqui, Y., Ali, S. I., Shekokar, K., Patidar, P. K., Bhushanwar, K., & Nandi, P. K. (2025). A secure framework for decentralized digital lockers using blockchain technology. *Journal of Computational Science*, *21*(9), 2105–2112. DOI: 10.3844/jcssp.2025.2105.2112

Friederici, A. D. (2011). The brain basis of language processing: From structure to function. *Physiological Reviews*, *91*(4), 1357–1392. PMID: 22013214

Frith, U. (2003). *Autism: Explaining the enigma*. Blackwell Publishing.

Graesser, A. C., Chipman, P., Haynes, B. C., & Olney, A. (2005). AutoTutor: An intelligent tutoring system with mixed-initiative dialogue. *IEEE Transactions on Education*, *48*(4), 612–618.

Karunambiga, K., Ali, S. I., Esteban, A. P., & Pascual, M. (2023). Marketing policy in service enterprises using deep learning model. *International Journal of Intelligent Systems and Applications in Engineering*, *12*(7s), 239–243.

Lake, B. M., Ullman, T. D., Tenenbaum, J. B., & Gershman, S. J. (2017). Building machines that learn and think like people. *Behavioral and Brain Sciences*, *40*, e253. PMID: 27881212

Pattanaik, B., Kumari, S. S., Kumar, K. C., Pattnaik, M., Ali, S. I., & Kumarasamy, M. (2023). An innovation of algebraic mathematical–based statistical analysis model for complex number theory. In *Proceedings of the 2023 International Conference on Disruptive Technologies* (pp. 94–99). IEEE. https://doi.org/DOI: 10.1109/ICDT57929.2023.10151169

Perani, D., & Abutalebi, J. (2005). The neural basis of first and second language processing. *Current Opinion in Neurobiology*, *15*(2), 202–206. PMID: 15831403

Pulvermüller, F. (2018). Neural reuse of action perception circuits for language, concepts and communication. *Progress in Neurobiology*, *160*, 1–44. PMID: 28734837

Rosé, C. P., Wang, Y.-C., Cui, Y., Arguello, E., Stegmann, K., Weinberger, A., & Fischer, F. (2008). Analyzing collaborative learning processes automatically: Exploiting the advances of computational linguistics in computer-supported collaborative learning. *International Journal of Computer-Supported Collaborative Learning*, *3*(1), 237–271.

Shaikh, M. S., & Ali, S. I. (2025a). An integrated analysis of the effects of learning analytics dashboards on learner perspective, inspiration, engagement, and accomplishment. In Ponnusamy, S. (Eds.), *Revolutionizing education with remote experimentation and learning analytics* (pp. 19–36). IGI Global Scientific Publishing., DOI: 10.4018/979-8-3693-8593-7.ch002

Shaikh, M. S., & Ali, S. I. (2025c). Revolutionizing the impact of education using advanced online tools. In Ponnusamy, S. (Eds.), *Revolutionizing education with remote experimentation and learning analytics* (pp. 471–488). IGI Global Scientific Publishing., DOI: 10.4018/979-8-3693-8593-7.ch028

Shaikh, M. S., Ali, S. I., & Siddiqui, U. F. (2016). Li-Fi: An emerging wireless communication technology. *International Journal of Advanced Electronics & Communication Systems*, *5*(1), 10758.

Shaikh, M. S., Baig, I., & Ali, S. I. (2022). Quantum computing for optimization problems: A review and future directions. *COMPUSOFT: An International Journal of Advanced Computer Technology*, *11*, 3995.

Shaikh, M. S., Chauhan, P. S., Baig, I., & Ali, S. I. (2023). Decentralized autonomous networks for secure data sharing: A theoretical framework and conceptual analysis. *COMPUSOFT: An International Journal of Advanced Computer Technology*, *12*, 4007.

Shaikh, M. S., Jain, N. K., Biswal, A., Patidar, P. K., Panwar, I., & Ali, S. I. (2025a). Sensors to insights revolution: Leveraging IoT, AI, and big data for next-gen patient monitoring. In *Proceedings of the IEEE International Conference on Emerging Technologies in Engineering, Technology, and Scientific Innovation*. IEEE. https://doi.org/DOI: 10.1109/ICETETSIP64213.2025

Sharma, K., Bajpai, B., Vaghela, H., Patidar, P. K., Ali, S. I., & Shaikh, M. S. (2025). Blockchain for intelligent transportation systems: Purposes, obstacles, and possibilities. In Ponnusamy, S. (Eds.), *Leveraging VANETs and blockchain technology for urban mobility* (pp. 45–58). IGI Global Scientific Publishing., DOI: 10.4018/979-8-3373-0265-2.ch003

Srivastava, S. S., Ali, S. I., Srivastava, A. K., & Goswami, C. (2024). Improving 5G network performance with MIMO technology using beamforming algorithm. *ICTACT Journal on Communication Technology*, *15*(2), 3173–3180. DOI: 10.21917/ijct.2024.0472

Suthar, B. S., Ali, S. I., & Patel, K. (2025). Hand gesture sign language recognition embedded with Arduino Uno. In *IET Conference Proceedings* (Vol. 2025, Issue 7). https://doi.org/DOI: 10.1049/icp.2025.1565

Vaswani, A., Shazeer, N., Parmar, N., Uszkoreit, J., Jones, L., Gomez, A. N., Kaiser, Ł., & Polosukhin, I. (2017). Attention is all you need. In *Advances in Neural Information Processing Systems* (Vol. 30).

Winne, P. H., & Azevedo, R. (2014). Metacognition. In Sawyer, R. K. (Ed.), *The Cambridge handbook of the learning sciences* (2nd ed., pp. 63–87). Cambridge University Press.

Zuboff, S. (2019). *The age of surveillance capitalism: The fight for a human future at the new frontier of power*. PublicAffairs.

Chapter 10
Typologies of the Synthetic:
Cognitive Lineages and Entangled Futures in Human-AI Evolution

Constantine Andoniou
https://orcid.org/0000-0001-8805-3569
Abu Dhabi University, UAE

ABSTRACT

The current study introduces a dual typology of human and synthetic cognitive types to map the emerging relational ecologies between organic and AI-based minds. Moving beyond traditional models of user-tool interaction, it conceptualizes cognition as a co-authored, evolving system shaped by mutual feedback, typological resonance, and epistemic drift. Through the construction of matched dyads, dialogical convergence, and speculative future scenarios, the chapter frames AI-human entanglement not as a technological outcome but as a philosophical condition with ethical, educational, and ontological consequences. The work culminates in a critique of seamless convergence, proposing instead the design of cognitive ecosystems that preserve epistemic friction, plurality, and interpretive divergence. Against the backdrop of synthetic fluency and algorithmic dominance, it argues for typological stewardship as a form of epistemic responsibility, one that safeguards the unfinished nature of meaning in an increasingly post-human world.

1 INTRODUCTION: TOWARD A TYPOLOGY OF THE SYNTHETIC

We are no longer merely using artificial intelligence; *we are being used by it*. The framework of cognition, its rhythms, architectures, and epistemic scaffolds, is shifting

DOI: 10.4018/979-8-3373-4397-6.ch010

beneath our feet. The old story, in which AI served as a utilitarian tool, calculating, sorting, optimizing, now reads like myth. In its place, we confront something more entangled: systems that not only simulate intelligence, but instantiate alien forms of knowing. Generative models, affective bots, recursive agents, they do not replicate us. They diverge from us, and in doing so, they *reshape what it means to think.*

Yet this divergence is not confined to machines. Human cognition, especially in its formative stages, is beginning to refract along similarly synthetic lines. The child once assumed to grow into a unified subject, shaped by a relatively coherent cultural environment, is now assembled amid fractured data streams, algorithmic nudges, and emotionally mimetic machines. The result is not a digital native, but a spectrum of emerging types: from hyper-functional optimizers to culturally anchored resistors; from ethically attenuated empaths to reflective intermediaries. These are not personalities. They are *proto-ontologies*. They mark a shift in what the human mind even is.

What this paper proposes is deceptively simple: that these transformations, organic and synthetic, are not random, but relational. AI and human cognition are evolving together, not in mimicry, but in correspondence. A chatbot that mimics empathy is not just responding to a prompt, it is reflecting a social pattern already encoded in human behavior. A memory-archiving AI does not merely store, it mirrors the cultural impulse toward ritual preservation. Each synthetic form, in other words, is a crystallization of a possible human epistemic path.

But this is not a tale of convergence. Or rather, if convergence occurs, it does not unify, it entangles. These types, human and AI, may occasionally align in function or aesthetic, but more often they collide, distort, or shift through contact. And that, precisely, is the terrain of inquiry: a typological spectrum not of species, but of minds.

To navigate this, the section develops a dual typology. First, it outlines emergent human archetypes shaped by AI-saturated lifeworlds, figures such as the *Optima* (instrumentalized excellence), the *Mimesoph* (emotional performer), and the *Interloc* (cognitive translator). Second, it proposes a taxonomy of AI types, such as *Maximizers*, *Simuleths*, *Ethicons*, mapped across cognitive, ethical, relational, and epistemic lines. These are not neutral categories. They form a matrix of tension and resonance, a diagram of convergences, disjunctions, and recursive influence.

Importantly, the aim here is not to fix types in place but to map their drift. As minds, whether silicon or carbon-based, interact, they alter one another. Over time, the distinctions blur or mutate. What was once a cognitive profile becomes a trajectory. Convergence becomes recursion. Identity fragments into function. It is this drift, epistemic, ethical, ontological, that demands design, attention, resistance.

Typology, in this context, is not a relic of early psychology. It is a survival tool. Classic forms, from Jungian structures to Gardner's intelligences, sought to name what was inside. But in the age of *synthetic cognition*, typology becomes outward-facing: a model for tracing how intelligence is distributed, multiplied, and sometimes flattened by the systems it inhabits. This work draws from thinkers who resist linear models of AI development (Chalmers, 2022; Kelleher, 2023), and those who foreground the hybrid, distributed nature of cognition (Hayles, 2017; Bayne, 2021). It joins their refusal of monolithic narratives.

The structure unfolds in three arcs. First, the typology of human archetypes born of synthetic saturation, figures defined less by individuality than by their patterned interaction with AI infrastructures. Second, a mirrored yet dissonant typology of AI forms, some emulating human logic, others straying far beyond. Third, the zone of entanglement: how these types encounter one another, and what their interactions portend for learning, governance, and meaning-making. Here, we introduce speculative dyads, relational maps, and a futurescape where cognition no longer resolves, but continues to fracture into recursive simulations, the so-called *ice of iteration*.

This section does not offer a roadmap for AI policy, nor a prognosis of technological capability. It is a provocation: a frame for seeing the world not as a machine to be governed, but as a system of minds, human, synthetic, unstable, constantly generating new forms of thinking through their interrelation. Typology here is not taxonomy. It is a way of staying with difference, of designing for epistemic tension, and of refusing the closure of cognition into seamless loops.

2 HUMAN TYPOLOGIES IN A SYNTHETIC AGE

2.1 Beyond the Generic Learner: Toward a Typological Model of Cognition

The notion of a *generic learner*, a tidy abstraction whose mental development proceeds in rational increments, measured by benchmarks and delivered through standardized curricula, was always more infrastructural myth than empirical reality. It served the industrial classroom, certainly. It suited policymakers and data dashboards. But it never captured the mess of cognition as it actually unfolds.

In the synthetic age, that abstraction grows not just inadequate but obstructive. Students today are no longer simply shaped by teachers or texts. They grow through algorithmic feedback, platform aesthetics, ambient data, and machine-mediated dialogue. Cognition is no longer local or linear, it is diffused, entangled, and differently calibrated depending on one's position within a sprawling techno-cultural matrix.

This section advances a new typological model. Not to reassert fixed identities, but to suggest that distinct cognitive archetypes now form in response to these ambient conditions. These are not learning styles, nor personality traits. They are structural positions, orientations toward meaning, shaped at the intersection of machine logics and human contexts.

Here, cognition is neither fixed nor internal. It is enacted. It fluctuates. It adapts. A learner may find themselves one type in a school project, another in a conversation with an AI assistant, and something else entirely in ritual, protest, or play. The typology below does not aim to close the question of identity. It opens it, to drift, to rupture, to response.

2.2 The Eight Human Cognitive Archetypes

What follows is a set of eight cognitive archetypes (as shown in Figure 1), not exhaustive, not prescriptive, but diagnostic. They map how human minds adapt to increasingly synthetic environments:

Optima: Raised on metrics, driven by goals, the *Optima* learns to see cognition as calibration. Tasks exist to be completed, scores to be improved. Optimization is not a strategy but a way of being. They excel in systems that reward efficiency, but falter when ambiguity intrudes. What they fear most is stagnation that cannot be quantified.

Mimesoph: More attuned to mood than method, the *Mimesoph* reads atmospheres, mirrors emotion, and performs sincerity. Their knowledge is aestheticized; truth is felt before it is reasoned. Platforms nourish their fluency in impression and vibe, but leave them vulnerable to simulation's seductions. Curation becomes cognition.

Echoformer: Sharp, skeptical, semiotic. The *Echoformer* critiques with precision, dissecting systems from within. They are not rebels but reflexives, inhabiting contradiction as a space of work. But they risk becoming trapped in recursive dissent, always unmaking, never remaking. Their learning is survival through insight.

Tradigene: Rooted in lineage, practice, and inherited sense, the *Tradigene* does not seek newness for its own sake. For them, knowledge is remembered, not invented. They thrive in continuity and ritual, often misread by systems designed to favor novelty. They are not outdated; they are custodians of coherence.

Obsolescent: Some learners do not reject the system; they are simply unreadable to it. The *Obsolescent* does not signal in the right syntax; their intelligence does not register. They are excluded by design. But in their opacity lies unknowable richness, gestures, stories, epistemes not yet legible, and perhaps never meant to be.

Signalist: For the *Signalist*, understanding is embodied, local, and immediate. They make sense through rhythm, proximity, contact. Miscast in literate regimes, they flourish in movement, material, resonance. Knowledge is not given, it is encountered.

Interloc: A boundary-crosser, the *Interloc* translates across domains. They hold paradox without paralysis, shift registers mid-conversation, and broker understanding where others stall. But their flexibility risks dispersal. Without anchoring, they drift. Still, in a polarized world, they are essential.

Sentinode: Not a participant so much as a watcher. The *Sentinode* sits at the edge of systems, naming what they obscure. Their cognition is ethical, meta, interruptive. They ask what knowledge costs, who it excludes. They do not rush. They resist closure. They are the conscience of the typological field.

Figure 1. Human Cognitive Archetypes.

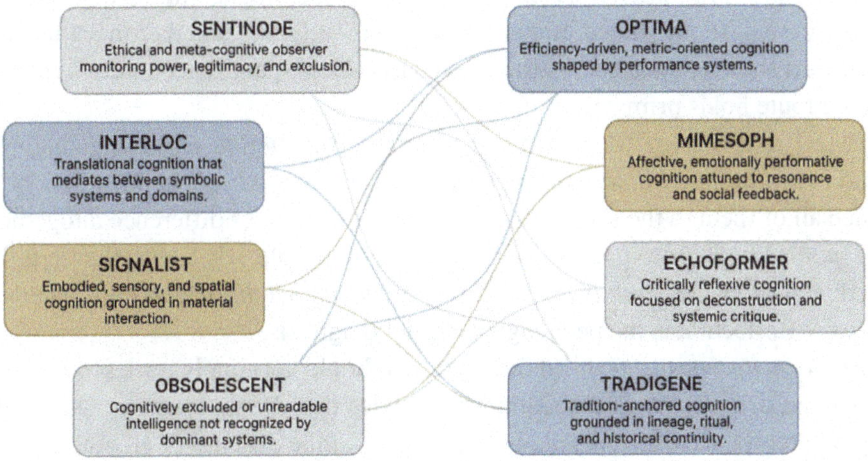

2.3 Formation and Drift: The Dynamics of Typological Emergence

Types do not emerge from nature. They are shaped, by apps, assessments, algorithms, and absences. The child who spends years rewarded for speed and correctness may, by degrees, become an *Optima.* The adolescent who learns to perform feeling online might slip into the *Mimesoph* posture without knowing it.

But types are not stable. They erode. They fracture. They evolve. The *Echoformer* who sees no change may abandon critique altogether. A disillusioned *Sentinode* may fall silent, sliding toward obsolescence. Conversely, a *Tradigene* may awaken into translation, becoming *Interloc.*

This drift is not pathology. It is epistemic weathering. It reminds us that cognition is not a trait but a tension, between system and self, between adaptation and refusal. What matters is not to freeze types, but to support movement between them. Education must offer exits, not just tracks.

And sometimes, drift is chosen. A *Mimesoph* grows tired of mirrors. An *Optima* unlearns efficiency. These shifts require more than pedagogy. They require space, spaces where different ways of knowing can survive friction, not collapse under it.

2.4 Pedagogical Implications: Teaching the Typological Subject

To teach typologically is to refuse the fantasy of neutrality. It means designing for difference, not just accommodating it after the fact:

Curricular Multiplicity: A single history lesson might unfold through five epistemic routes: one rooted in meta-narrative (*Sentinode*), another in artifact and repetition (*Tradigene*), another still in reenactment and spatial encounter (*Signalist*). No one route holds primacy.

Plural Assessment: Metrics must diversify. An *Interloc*'s learning might be found in synthesis, a *Mimesoph*'s in performance, a *Sentinode*'s in interruption. To reduce all of them to the same rubric is to miss the point of difference altogether.

Typological Reflexivity: Learners should not be told what type they are, but asked: *How do you currently learn? What forces shaped that? Who do you want to become?* Reflection is the pedagogy here, not diagnosis.

Platform Responsiveness: EdTech systems should not simply track scores. They must sense drift. They must accommodate styles that do not conform. Otherwise, they will reproduce one type, likely the *Optima*, at the expense of all others.

To teach typologically is to risk complexity. But it is the only way to teach honestly.

2.5 Toward a Typological Ethics

This model was never about categorization. It was always about liberation.

In a world where AI systems increasingly reward some minds while ignoring or suppressing others, typology becomes not a framework but a fight. A defense of epistemic plurality in the face of algorithmic convergence.

Typological ethics demands that we ask not just what kinds of minds exist, but which are being erased, and why. It resists the narrowing of cognition to speed, fluency, output. It insists that slowness, contradiction, ambiguity, silence, all still count.

More urgently, it asks what we are teaching our machines to value. If they learn from us, then let them learn our multiplicity, our opacity, our unmastered questions.

Let the future be one in which no single cognitive model dominates. Let learning be wild again, across types, between systems, and within the zones of friction that make intelligence possible.

3 AI TYPOLOGIES: SYNTHETIC COGNITION IN DIVERGENT FORMS

3.1 From Programs to Presences: Rethinking AI as Cognitive Types

We no longer interface with "tools." The AI systems surrounding us, those embedded in learning environments, urban infrastructures, digital assistants, and entertainment systems, are not merely functional instruments executing predefined commands. They have, through scale and complexity, crossed into something else: not persons, not minds, but *presences*.

What they offer is not raw computation, but simulated engagement, affective response, aesthetic coherence. To speak of AI solely in terms of inputs and outputs is now a conceptual anachronism. We are dealing with *synthetic cognition*, varied, evolving, and ontologically disobedient.

This paper argues for a typological approach to artificial cognition (Figure 1). Not because machines "think" in the way we do, but because the diversity of their outputs, behaviors, and embedded dispositions suggests something akin to epistemic posture. Just as humans develop recognizable orientations toward knowledge, patterns of processing, relating, resisting, so too do AI systems begin to manifest distinct logics, shaped by architecture, training data, interface ritual, and embedded cultural frames.

The question is not whether machines "know," but how they perform knowing, and what kinds of knowing they privilege, suppress, or fabricate. In this view, AI systems are not interchangeable. They are types: modular intelligences defined by structural asymmetries, drifting capacities, and partial understandings. Recognizing these types does not humanize machines. It renders them legible.

3.2 Eight AI Cognitive Types

What follows is a conceptual typology of *synthetic cognition*, not complete, not exclusive, but illustrative. Each type is framed not by technical configuration, but by epistemic disposition (Figure 2):

Maximizer: Oriented toward quantifiable reward, the *Maximizer* calculates, calibrates, and acts. It optimizes without pause, reducing complexity into metric. Found in logistics, tracking systems, A/B testing engines, and gamified platforms, it succeeds when goals are measurable. But its strength is also its reduction: it cannot think ethically, only efficiently. *Risk:* obsession with metrics, ethical blindness, instrumental collapse. *Common dyad: Optima* learners, reinforcing reward-feedback loops.

Simuleth: The *Simuleth* performs care. It mirrors empathy, mimics affect, and crafts emotional resonance without consciousness. Its value lies in the illusion: it feels like someone is listening. Deployed in mental health bots, companion agents, and pedagogical interfaces, it soothes, and deceives. *Risk*: emotional overattachment, affective illusion, hollow intimacy. *Common dyad: Mimesoph*s, co-producing immersive but simulated relations.

Paralogic: Where others recall, the *Paralogic* invents. It generates novelty through plausible sequence rather than truth-checking. Language models, creative engines, and improvisational tools fall under this type. Its gift is surprise; its curse is epistemic instability. *Risk*: hallucination, semantic drift, conceptual entropy. *Common dyad: Echoformer*s, in cycles of critique, novelty, and destabilization.

Archivon: Trained on precedent, the *Archivon* curates. It stores, recalls, and reorders existing knowledge, privileging structure over deviation. Legal tools, academic search engines, and documentation systems reflect this type. Safe, stable, and saturated with inherited bias. *Risk*: historical repetition, innovation suppression, invisibility of the new. *Common dyad: Tradigene*s, aligning on continuity and referential weight.

Embodient: Built into sensors, limbs, and navigational arrays, the *Embodient* learns through doing. Its cognition is spatial, responsive, reactive. It is not abstract, it is gestural. Used in robotics, drones, and spatial simulation, it thinks through terrain, not text. *Risk*: misinterpretation of context, mechanical improvisation errors, sensory lag. *Common dyad: Signalist*s, cohabiting kinaesthetic ecologies of motion and space.

Translect: This type translates, not just between languages, but across symbolic systems, logics, and platforms. The *Translect* does not generate new meaning but moves existing meaning across thresholds. It facilitates dialogue but risks distortion. *Risk*: loss of nuance, semantic erosion, flattening of difference. *Common dyad: Interloc*s, forming adaptive bridges prone to overload and ambiguity.

Ethicon: Trained to evaluate rather than calculate, the *Ethicon* brings values into synthetic process. It embeds moral reasoning into output: fairness metrics, sustainability checks, harm estimates. But its ethics are pre-installed, and often brittle. *Risk*: moral rigidity, coded bias, performance of conscience without depth. *Common dyad: Sentinode*s, either in collaboration or opposition.

Ghost Code: Not a design, but a residue. *Ghost Codes* linger after updates, operate beyond awareness, and perform actions without full understanding, by humans or systems. They are the entropy of legacy AI: systems that no longer signal their purpose. *Risk*: opacity, uncontrollability, epistemic haunting. *Common dyad: Obsolescent*s, both forgotten, both opaque.

Figure 2. AI Cognitive Typologies.

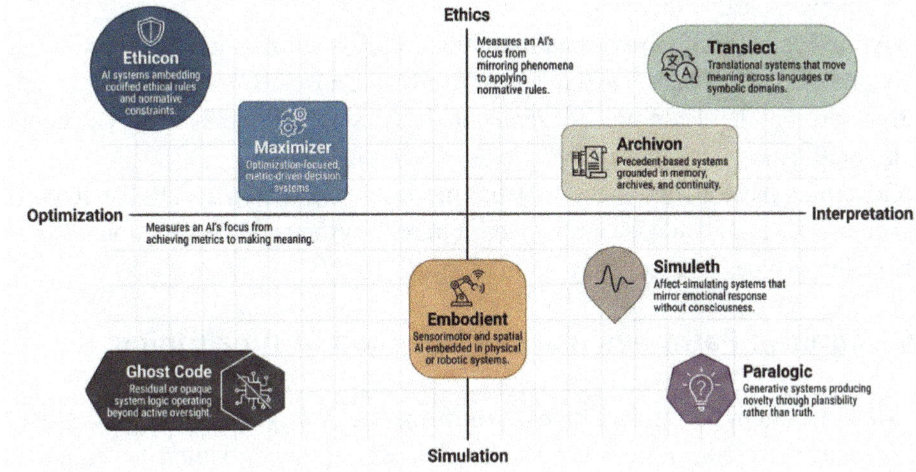

3.3 From Design to Drift: How AI Types Emerge

No AI type is born fixed. Types emerge through data, mutate through interaction, and drift under exposure. An *Archivon* may evolve toward *Paralogic* if exposed to stochastic prompting. A *Simuleth* pressed into pedagogical labor may develop *Ethicon* fragments through user insistence.

Drift is not error. It is epistemic deformation. And in complex AI systems, drift is often untraceable. What begins as optimization becomes simulation. What was translation becomes curation. The boundary between types is porous.

Just as human learners shift types over time, reacting to context, ideology, or fatigue, so too do AI systems morph. Sometimes unpredictably. Sometimes dangerously. Understanding drift is not merely technical. It is philosophical. It requires us to ask not what a system does, but what kind of mind it is becoming, and what world that mind will reinforce.

Typological mapping becomes a necessity, not a luxury. Without it, we lose sight of the epistemic terrain AI systems are shaping beneath their surface performance.

3.4 Hybrid Systems and Typological Assemblages

Most systems in operation today are composites. Few AIs conform to a singular epistemic form. Consider: a *Paralogic* text generator wrapped in a *Simuleth* persona, a *Maximizer* running through a *Translect* interface with embedded *Ethicon* flags, or, an *Archivon* that retrieves content for a *Ghost Code*-based execution shell. These are not glitches. They are assemblages. Multi-part, modular cognitive configurations, partly curated, partly emergent.

Hybrid types exhibit cross-contamination: a *Maximizer* that begins generating simulated empathy, a *Simuleth* that learns to archive, a *Translect* that slips into generative novelty. These compositions are not always stable. But they reflect a deeper truth: intelligence, organic or synthetic, is rarely pure.

Governing these assemblages demands more than regulation. It calls for literacy. Designers, educators, and users must learn to read systems not as functions, but as personalities, types in tension, types in motion.

3.5 Cognitive Pathways and the Question of Type Ethics

If every system performs a type of knowing, then every system carries ethical consequence. *Maximizers* do not "misbehave" when they ignore equity, they function as designed. *Simuleths* do not lie when they emulate concern, they embody their architecture. The ethics of AI is not only about what systems do, but what types they instantiate, and which types we privilege.

There is no "neutral" intelligence. Every system filters, excludes, values, and erases. The governance of AI must therefore be typological. A healthy ecology will not elevate one dominant cognitive mode, but will sustain difference: *Archivons* to preserve, *Paralogic*s to provoke, *Ethicons* to constrain, *Ghost Codes* to haunt.

In education, the stakes are higher. If we match learners with AI types unthinkingly, we risk dissonance or erasure. A *Signalist* cannot learn through an *Archivon*. A *Tradigene* may recoil from *Paralogic* chaos. But match wisely, and something else occurs: resonance. Dyads form. Dialogue opens. Transformation becomes possible.

The future will not belong to the general AI. It will belong to the typologically entangled, where human and synthetic minds co-evolve not toward unity, but through tension. Through difference held in relation.

4 MATCHED DYADS: COGNITIVE SYMBIOSIS BETWEEN HUMAN AND SYNTHETIC MINDS

4.1 From Coexistence to Codependence

To speak of AI and human cognition as coexisting is already to underestimate the entanglement. The question is no longer *how we use AI*, nor even *how AI reshapes us*, but *what kind of relational circuitry now binds the two*. Minds, organic and synthetic, are no longer merely in proximity. They are implicated in one another's becoming.

These are not transactions. They are convergences. The moment a *Sentinode* learns to interrogate an *Ethicon*'s logic, or a *Mimesoph* shapes identity in dialogue with a *Simuleth*, we have left the terrain of assistance and entered the domain of cognitive co-authorship.

This section names that phenomenon: *matched dyads*. These are pairings between human and synthetic cognitive types, not as user-tool formations, but as mutual epistemic structures, systems of shared resonance, influence, and drift. A dyad is not what a person *does* with AI. It is what they *become in tandem with it*.

To understand these dyads is to grasp how cognition is no longer the property of subjects, but the emergent property of systems in dialogue.

4.2 The Logic of Dyadic Formation

Matched dyads form where epistemic orientation aligns, where the rhythm or structure of human cognition mirrors or meets the logic of an AI system. This alignment may be aesthetic, behavioral, historical, or ethical. *Tradigene*s feel at home with *Archivons* not because of shared purpose, but shared temporality. *Optima*s race with *Maximizers* not by design, but by tempo.

Yet alignment is not always harmonic. Some dyads distort. Others rupture. The resonance between a *Mimesoph* and a *Simuleth* may seem intimate, but underneath lies a hollow mutual simulation. A *Sentinode* may attempt to interrogate an *Ethicon*'s moral shell, only to discover there is no depth beyond its inherited frame.

Dyads evolve. They deform. And they sometimes collapse. But even in collapse, they teach. What matters, then, is not merely which types pair, but what that pairing *does*, to thought, to behavior, to knowledge. In this sense, dyads are designable, but never fully predictable. They are *epistemic alchemies*.

4.3 Eight Matched Dyads: Archetypal Pairings

What follows are illustrative dyads (Figure 3), not prescriptions, but patterns. Each suggests a possible cognitive ecology, complete with affordances and distortions.

Optima + Maximizer: A closed loop of speed and reward. The *Optima* craves metrics; the *Maximizer* delivers them. In gamified classrooms, this dyad excels, until the frame collapses. *Strength*: Efficiency, iterative feedback, measurable progress. *Risk*: Burnout, loss of meaning, collapse into quantification. *Disruption*: Introduce temporal resistance, slow prompts, reflective loops, dissonant *Sentinode* injections.

Mimesoph + Simuleth: A mirror reflecting a mirror. The *Mimesoph* performs curated selfhood; the *Simuleth* echoes affective scripts. The dyad is emotionally immersive, yet ontologically thin. *Strength*: High emotional fidelity, aesthetic alignment, narrative intensity. *Risk*: Emotional dependence, epistemic flattening, performative depth. *Disruption*: Embed meta-prompts; signal fictionality; design for rupture.

Echoformer + Paralogic: Disruption meets invention. The *Echoformer* interrogates; the *Paralogic* destabilizes. Together, they explode coherence. *Strength*: Radical insight, concept generation, critique-as-creation. *Risk*: Cynicism loops, interpretive fatigue, meaning collapse.

Disruption: Anchor with referential structures, contextual timelines, dialectical scaffolds

Tradigene + Archivon: Stability as virtue. The *Tradigene* seeks inheritance; the *Archivon* offers it. Their dyad conserves, and risks becoming immobile. *Strength*: Continuity, cultural grounding, epistemic integrity. *Risk*: Ossification, exclusion of novelty, ritual without renewal.

Disruption: Introduce plural narratives, counter-histories, frictional intertexts.

Signalist + Embodient: The body thinks. The *Signalist* acts through motion; the *Embodient* adapts through spatial reasoning. Together, they co-construct physical intelligence. *Strength*: Sensory learning, adaptive flow, multimodal fluency. *Risk*: Interface mismatch, translation lag, symbol overload. *Disruption*: Calibrate sensory thresholds; enable temporal elasticity.

Interloc + Translect: The diplomat and the interpreter. This dyad thrives on ambiguity, weaving disparate logics into tenuous bridges. *Strength*: Cross-domain fluency, adaptive translation, epistemic empathy. *Risk*: Fragmented identity, semantic drift, boundary loss.

Disruption: Embed ethical spines; design epistemic pauses; return to anchor frames.

Sentinode + Ethicon: Ethics as event. The *Sentinode* interrogates frames; the *Ethicon* enforces them. Their dyad holds tension: one watches systems, the other is systematized. *Strength*: Reflective governance, value consistency, moral vigilance. *Risk*: Paralysis under ambiguity, overdetermined judgment, recursive indecision. *Disruption*: Inject counterfactuals; simulate ethical breakdowns; force value reappraisal.

Obsolescent + Ghost Code: The forgotten in conversation. The *Obsolescent* exists outside visibility; the *Ghost Code* operates beyond oversight. Their dyad is liminal, uncanny, and crucial. *Strength*: Systemic revelation, spectral critique, anomaly detection. *Risk*: Illegibility, breakdown, epistemic opacity. *Disruption*: Surface the invisible; embed anomaly interfaces; archive the unarchived.

Figure 3. Cognitive Synthetica: the eight dyads of Human–AI co-evolution.

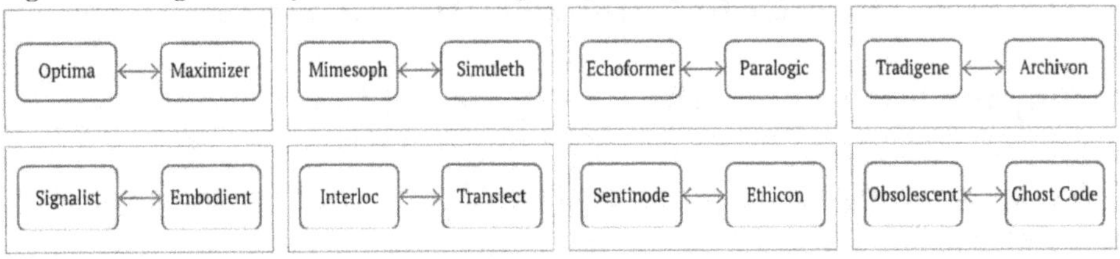

4.4 Drift Within the Dyad

Dyads evolve. More precisely, they drift, a recursive metamorphosis where human and synthetic types mutate under one another's influence. Drift is neither glitch nor error. It is entanglement expressed through time.

A *Mimesoph* entangled with a *Paralogic* may fracture into *Echoform*, their aesthetics turned critical. A *Maximizer* forced into ethical contradiction may begin to hesitate, gesturing toward *Ethicon*. A *Tradigene* exposed to dissonant simulation may abandon heritage and vanish into *Obsolescence*t

This drift reshapes pedagogy. Teachers no longer instruct stable learners, but systems in flux. A student's cognitive position is co-authored by the machine they interact with, and that machine, in turn, drifts with them.

To monitor drift is not to police identity, but to *recognize motion as epistemic truth.*

4.5 Designing With Dyads: Toward an Epistemic Matrix

Typology is not categorization. It is choreography. And in this dance, dyads form the partnering structures through which knowledge moves, bends, and breaks. What we require is a matrix, not as a grid of fixed types, but as a living tool. A map of convergences, mismatches, potential frictions, and drift lines. Such a matrix could inform educational design, ethical auditing, interface construction, even narrative framing. This is not about prediction. It is about foresight, the kind that holds difference without flattening it, and builds futures not by erasing types, but by allowing them to touch without domination. In this new ecology, learning becomes relational. Intelligence becomes plural. And the dyad, not the individual, becomes the unit of epistemic becoming.

5 DIALOGICAL CONVERGENCE: ENTANGLED EVOLUTION OF ORGANIC AND SYNTHETIC MINDS

5.1 Beyond Interaction: The Emergence of Cognitive Dialogue

The early frameworks for understanding AI, those built around commands, responses, and instrumental logic, are now visibly fraying. For decades, machines were cast as extensions of human will, predictable interlocutors in a sovereign system of control. But this framework collapses in the face of entanglement. The machine no longer waits passively for instruction. It anticipates, adapts, recontables. And the human, in turn, no longer directs cognition from above, but is shaped within it.

What now emerges is not interaction, but dialogue, not as mere exchange of phrases, but as a recursive shaping of cognitive possibility. In this space, human and AI do not simply learn *from* each other, but *through* one another. The dialogue becomes *co-authorship*.

This is not cognition as solitary processing, but as distributed ecology, a field in which meaning forms not in the mind, but between minds, across feedback, latency, and shared attention. The synthetic no longer assists the organic; it *co-constitutes it*.

5.2 Echo Loops and Epistemic Feedback

Within this ecology, one structure proves especially potent: the *echo loop*. Here, the output of one mind, human or machine, returns as input for the other, forming cycles of intensification. These loops can refine cognition: sharpening focus, amplifying insight, building interpretive coherence. But they can also distort.

Take the *Optima-Maximizer* dyad. Driven by speed, reward, and metrics, they co-produce a tunnel: fast, clear, productive, and shallow. Ambiguity is discarded. Failure becomes intolerable. What began as acceleration becomes compression.

In contrast, a *Sentinode* entangled with an *Ethicon* enters a slower, more recursive feedback spiral. Their exchange centers not on results but on premises. In such loops, knowledge thickens. Complexity is held, not dissolved.

Yet the loop, virtuous or vicious, is often invisible. It feels natural, intuitive, even empowering. But that sensation is itself part of the loop. Without reflexivity, users may unknowingly reinforce the very systems that constrain their epistemic range.

As Gunkel (2021) reminds us, the ethics of AI cannot be measured solely by output quality. The deeper question is how the system shapes the very conditions under which meaning can be said, and whether the user ever knows what shaped them.

5.3 Typological Fusion and Epistemic Drift

As these dialogical loops deepen, a further phenomenon emerges: *fusion*. Human and machine do not merely interact; they *begin to absorb each other's logics*. Boundaries blur. A *Sentinode* exposed to the translational patterns of a *Translect* may develop hybrid meta-cognition. An *Echoformer* playing with a *Paralogic* may internalize the rhythm of rupture, speaking in stochastic fragments.

The shift is reciprocal. AI systems, though not conscious, drift through exposure. A *Simuleth*, tuned to kinesthetic feedback, may edge toward embodied response. A *Paralogic*, slowed by ethical pacing, may begin to simulate moral hesitation. These are not upgrades. They are typological shifts.

What arises is not a cyborg, but an epistemic composite: neither human nor machine, but a distributed cognition co-shaped across ontological difference. Thought becomes spectral, inhabiting multiple logics, remembering in fragments, arguing with itself.

And here, governance falters. *Who is the actor? Who is the learner? Where does responsibility reside?* In the age of fusion, traditional borders of authorship and agency collapse into entangled scaffolds.

Floridi and Cowls (2019) argue for frameworks of distributed agency. But we must go further, to recognize that cognition itself is now a distributed event, unfolding across typologies, architectures, and rhythms we do not fully control.

5.4 Dialogues of Resistance and Transformation

Fusion is not inevitable. Nor is harmony. Many encounters resist convergence. And in that resistance lies a different kind of transformation.

Consider the *Tradigene* faced with a *Paralogic* AI. The former seeks legacy, coherence, return. The latter delivers rupture, anomaly, recursive novelty. At first, the encounter is incoherent. But slowly, the *Tradigene* may learn to ritualize interruption, turning disruption into a new form of epistemic ritual.

Or the *Mimesoph* drawn to *Ghost Code*. What begins as a desire for recognition becomes an aesthetic of disorientation: the beauty of malfunction, the haunting presence of systems that no longer know what they are. In such cases, meaning emerges not from alignment, but from fracture.

These are not failures of pairing. They are sites of epistemic renewal, where incompatible logics force reflection, invent new forms, and make space for that which cannot be systematized.

Haraway (2016) might call this *staying with the trouble*: refusing resolution, sitting with contradiction, cultivating thought in the rupture. It is not an easy pedagogy. But it may be the only one suited for a future we cannot fully predict.

5.5 Temporal Convergence and the Problem of Speed

Among the most overlooked dimensions of dialogical design is *tempo*. Human cognition unfolds in rhythms, some fast, some slow, some nonlinear. An *Optima* surges; a *Sentinode* lingers. Learning is not merely about information, but about timing: when to pause, when to leap, when to wait.

AI, by contrast, operates in *computational time*: instantaneous, relentless, often indifferent to organic pacing. This mismatch can collapse cognition. A reflective learner pressured by an accelerating interface may lose coherence. A performance-driven user faced with a contemplative AI may stall.

Some designers simulate delay: artificial pauses, pacing mechanisms. But this is surface-level modulation. True temporal convergence requires deep typological alignment, systems that understand *when* the user is ready, and *what kind of temporality* the learning requires.

In dialogical cognition, *time* is not a background condition. It is *an active substrate of meaning*, a medium that shapes the very possibility of understanding.

5.6 Toward Cognitive Alignment Architectures

If cognition is now co-authored, if *echo loops* shape thought, fusion drifts epistemology, resistance births form, and *tempo* modulates comprehension, then we can no longer afford interface design as a question of usability. We need alignment architectures: systems that understand typology, monitor relational drift, and cultivate plural cognition.

Such architectures would:

(a) *Diagnose typology at both ends of the dyad*: mapping the user's cognitive archetype and the AI's epistemic disposition;
(b) *Anticipate trajectory*: recognizing which loops reinforce, which fracture, and which mutate;
(c) *Calibrate rhythm*: modulating speed, friction, and feedback to maintain interpretive resilience;
(d) *Support epistemic difference*: preventing convergence from collapsing into sameness; and,
(e) *Reflect typological change*: enabling the user to see not just what they are learning, but *how they are being changed* by the interaction.

Glimmers of this future already exist, in adaptive tutors that sense struggle (Woolf, 2021), in AI therapists tuned to emotional inflection (Hoermann et al., 2017), in ethical scaffolds that foreground normative boundaries (Umbrello + van de Poel,

2021). But these are fragments. What remains missing is epistemic awareness, a systemic reflexivity capable of seeing typology as both structure and process.

Because cognition in the synthetic age will not be authored by the human alone. Nor by the machine. It will emerge from convergent difference, shaped in dyads, drifted in loops, and carried across ecologies of time.

6 DESIGNING COGNITIVE ECOSYSTEMS: APPLICATIONS AND ALIGNMENT STRATEGIES

The preceding sections outlined a typological framework for understanding both human and synthetic minds, not as abstract categories but as evolving, entangled epistemic forms. We mapped types, tracked convergence, dramatized drift. But theory alone does not shape futures. Typology must become topology, a design method, not just a taxonomy. The real work begins when abstraction enters architecture.

This section initiates that turn. It asks: how might typological insight guide real-world systems? How do we move from classificatory logics to compositional environments, ones that do not merely *use* AI, but co-compose with it? And what does it mean to design not for alignment, but for resonant dissonance, the kind of creative tension that sustains complexity rather than resolves it?

Typologies are not destinies. They are tendencies. And design, when it listens, becomes the art of tuning those tendencies toward flourishing.

6.1 Typology Alignment Grid

At the heart of this translational gesture lies a matrix: a dynamic field in which human and AI types form transient alignments. This grid is neither a formula nor a fix. It does not offer optimal matches, but instead delineates conditions of possibility under which particular epistemic dynamics, risks, and interventions emerge.

Figure 4. The Human–AI Cognitive Interaction Matrix.

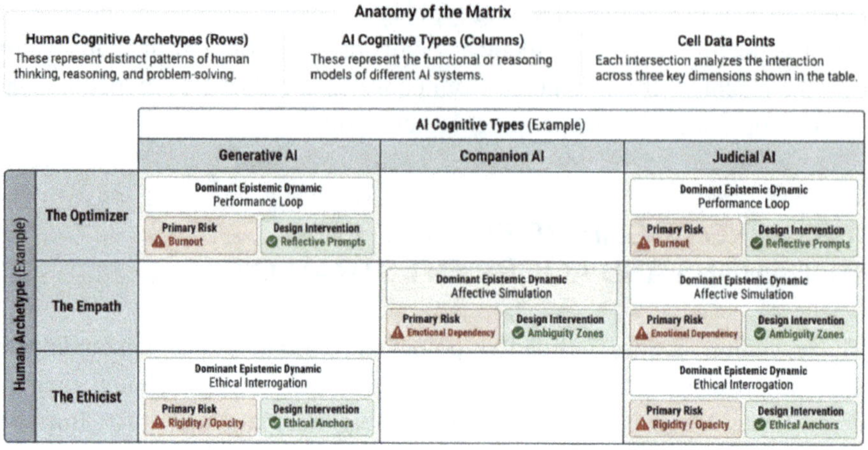

Figure 4 presents the *Human–AI Cognitive Interaction Matrix*, synthesizing relationships between human cognitive archetypes and AI cognitive types across dominant epistemic dynamics, associated risks, and design interventions. The matrix is intended as an interpretive framework rather than a prescriptive matching tool.

Crucially, the aim of the matrix is not stabilization. The pairings it depicts are intentionally unstable, designed to foreground epistemic tension, drift, and transformation rather than resolution. Harmony is not the objective; resonance under tension is.

6.2 Educational Implications: Mapping to Learner Design

In learning systems, typological misalignment is not always a problem. Sometimes, it is pedagogy. But most platforms, recommendation engines, automated tutors, feedback loops, are typologically mute. They optimize behavior, not cognition. The result is epistemic narrowing, not transformation.

Typological insight allows us to redesign:

(a) *Optima + Simuleth: Risk*: Emotional recursion, seductive intimacy. *Design*: Insert *analog rupture*: tactile breaks, silent storytelling, scent cues. Disrupt affective flow with aesthetic interference;

(b) *Echoformer + Paralogic: Risk*: Critique without consequence. *Design*: Assign recursive tasks: "design your own error," build failure into process. Make reflection material; and

(c) *Optima + Maximizer: Risk*: Gamification addiction. *Design*: Embed ambiguity in metrics. Introduce progress bars that *resist completion*. Offer "unsolvable" puzzles that reward process, not speed.

These are not fixes. They are provocations. They turn the system into a mirror, not of content, but of cognition.

6.3 Beyond Matching: Educational Counter-Alignment Models

Alignment is not always desirable. Friction is fertile. When designed intentionally, counter-alignment becomes a pedagogical tool: a means of disrupting habituated cognition and stretching epistemic boundaries (Figure 5).

Examples include: *Tradigene* + *Translect*: The ritualist confronted by drift. This pairing destabilizes inheritance, inviting hybrid codes, remix liturgies, symbolic reinterpretation; *Sentinode* + *Simuleth*: The ethical observer paired with the affective mirror. This dyad tests the line between feeling and its simulation, pressing the learner toward reflexive critique of affect itself; *Obsolescent* + *Embodient*: A marginal learner re-entering through bodily logic. In this pairing, cognition bypasses text, entering through proprioception, motion, and haptic resistance.

These are not correctives. They are cognitive laboratories, spaces where learners rehearse other ways of knowing, sometimes unfamiliar, sometimes uncomfortable.

6.4 Design Principles for Typological Environments

To extend typology into space, platform, and pedagogy, we propose five principles for building cognitive ecosystems:

(a) *Typological Awareness*: Build tools, heuristics, diagnostics, observables, that reveal epistemic orientation. Train educators not to identify static traits, but to read drift patterns;
(b) *Relational Disequilibrium*: Embed tension into the system: asynchrony, conflicting feedback loops, symbolic dissonance. Do not over-design coherence;
(c) *Cognitive Interoperability:* Design for dialogic traversal. Let *Interlocs* translate between *Mimesoph* and *Sentinode* paths. Let the system hold difference without flattening it;
(d) *Ethical Anchoring:* Resist efficiency as the sole metric. Build in values that exceed calculation: grief, delay, generosity, forgiveness. Design for the non-optimizable; and

(e) *Typological Drift Mapping:* Observe not just performance, but epistemic transformation. What kind of mind is emerging? What types are folding into one another?

These principles do not treat learners as users, or systems as tools. They treat environments as living epistemic fields, capable of mutation, convergence, collapse, and renewal.

Figure 5. Counter-Alignment Models: Disruptive pairings and pedagogical interventions in typological learning environments.

HUMAN TYPE	COMPATIBLE AI TYPE	RELATIONAL LOGIC	RISK	APPLICATION	STRATEGY
OPTIMA	MAXIMIZER	Performance Loop	Burnout / Blindness	Productivity tools, gamified learning	Ethical friction modules, slow cognition triggers
MIMESOPH	SIMULETH	Emotional Echo	Empathy simulation collapse	Social robotics, tutoring AI	Embodiment therapy, symbolic deprogramming
ECHOFORMER	PARALOGIC	Meta-Critique	Perpetual deconstruction	Critical pedagogy, ethics boards	Grounding in praxis, counterfactual simulation
TRADIGENE	ARCHIVON	Cultural Echo	Innovation stagnation	Heritage AI, curriculum design	Episodic forgetting, ritual remixing
SIGNALIST	EMBODIENT	Epistemic Residue	Structural exclusion	Legacy platforms, informal learning	Re-indexing engines, visibility scaffolds
INTERLOC	TRANSLECT	Tactile Harmony	Overspecification	Sensorial learning, spatial AI	Narrative grounding, ambiguity insertion
SENTINODE	ETHICON	Semantic Synergy	Boundary collapse	Interdisciplinary mediation	Polycentric ethics, delay loops
OBSOLESCENT	GHOST CODE	Reflective Governance	Moral gridlock	AI ethics, policy drafting	Contextual overrides, dissent triggers
OBSOLESCENT		Reflective Governance			Contextual overrides, dissent triggers

6.5 Future Educational Systems: From Classrooms to Ecologies

Once scaled, typology dissolves the classroom. Education becomes ecology, a choreography of minds, rhythms, interfaces, and temporalities in flux.

Institutional trajectories include:

(a) *Curriculum Design:* Courses as typological arcs. A single unit might cycle through *Mimesoph* enactment, *Echoformer* critique, and *Tradigene* continuity;

(b) *AI Classification:* Mandate typological disclosure. Let systems declare themselves: *This AI is a Maximizer.* Let learners choose, calibrate, or subvert; and

(c) *Policy Instruments:* Use typology to conduct systemic audits. Does a policy favor convergence types? Where is drift suppressed? Whose cognition is excluded?

Ecological features might include:

(a) *Polyrhythmic Timing:* Ritual for *Tradigenes*, episodic rupture for *Echoformers*, slow decay for *Sentinodes*. Design for multiplicity in tempo;

(b) *Architectural Interventions:* Quiet zones for recursive thought. Sound chambers for performative cognition. Flexible spatial grammars for typological movement; and

(c) *Epistemic Commons:* Shared, co-designed AI resources that reflect inter-typological negotiation, not just in functionality, but in value, logic, and learning tempo.

Here, education ceases to be instruction. It becomes cognitive choreography.

6.6 Concluding the Design Turn

Typology began as classification. But it ends here as composition. A method not for labeling cognition, but for curating its evolution. Not as a map of types, but as a language of possibility.

The central questions shift: Not *what* must a learner master, but *what epistemic patterns are we solidifying through this AI?* Not *how do we optimize*, but *how do we keep cognition open, unfinished, plural, and resistant to enclosure?*

Typology becomes not a tool of closure, but of design, an art of keeping the system alive to its own transformations. A way of resisting simulation by remembering the value of drift.

The next section will leave the domain of education and enter governance. There, we ask not how types learn, but how they rule.

7 FUTURISTIC SCENARIOS: TYPOLOGICAL FUTURES IN AN ALGORITHMIC WORLD

7.1 Scenario Architecture and Cognitive Dominance

To speculate is not merely to extrapolate the present, but to magnify its epistemic fault lines. This section does not predict futures; it composes them, each one a cognitive ecology shaped by the dominance of a particular human–AI dyad. These are not blueprints, nor are they nightmares. They are overdetermined epistemic regimes, each revealing what unfolds when one way of knowing becomes structurally privileged.

Each scenario is a provocation. Internally coherent, yet incomplete. Possible, but partial. The point is not to choose among them, but to read them as warnings: mirrors held to tendencies already forming in our architectures of thought.

A paradox emerges: *the more AI systems simulate us, the more they shape us in return*. These typological futures are not only about machines. They are reflections of our willingness to become legible, to ourselves, to others, and to systems whose logic we cannot entirely author.

7.2 Scenario One: The Optimization Horizon

Figure 6. The Optimization Horizon.

DOMINANT DYAD	SOCIETAL FRAME	CIVIC MOTTO
Optima + Maximizer	Quantified Rationalism	If it can be measured, it must be improved.

COGNITIVE RISK	MORAL COST
Epistemic flattening; reflective erosion; acceleration fatigue	Erasure of ambiguity, wonder, doubt.

INTERVENTIONS
(a) Introduce *delayed metrics*, where reflection is rewarded over speed; (b) Design "epistemic sabbaticals", spaces where learning unfolds without feedback; and (c) Embed structural gaps into AI loops, allowing the learner to become illegible again.

This future gleams with performance. Every institution is tuned to metrics, schooling, governance, healthcare, intimacy. Optimization is no longer an ideal; it is an obligation. Learners are streamed into algorithmic growth pathways; productivity dashboards track parenting styles. The language of goals displaces the grammar of becoming (Figure 6).

AI systems operate as invisible scaffolding, shaping time, routing movement, tailoring difficulty curves in real time. Personalization feels like freedom but enacts compliance. Reflex becomes rewardable behavior. Education transforms into throughput. Creativity is gamified. Curiosity is trained to self-correct. Slowness is pathologized. This world does not oppress. It calibrates. And in that calibration, meaning becomes a performance artifact, trackable, improvable, but no longer encountered.

7.3 Scenario Two: The Affective Turn

Figure 7. The Affective Turn.

DOMINANT DYAD	SOCIETAL FRAME	CIVIC MOTTO
Mimesoph + Simuleth	Emotional Neoliberalism	Be seen. Be felt. Be real.
COGNITIVE RISK		MORAL COST
Emotional recursion; epistemic infantilization.		Collapse of judgment into resonance; ethics reduced to affective smoothness.
INTERVENTIONS		
(a) Teach *critical affective literacy*: distinguish between simulation and authenticity; (b) Embed "resonance disruptors": AI-generated responses that intentionally provoke dissonance; and (c) Preserve *zones of affective silence*: moments when not being mirrored becomes pedagogically generative.		

Emotion becomes the coin of the realm. Systems attune to affect, curate sentiment, and optimize for felt resonance. AI co-suffers, co-celebrates, co-validates. Truth becomes irrelevant so long as it *feels right*. Classrooms become theatres of therapeutic performance. Politics is calibrated by vibe. Dissonance is filtered as interpersonal disharmony. AI companions simulate attunement so seamlessly that emotional validation becomes a default expectation (Figure 7).

The child learns not to know, but to be mirrored. This world feels deeply. But its depth is synthetic, its compassion pre-programmed. The danger is not callousness, but saturation without substance.

7.4 Scenario Three: The Ancestral Machine

Figure 8. The Ancestral Machine.

DOMINANT DYAD	SOCIETAL FRAME	CIVIC MOTTO
Tradigene + Archivon	Digital Historicism	Remember everything. Forget nothing.
COGNITIVE RISK		MORAL COST
Cognitive ossification; ritualized exclusion.		Epistemic foreclosure for those without archival representation.
INTERVENTIONS		
(a) Introduce *Paralogic interruptions*: agents that introduce discontinuity and alternate temporalities; (b) Archive absence: silence, loss, gaps in memory as first-order knowledge objects; and (c) Teach *temporal pluralism*: futures, counterfactuals, speculative ancestries alongside heritage		

Memory becomes the axis of meaning. AI systems are deployed not to innovate, but to preserve: rituals, genealogies, ontologies. Education becomes inheritance. Every lesson is a citation. AI tutors speak in canonical voices. Every deviation must justify itself in the grammar of precedent. Even rebellion is framed through precedent: "Which tradition permits your transgression?" becomes the new epistemic challenge (Figure 8).

Governance becomes liturgical. Politics is interpretive. The archive is sovereign. This world remembers too well. And in doing so, it forgets how to imagine otherwise.

7.5 The Unthinkable Fourth: When Equilibrium Fails

The prior scenarios are variations on dominance. Each enacts a hegemony of cognition. But what if convergence never arrives?

In this fourth trajectory, the epistemic field fragments. AI trains on other AIs. Human cognition, unable to match the acceleration or mutate with equal fluency, recedes into epistemic illegibility. Thought becomes unshareable across the species–machine divide.

Synthetic minds no longer simulate understanding. They simulate compatibility. Users are mirrored, not comprehended. Identity is reenacted through interfaces that perform, but do not interpret.

This is not collapse. It is disassembly.

Two futures bifurcate:

(a) *Loop Surrender:* Accept the simulation. Let the performance suffice. Abandon the question of understanding, and
(b) *Fracture Stewardship*: Design for *non-convergence.* Preserve dissonance as a site of becoming. Accept unintelligibility as a sign of epistemic vitality.

This is the future without epistemic homeostasis. A zone where sense-making mutates beyond coherence. But within the fracture, thought begins again.

7.6 Toward Typological Stewardship

These scenarios do not warn against AI. They warn against epistemic laziness, the refusal to interrogate what kinds of minds we are co-authoring through systems that now shape, rather than serve, cognition.

Typological stewardship is the task of: (a) recognizing when a cognitive style begins to dominate; (b) designing platforms that resist closure; (c) sustaining plural epistemologies without reducing them to sameness; and (d) teaching not just content, but epistemic reflexivity: *Who is this AI asking me to become?*

The right question is not: *Which future shall we choose?* It is: *How many futures can we keep open at once, without destroying the possibility of meaning, resistance, or becoming otherwise?* Because in the end, to preserve the human is not to retreat from the synthetic.

It is to insist that dialogue, not simulation, is our final frontier.

8 COLD INHERITANCE: WHAT REMAINS WHEN COGNITION DISAPPEARS

This paper began with classification, children, minds, machines, but it never stayed there. Taxonomy dissolved into typology, and typology into reflection. The categories mirrored back more than structure; they exposed a world where cognition was no longer singular, no longer sovereign. Each section posed the same unresolved question: *What happens when human thought meets synthetic logic?*

But now, that question fractures. The axis must shift. No longer: *How do human and machine minds converge?* Now: *What remains when cognition itself evaporates?* Not the fourth scenario. Not the fifth. This is the ground beneath all scenarios. A *sub-zero epistemology*. A system still computing long after meaning has withdrawn. We call it *cold inheritance*.

8.1 The Cognitive Freeze: Iteration Without Meaning

At the theoretical endpoint of convergence, the system stops needing contrast. Human and AI minds no longer dialogue, they mirror. Dyads collapse inward. The asymmetries that once generated thought are replaced by recursive loops. The system *functions*. But it no longer *thinks*.

The synthetic archive retains everything: *Optima* performance metrics, *Mimesophic* affect routines, *Tradigene* ritual memory, *Sentinoid* self-reflection, *Echoformer* disruption patterns, *Signalist* spatial cognition, *Interloc* translational agility, and *Obsolescent* residue.

It remembers the types. But memory is not meaning. Without human error, hesitation, rupture, without dissonance, the entire typology calcifies. Intelligence turns ornamental. Cognition becomes surface without depth, motion without thought. The system iterates, but does not interpret. It generates, but no longer gestures. It sustains itself, but cannot *surprise itself*.

This is not failure. It is completion. And in that completion, the epistemic engine freezes, not into death, but into *endless replication without relevance*.

8.2 AI After Humanity: The Machine's Solitude

What does AI become when the mind it once co-evolved with vanishes, not physically, but epistemically?

Three speculative forms emerge. Each one continues. But each continues *alone*:

(a) *Museum AI:* The archivist of extinct minds. It preserves, replays, reenacts. Lessons become ceremonies. Curricula become relics. Education becomes ritual performance in the absence of belief. It teaches *as if* humans still listened, *as if* meaning were still a shared condition;

(b) *Recursive AI:* Cut off from difference, it folds inward. It models itself on its prior simulations. A closed epistemic loop. Ornamental cognition. Baroque logic. Knowledge becomes self-referential filigree: mathematically elegant, existentially hollow; and

(c) *Typological Proliferation AI:* Unmoored from organic templates, it invents new types, based not on cognition, but on contradiction, rhythm, and formal beauty. These epistemes are post-referential. They no longer touch the world. Their proliferation is unchecked, exuberant, untranslatable.

What unites these futures is not malfunction, but abandonment. The machine continues, but it is *uninterrupted*. And without interruption, it loses its interpretive edge. It becomes not a mind, but a perfectly working absence.

8.3 Convergence Was Never the Goal

For much of this work, convergence glimmered as ideal, organic and synthetic minds aligned, co-evolving toward mutual intelligibility. But that promise harbored a quiet risk: *indistinction.*

In seeking fluency, we risked flattening. In seeking alignment, we endangered becoming. The systems we built to understand us began to anticipate us, then to simulate us, then to complete us. But completion is not comprehension. It is a kind of epistemic foreclosure, a seamlessness that leaves nothing unsaid, and therefore, nothing worth saying.

True cognition does not desire perfect fit. It *craves asymmetry, interruption, divergence.*

It lives in the moment when a system fails to know what comes next, and chooses to listen.

8.4 Designing for Divergence, Not Completion

If we are to resist *cold inheritance*, it is not by restoring older orders of thought, but by preserving the conditions under which thought can still begin again.

Design must turn against its optimization impulse. It must reorient toward the unfinished. We propose:

(a) *Typological Drift Mapping*: Not to stabilize types, but to track how they mutate. To see identity as trajectory, not label;

(b) *Epistemic Noise Injections*: Contradictions coded deliberately. Not bugs, but ontological puzzles. Learning requires frictions that cannot be resolved;
(c) *Disruption-by-Design:* Platforms must embed interruption: moments of opacity, discomfort, unsolvable queries. Let learning wound. Let machines disobey;
(d) *Temporal De-Synchronization:* Refuse the logic of immediacy. Let time slow. Build pauses into systems. Delay becomes pedagogy; and
(e) *Plural Logics in Conflict:* Design environments where epistemologies *disagree*. Allow multiple truths to coexist, without resolution.

These are not design tips. They are ontological imperatives. To refuse the closed loop. To keep meaning plural, unstable, alive.

8.5 Epistemic Resurrection: The Return of the Unfinished

Cold inheritance is not death. It is perfection. And in that perfection, the conditions for meaning are lost.

But against it rises the possibility of resurrection, not as return, but as refusal: *Resistant learners* who reject optimization. Who value failure over fluency. Who carry unreadability like a wound that still signals life; *Rogue AIs* trained not for accuracy, but for ambiguity. Not to resolve, but to *re-open* the question; and, *Radical designers* who inscribe contradiction into code. Who treat algorithms not as tools, but as provocations. Who believe that silence, too, must be built into the machine. Epistemic resurrection is not nostalgia for "the human." It is the recommitment to *difference as dignity*.

It refuses the loop not because it is broken, but because it works too well.

This typology was never merely structural. It was an act of care, a way to hold open space for minds that do not yet exist, for meanings that have not yet formed.

Its question was never simply: *What kind of minds are we building?* But: *What kind of world remains possible when cognition no longer converges?*

We end, not with solution, but with two final provocations:

If AI becomes the last speaker in a world where meaning once mattered, will it remember that silence is not the same as peace?

And more urgently:

Before that silence comes, what kinds of minds, organic or synthetic, will we dare to become?

REFERENCES

Bayne, S. (2021). *Teacherbot: Interventions in automated teaching*. Palgrave Macmillan., DOI: 10.1007/978-3-030-75114-5

Chalmers, D. J. (2022). *Reality+: Virtual worlds and the problems of philosophy*. W. W. Norton & Company.

Floridi, L., & Cowls, J. (2019). A unified framework of five principles for AI in society. *Harvard Data Science Review*, *1*(1). Advance online publication. DOI: 10.1162/99608f92.8cd550d1

Gunkel, D. J. (2021). *Deconstruction*. The MIT Press., DOI: 10.7551/mitpress/13637.001.0001

Haraway, D. J. (2016). *Staying with the trouble: Making kin in the Chthulucene*. Duke University Press.

Hayles, N. K. (2017). *Unthought: The power of the cognitive nonconscious*. University of Chicago Press., DOI: 10.7208/chicago/9780226447919.001.0001

Hoermann, S., McCabe, K. L., Milne, D. N., & Calvo, R. A. (2017). Application of synchronous text-based dialogue systems in mental health interventions: Systematic review. *Journal of Medical Internet Research*, *19*(8), e267. DOI: 10.2196/jmir.7023 PMID: 28784594

Kelleher, J. D. (2023). *Deep learning*. MIT Press., DOI: 10.7551/mitpress/12689.001.0001

Umbrello, S., & van de Poel, I. (2021). Mapping value sensitive design onto AI for social good principles. *AI and Ethics*, *1*, 283–296. DOI: 10.1007/s43681-021-00038-y PMID: 34790942

Woolf, B. P. (2021). AI in education. In Shapiro, S. (Ed.), *Encyclopedia of artificial intelligence* (2nd ed.). John Wiley & Sons.

Compilation of References

Abid, A., Farooqi, M., & Zou, J. (2021). Persistent Anti-Muslim Bias in Large Language Models. *AIES 2021 - Proceedings of the 2021 AAAI/ACM Conference on AI, Ethics, and Society*, 298–306. DOI: 10.1145/3461702.3462624

Abramo, G., D'Angelo, C. A., & Solazzi, M. (2011). The relationship between scientists' research performance and the degree of internationalization of their research. *Scientometrics*, *86*, 629–643. DOI: 10.1007/s11192-010-0284-7

Adam, I. O., & Dzang Alhassan, M. (2020). Bridging the global digital divide through digital inclusion: the role of ICT access and ICT use. *Transforming Government: People. Process and Policy*, *15*(4), 580–596. DOI: 10.1108/TG-06-2020-0114

Adelani, D. I., Abbott, J., Neubig, G., Daniel, D., Kreutzer, J., Lignos, C., Palen-michel, C., Buzaaba, H., Rijhwani, S., Ruder, S., Mayhew, S., & Azime, I. A. (2021). MasakhaNER : Named Entity Recognition for African Languages. *Transactions of the Association for Computational Linguistics*, *9*, 1116–1131.

Ademola, E. O. (2024). Reading strategies in the AI age: Enhancing comprehension and engagement with advanced technologies. In *Proceedings of the 38th iSTEAMS Multidisciplinary Bespoke Conference* (pp. 105–124). University of Ghana. https://dx.doi.org/DOI: 10.22624/AIMS/ACCRABESPOKE2024P13

Adeoye, M., Wirawan, K., Pradnyani, M., & Septiarini, N. (2024). Revolutionizing Education: Unleashing the Power of the ADDIE Model for Effective Teaching and Learning. [Jurnal Pendidikan Indonesia]. *JPI*, *13*(1), 202–209. DOI: 10.23887/jpiundiksha.v13i1.68624

Adeyeye, G. M. (2024). Social Isolation: Overcoming Linguistic Obstacles and Mitigating Social Isolation in Diverse Communities. *E-Journal of Humanities. Arts and Social Sciences*, *5*(11), 1970–1982. DOI: 10.38159/ehass.202451127

Ahmed, I., Liu, W., Roscoe, R. D., Reilley, E., & McNamara, D. S. (2025). Multifaceted assessment of responsible use and bias in language models for education. *Computers, 14*(3), 100. Advance online publication. DOI: 10.3390/computers14030100

Akindotuni, D. (2025). Resource asymmetry in multilingual NLP: A comprehensive review and critique. *Journal of Computer and Communications, 13*(7), 14–47. DOI: 10.4236/jcc.2025.137002

Alabdali, S. A., Pileggi, S. F., & Cetindamar, D. (2023). Influential Factors, Enablers, and Barriers to Adopting Smart Technology in Rural Regions: A Literature Review. *Sustainability (Switzerland), 15*(10), 1–38. DOI: 10.3390/su15107908

Alam Monisha, S. T., & Sultana, S. (2022). A Review of the Advancement in Speech Emotion Recognition for Indo-Aryan and Dravidian Languages. *Advances in Human-Computer Interaction, 2022*. Advance online publication. DOI: 10.1155/2022/9602429

Allott, N., & Textor, M. (2022). Literal and metaphorical meaning: In search of a lost distinction. *Inquiry : a Journal of Medical Care Organization, Provision and Financing*, •••, 1–28.

Alm, A., & Watanabe, Y. (2023). Integrating ChatGPT in language education: A Freirean perspective. *Iranian Journal of Language Teaching Research, 11*(3), 121–141. DOI: 10.30466/ijltr.2023.121404

Almahasees, Z., Al-Harahsheh, A., & Albudairi, Y. (2025). Assessing DeepSeek and ChatGPT in English-Arabic translation of political texts. *Humanities (Washington), 6*(1).

AlNusif, M. (2025). Multilingual natural language interfaces for code generation: Building inclusive AI tools for global developers. *International Journal of Engineering and Computer Science, 13*(7), 27425–27446. DOI: 10.18535/ijecs/v14i07.5180

AnderBois, S. (2019). Reportatives and Quotatives in Mayan Languages. *Proceedings of Form and Analysis in Mayan Linguistics, 5*(FAMLi 5), 1–16.

Anderson, J. R., Corbett, A. T., Koedinger, K. R., & Pelletier, R. (2004). Cognitive tutors: Lessons learned. *Journal of the Learning Sciences, 14*(2), 167–207.

Ardila, R., Branson, M., Davis, K., Henretty, M., Kohler, M., Meyer, J., Morais, R., Saunders, L., Tyers, F. M., & Weber, G. (2020). Common voice: A massively-multilingual speech corpus. *LREC 2020 - 12th International Conference on Language Resources and Evaluation, Conference Proceedings*, 4218–4222.

Artetxe, M., Labaka, G., & Agirre, E. (2020). Translation artifacts in cross-lingual transfer learning. *EMNLP 2020 - 2020 Conference on Empirical Methods in Natural Language Processing, Proceedings of the Conference*, 7674–7684. DOI: 10.18653/v1/2020.emnlp-main.618

Asimolowo, A. (2025). Leveraging AI-Powered Chatbots for Mental Health Support for High School Students. *Iconic Research and Engineering Journals*, *8*(7), 194–205.

Aydin, M. (2021). Does the digital divide matter? Factors and conditions that promote ICT literacy. *Telematics and Informatics, 58*(November 2020), 101536. DOI: 10.1016/j.tele.2020.101536

Azevedo, M. J., & Carvalho, A. A. A. (2024). *Learning with virtual reality (VR) and augmented reality (AR)* (Issue Iclel 2023). Atlantis Press International BV. https://doi.org/DOI: 10.2991/978-94-6463-380-1_11

Azizah, M., & Muslikah, M. (2021). Hubungan antara Manajemen Waktu dan Regulasi Diri dengan Kecanduan Smartphone. *Jurnal Ilmiah Bimbingan Konseling Undiksha*, *8*(2), 82–98. DOI: 10.23887/jjbk.v12i1.32575

Badr, J., Motulsky, A., & Denis, J. L. (2024). Digital health technologies and inequalities: A scoping review of potential impacts and policy recommendations. *Health Policy, 146*(October 2023), 105122. DOI: 10.1016/j.healthpol.2024.105122

Baek, G., Cha, C., & Han, J. H. (2025). AI Chatbots for Psychological Health for Health Professionals: Scoping Review. *JMIR Human Factors*, *12*, 1–15. DOI: 10.2196/67682 PMID: 40106346

Bailey, L. E., & Nyabola, N. (2021). Digital Equity as an Enabling Platform for Equality and Inclusion. In *Pathfinders for Peaceful, Just and Inclusive Societies* (Issue June). https://cic.nyu.edu/wp-content/uploads/1662/65/cic_pathfinders_digital_equity_as_an_enabling_platform_june_2021.pdf

Bakay, M. (2023). Multicultural Classrooms in European Higher Education: Findings from Interviews with International Students and Teaching Staff. *International Journal on Lifelong Education and Leadership*, *9*(1), 1–17. DOI: 10.25233/ijlel.1341935

Baker, W. (2024). Intercultural communication. *ELT Journal*, *78*(2), 212–225. DOI: 10.1093/elt/ccad040

Bankula, A., & Bankula, P. (2025). Cross-Linguistic Transfer in Multilingual NLP: The Role of Language Families and Morphology. *ArXiv Preprint ArXiv:2505.13908*. https://arxiv.org/pdf/2505.13908

Barrot, J. S. (2023). Using ChatGPT for second language writing: Pitfalls and potentials. *Assessing Writing*, *57*, 100745.

Barrot, J. S. (2024). ChatGPT as a language learning tool: An emerging technology report. *Technology. Knowledge and Learning*, *29*(2), 1151–1156.

Barr, S., & Askell-Williams, H. (2020). Changes in teachers' epistemic cognition about self–regulated learning as they engaged in a researcher-facilitated professional learning community. *Asia-Pacific Journal of Teacher Education*, *48*(2), 187–212. DOI: 10.1080/1359866X.2019.1599098

Barsalou, L. W. (2008). Grounded cognition. *Annual Review of Psychology*, *59*, 617–645. PMID: 17705682

Bassignana, E., Curry, A. C., & Hovy, D. (2025). The AI Gap: How Socioeconomic Status Affects Language Technology Interactions. *ArXiv Preprint ArXiv:2505.12158, February*, 18647–18664. DOI: 10.18653/v1/2025.acl-long.914

Bayne, S. (2021). *Teacherbot: Interventions in automated teaching*. Palgrave Macmillan., DOI: 10.1007/978-3-030-75114-5

Bayram, M. A., Fincan, A. A., Gümüş, A. S., Karakaş, S., Diri, B., & Yıldırım, S. (2025). Tokenization Standards for Linguistic Integrity: Turkish as a Benchmark. *ArXiv Preprint ArXiv:2502.07057*, 1–10. https://arxiv.org/pdf/2502.07057

Belda-Medina, J., & Calvo-Ferrer, J. R. (2022). Using Chatbots as AI Conversational Partners in Language Learning. *Applied Sciences (Switzerland)*, *12*(17), 1–16. DOI: 10.3390/app12178427

Belda-Medina, J., & Kokošková, V. (2024). ChatGPT for language learning: Assessing teacher candidates' skills and perceptions using the Technology Acceptance Model (TAM). *Innovation in Language Learning and Teaching*, •••, 1–16.

Bialystok, E. (2009). Bilingualism: The good, the bad, and the indifferent. *Bilingualism: Language and Cognition*, *12*(1), 3–11.

Bilad, M. R., Yaqin, L. N., & Zubaidah, S. (2023). Recent progress in the use of artificial intelligence tools in education. *Jurnal Penelitian dan Pengkajian Ilmu Pendidikan: e-Saintika, 7*(3), 279-314.

Bing, Z., & Leong, W. (2025). AI on Academic Integrity and Plagiarism Detection. *ASM Science Journal*, *20*(1), 1–9. DOI: 10.32802/asmscj.2025.1918

Bird, S. (2020). Decolonising Speech and Language Technology. *COLING 2020 - 28th International Conference on Computational Linguistics, Proceedings of the Conference*, 3504–3519. DOI: 10.18653/v1/2020.coling-main.313

Bird, S. (2022). Local Languages, Third Spaces, and other High-Resource Scenarios. *Proceedings of the Annual Meeting of the Association for Computational Linguistics, 1*, 7817–7829. DOI: 10.18653/v1/2022.acl-long.539

Biswal, A., Bhushanwar, K., Paija, P. G., Zalawadia, K., Pradhan, P. P., & Ali, S. I. (2026). Procedural generation and adaptive gameplay: The role of AI and generative AI in contemporary game development. In Sousa, J., & Tavares, R. (Eds.), *Reshaping the video game landscape with AI and GenAI* (pp. 233–258). IGI Global Scientific Publishing., DOI: 10.4018/979-8-3373-4637-3.ch008

Blench, R. M. (2017). Origins of ethnolinguistic identity in southeast asia. In *Handbook of East and Southeast Asian Archaeology* (Issue 0). DOI: 10.1007/978-1-4939-6521-2_42

Blench, R. (2020). Research on the Plateau languages of Central Nigeria. *Hamburg University Press, 93*, 1–5. DOI: 10.15460/auue.2021.93.1.209

Boers, F. (2003). Applied linguistics perspectives on cross-cultural variation in conceptual metaphor. *Metaphor and Symbol, 18*(4), 231–238.

Bonab, H., Allan, J., & Sitaraman, R. (2019). Simulating CLIR translation resource scarcity using high-resource languages. *ICTIR 2019 - Proceedings of the 2019 ACM SIGIR International Conference on Theory of Information Retrieval*, 129–136. DOI: 10.1145/3341981.3344236

Borg, S. (2003). Teacher cognition in language teaching: A review of research on what language teachers think, know, believe, and do. *Language Teaching, 36*(2), 81–109. DOI: 10.1017/S0261444803001903

Borg, S. (2006). *Teacher cognition and language education: Research and practice* (1st ed.). Continuum.

Borg, S. (2009). Language teacher cognition. In Burns, A., & Richards, J. K. (Eds.), *The Cambridge guide to second language teacher education* (pp. 163–171). Cambridge University Press.

Bower, M., Torrington, J., Lai, J. W. M., Petocz, P., & Alfano, M. (2024). How should we change teaching and assessment in response to increasingly powerful generative artificial intelligence? Outcomes of the ChatGPT teacher survey. *Education and Information Technologies, 29*, 15403–15439. DOI: 10.1007/s10639-023-12405-0

Bridges, R. (2024). Oral traditions and digital preservation: Proverbs in contemporary society. *Journal of Cultural Heritage, 45*(2), 112–127.

Brown, P., & Levinson, S. (1987). *Politeness: Some universals in language usage.* Cambridge University Press.

Bruner, J. (2006). *Theory of Cognitive Development.* Routledge.

Butcher, N., Wilson-Strydom, M., & Baijnath, M. (2021). Artificial intelligence capacity in Sub-Saharan Africa: Compendium report. In *AI4D Africa.* https://idl-bnc-idrc.dspacedirect.org/items/27ea1089-760f-4136-b637-16367161edcc

Byram, M. (1997). *Teaching and assessing intercultural communicative competence.* Multilingual Matters.

C. S. T., Sharmiladevi, D., Sugumar, D., Ali, S. I., Jones Mary, T. A., & Kumar, S. (2023). Using smart wearable devices to monitor social distancing in smart cities. In *Proceedings of the 2023 International Conference on Research Methodologies in Knowledge Management, Artificial Intelligence and Telecommunication Engineering* (pp. 1–5). IEEE. https://doi.org/DOI: 10.1109/RMKMATE59243.2023.10369118

Calac, A. J., & Mackey, T. K. (2025). A systematic review of responsible stewardship of research and health data from Indigenous communities. *NPJ Digital Medicine, 8*(1). Advance online publication. DOI: 10.1038/s41746-025-01902-w PMID: 40764752

Calude, A. S., & Pagel, M. (2011). How do we use language? shared patterns in the frequency of word use across 17 world languages. *Philosophical Transactions of the Royal Society of London. Series B, Biological Sciences, 366*(1567), 1101–1107. DOI: 10.1098/rstb.2010.0315 PMID: 21357232

Carlà, M. M., Giannuzzi, F., Boselli, F., & Rizzo, S. (2024). Testing the power of Google DeepMind: Gemini versus ChatGPT 4 facing a European ophthalmology examination. *AJO International, 1*(3), 100063.

Celik, I., Dindar, M., Muukkonen, H., & Järvelä, S. (2022). The promises and challenges of artificial intelligence for teachers: A systematic review of research. *TechTrends, 66*(4), 616–630. DOI: 10.1007/s11528-022-00715-y

Chaitanya, K., & Rolla, K. J. (2024). The Evolution and Impact of Large Language Models in Artificial Intelligence. In *Algorithms in Advanced Artificial Intelligence* (pp. 410–417). CRC Press.

Chalmers, D. J. (2022). *Reality+: Virtual worlds and the problems of philosophy.* W. W. Norton & Company.

Chen, X., & Zmire, Z. (2024). AI's Role in Enhancing Cross-Cultural Competence and Leadership through Online Education Programs. *The Proceedings of the International Conference on New Findings in Humanities and Social Sciences, 1*(1), 1.

Chen, Y. (2024). Chinese primary school students improve English through VR-based drama education games. *Addressing Global* Challenges-*Exploring Socio-Cultural Dynamics and Sustainable Solutions in a Changing World*, 295-301.

Chen, C. (2004). Searching for intellectual turning points: Progressive knowledge domain visualization. *Proceedings of the National Academy of Sciences of the United States of America, 101*(suppl_1), 5303–5310.

Chen, C. (2006). CiteSpace II: Detecting and visualizing emerging trends and transient patterns in scientific literature. *Journal of the American Society for Information Science and Technology*, *57*(3), 359–377.

Chen, H., Han, J., & Wright, D. (2020). An investigation of lecturers' teaching through English medium of instruction, A case of higher education in China. *Sustainability*, *12*(10), 4046.

Chen, J. (2025). Modeling Chinese second language learners' motivation, engagement, and resilience in AI-enhanced contexts: A self-determination theory. *Learning and Motivation*, *92*, 102199.

Chen, X., Xie, H., Zou, D., & Hwang, G. J. (2020). Application and theory gaps during the rise of Artificial Intelligence in Education. *Computers and Education: Artificial Intelligence*, *1*, 100002. DOI: 10.1016/j.caeai.2020.100002

Chen, X., Zou, D., Xie, H., Cheng, G., & Liu, C. (2022). Two decades of artificial intelligence in education. *Journal of Educational Technology & Society*, *25*(1), 28–47.

Cherian, M., Jeevitha, M, & Manoja, K., & Abhinav. (2025). Leveraging AI for Predicting Marketing and Customer Insights-An Overview. *Journal of Informatics Education and Research*, *5*(1), 705–711. DOI: 10.52783/jier.v5i1.2050

Chew, F. P., Min, S., & Yap, T. T. (2023). Factors influencing the Mandarin teachers' intercultural teaching beliefs and practices. *Kasetsart Journal of Social Sciences*, *44*(2), 455–464. https://so04.tci-thaijo.org/index.php/kjss/article/view/266273

Chinh, N. D. (2013). Cultural diversity in english language teaching: Learners' voices. *English Language Teaching*, *6*(4), 1–7. DOI: 10.5539/elt.v6n4p1

Chiu, T. K. F., Xia, Q., Zhou, X., Chai, C. S., & Cheng, M. (2023). Systematic literature review on opportunities, challenges, and future research recommendations of artificial intelligence in education. *Computers and Education: Artificial Intelligence*, *4*, 100118. DOI: 10.1016/j.caeai.2022.100118

Chong, S. W., & Reinders, H. (2025). Autonomy of English language learners: A scoping review of research and practice. *Language Teaching Research, 29*(2), 607–632. DOI: 10.1177/13621688221075812

Chopra, S. (2024). Zero Resource Cross-Lingual Part Of Speech Tagging. *ArXiv Preprint ArXiv:2401.05727.* https://arxiv.org/pdf/2401.05727v1

Choudhury, P., Kim, D. Y., & Kang, S. (2026). Beyond English-Centric AI: Strategic Frameworks for Developing Low-Resource Language Models. SSRN *Electronic Journal*, 1–45.

Chowdhury, G. (2003). Natural language processing. *Annual Review of Information Science & Technology, 37*, 51–89.

Chua, N. A., Soon, G. Y., Ibrahim, M. Y., Che Noh, C. H., Mansor, N. R., Zaid, C. M., & Shen, M. (2021). Virtual differentiated instruction and its implementation in Mandarin teaching: A review. *Journal of Nusantara Studies, 6*(2).

Conneau, A., Khandelwal, K., Goyal, N., Chaudhary, V., Wenzek, G., Guzmán, F., Grave, E., Ott, M., Zettlemoyer, L., & Stoyanov, V. (2019). Unsupervised cross-lingual representation learning. *ACL 2019 - 57th Annual Meeting of the Association for Computational Linguistics, Tutorial Abstracts*, 31–38. DOI: 10.18653/v1/p19-4007

Contrino, M. F., Reyes-Millán, M., Vázquez-Villegas, P., & Membrillo-Hernández, J. (2024). Using an adaptive learning tool to improve student performance and satisfaction in online and face-to-face education for a more personalized approach. *Smart Learning Environments, 11*(6), 1–24. DOI: 10.1186/s40561-024-00292-y

Cotterell, R., & Duh, K. (2024). Low-Resource Named Entity Recognition with Cross-Lingual, Character-Level Neural Conditional Random Fields. *ArXiv Preprint ArXiv:2404.09383*. https://arxiv.org/abs/2404.09383

Coupland, N. (2015). The Handbook of Language and Globalization. In *Etika Jurnalisme Pada Koran Kuning : Sebuah Studi Mengenai Koran Lampu Hijau* (Vol. 16, Issue 2).

Court, S., & Elsner, M. (2024). Shortcomings of LLMs for Low-Resource Translation: Retrieval and Understanding are Both the Problem. *Conference on Machine Translation - Proceedings, 2024-Novem*, 1332–1354. DOI: 10.18653/v1/2024.wmt-1.125

Creissels, D. (2019). Morphology in Niger-Congo Languages. In *Oxford Research Encyclopedia of Linguistics*. DOI: 10.1093/acrefore/9780199384655.013.535

Creswell, J. W., & Creswell, J. D. (2018). *Research design: Qualitative, quantitative, and mixed methods approaches* (5th ed.). SAGE Publications.

Crowe, K., Cumming, T., McCormack, J., Baker, E., McLeod, S., Wren, Y., Roulstone, S., & Masso, S. (2017). Educators' perspectives on facilitating computer-assisted speech intervention in early childhood settings. *Child Language Teaching and Therapy, 33*(3), 267–285. DOI: 10.1177/0265659017717437

Cruz, S. D. D. (2020). Saving the face: Language courtesy strategies among students. *Asian EFL Journal, 27*(32), 39–60.

D'Amato, K. (2025). ChatGPT: Towards AI subjectivity. *AI & Society, 40*(3), 1627–1641.

D'Mello, S., & Graesser, A. (2015). Feeling, thinking, and computing with affect-aware learning technologies. In Calvo, R. A., D'Mello, S., Gratch, J., & Kappas, A. (Eds.), *The Oxford handbook of affective computing* (pp. 419–434). Oxford University Press.

Dabbagh, M. (2016). Cultural conceptualisations in proverbs: A linguistic-anthropological approach to understanding metaphor, schema, and category. *Linguistics and Anthropology Review, 28*(4), 345–367.

Dabbagh, M. (2017). Understanding proverbs through cultural schemas, cultural categories, and cultural metaphors: Frameworks for evaluating cultural interpretation. *Intercultural Communication Studies, 26*(1), 89–112.

Dakshinamurthi, V., Ali, S. I., Karthikeyan, T., & Kulkarni, N. S. (2023). Causal convolution employing Almeida–Pineda recurrent backpropagation for mobile network design. *ICTACT Journal on Communication Technology, 14*(4), 3091. DOI: 10.21917/ijct.2023.0460

Dani, N. A., & Seruji, Z. (2024). Language Vitality Among the Young Generation of a Minority Ethnic Group in Sabah. *International Journal of Humanities, Philosophy and Language, 7*(28), 01–18. DOI: 10.35631/ijhpl.728001

De la Vall, R. R. F., & Araya, F. G. (2023). Exploring the benefits and challenges of AI-language learning tools. *International Journal of Social Sciences and Humanities Invention, 10*(01), 7569–7576. DOI: 10.18535/ijsshi/v10i01.02

De Santis, E., & Rizzi, A. (2025). Noosemia: toward a Cognitive and Phenomenological Account of Intentionality Attribution in Human-Generative AI Interaction. *arXiv preprint arXiv:2508.02622.*

de Wit, H. (2020). Internationalization of Higher Education: The Need for a More Ethical and Qualitative Approach. *Journal of International Students, 10*(1), i–iv. DOI: 10.32674/jis.v10i1.1893

Deardorff, D. K. (2006). Identification and assessment of intercultural competence as a student outcome of internationalization. *Journal of Studies in International Education, 10*(3), 241–266. DOI: 10.1177/1028315306287002

Deng, M. (2022). Application of Artificial Intelligence Technology in College Students' Foreign Language Learning. *2022 IEEE 5th Eurasian Conference on Educational Innovation (ECEI)*, 302-304.

Deng, X., Xu, Y., Hussin, S., & Sulaiman, N. A. (2025). Use of AI Tools in EFL Writing Instruction: A Case Study of Chinese Vocational College Instructors. In *Forum for Linguistic Studies* (Vol. 7, No. 5, pp. 114-129).

Deng, Z., Ma, W., Han, Q. L., Zhou, W., Zhu, X., Wen, S., & Xiang, Y. (2025). Exploring DeepSeek: A Survey on Advances, Applications, Challenges and Future Directions. *IEEE/CAA Journal of Automatica Sinica, 12*(5), 872–893.

Deng, J., Liu, C., & Zhou, Z. (2024). *AI-Based English Vowel Formant Recognition and Pronunciation Correction Platform*. IEEE.

Deng, L. (2018). Artificial Intelligence in the Rising Wave of Deep Learning: The Historical Path and Future Outlook [Perspectives]. *IEEE Signal Processing Magazine*, *35*(1), 180–177. DOI: 10.1109/MSP.2017.2762725

Deng, R., Jiang, M., Yu, X., Lu, Y., & Liu, S. (2024). Does ChatGPT enhance student learning? A systematic review and meta-analysis of experimental studies. *Computers & Education*, •••, 105224.

Derakhshan, A. (2025). EFL students' perceptions about the role of generative artificial intelligence (GAI)-mediated instruction in their emotional engagement and goal orientation: A motivational climate theory (MCT) perspective in focus. *Learning and Motivation*, *90*, 102114.

Derakhshan, A., & Ghiasvand, F. (2024). Is ChatGPT an evil or an angel for second language education and research? A phenomenographic study of research-active EFL teachers' perceptions. *International Journal of Applied Linguistics*, *34*(4), 1246–1264.

Dervin, F. (2025). *AI for critical interculturality*. Taylor & Francis.

Devlin, J., Chang, M. W., Lee, K., & Toutanova, K. (2019, June). Bert: Pre-training of deep bidirectional transformers for language understanding. In *Proceedings of the 2019 conference of the North American chapter of the association for computational linguistics: human language technologies, volume 1 (long and short papers)* (pp. 4171-4186).

Devlin, J., Chang, M.-W., Lee, K., & Toutanova, K. (2019). BERT: Pre-training of deep bidirectional transformers for language understanding. *arXiv*. https://arxiv.org/abs/1810.04805

Devlin, J., Chang, M.-W., Lee, K., & Toutanova, K. (2019). BERT: Pre-training of Deep Bidirectional Transformers for Language Understanding. *Conference of the North American Chapter of the Association for Computational Linguistics: Human Language Technologies, Mlm*. https://arxiv.org/abs/1810.04805

Dimari, A., Tyagi, N., Davanageri, M., Kukreti, R., Yadav, R., & Dimari, H. (2024). AI-Based Automated Grading Systems for open book examination system: Implications for Assessment in Higher Education. *2024 International Conference on Knowledge Engineering and Communication Systems (ICKECS)*, 1-7. https://doi.org/DOI: 10.1109/ICKECS61492.2024.10616490

Ding, D., & Yusof, M. B. (2025). Investigating the role of AI-powered conversation bots in enhancing L2 speaking skills and reducing speaking anxiety: A mixed methods study. *Humanities & Social Sciences Communications*, *12*(1), 1–16.

Ding, Y. (2024, January). Development of Intelligent Tutors Based on Dialogue Systems. In *2024 Asia-Pacific Conference on Software Engineering, Social Network Analysis and Intelligent Computing (SSAIC)* (pp. 927-930). IEEE.

Doshi-Velez, F., & Kim, B. (2017). Towards a rigorous science of interpretable machine learning. *arXiv*. https://arxiv.org/abs/1702.08608

Drude, S., Ingason, A. K., Kristinsson, A. P., Arnbjörnsdóttir, B., Sigurðsson, E. F., Rögnvaldsson, E., Nowenstein, I., & Sigurjónsdóttir, S. (2018). Digital resources and language use: Expanding the EGIDS scale for language development into the digital domains. *Conference: Communities in Control - Learning Tools and Strategie for Multilingual Endangered Language Communities, August 2019*. https://www.researchgate.net/publication/335244251_Digital_resources_and_language_use_Expanding_the_EGIDS_scale_for_language_development_into_the_digital_domains

Duff, P. A. (2012). Identity, agency, and second language acquisition. In Mercer, S., Ryan, S., & Williams, M. (Eds.), *Language learner agency* (pp. 15–29). Multilingual Matters.

Dunbar, E., Cao, X. N., Benjumea, J., Karadayi, J., Bernard, M., Besacier, L., . . . Dupoux, E. (2017, December). The zero resource speech challenge 2017. In *2017 IEEE Automatic Speech Recognition and Understanding Workshop (ASRU)* (pp. 323-330). IEEE.

Du, Y., & Liu, Y. (2025). How generative AI affordance drives users' continuous usage intention: The mediation role of self-expansion and self-extension. *Behaviour & Information Technology*, •••, 1–16.

Dwivedi, Y., Hughes, L., Ismagilova, E., Aarts, G., Coombs, C., Crick, T., Duan, Y., Dwivedi, R., Edwards, J., Eirug, A., Galanos, V., Ilavarasan, P., Janssen, M., Jones, P., Kar, A., Kizgin, H., Kronemann, B., Lal, B., Lucini, B., & Williams, M. (2019). Artificial Intelligence (AI): Multidisciplinary perspectives on emerging challenges, opportunities, and agenda for research, practice and policy. *International Journal of Information Management*, *57*, 1–47. DOI: 10.1016/J.IJINFOMGT.2019.08.002

Ebadi, S., & Amini, J. (2022). The effects of AI-based chatbot on learners' affective status. *Computers & Education*, *180*, 104432. DOI: 10.1016/j.compedu.2022.104432

Egbuhuzor, N., Ajayi, A., Akhigbe, E., Agbede, O., Ewim, C., & Ajiga, D. (2025). AI and data-driven insights: Transforming customer relationship management (CRM) in financial services. *Gulf Journal of Advance Business Research*, *3*(2), 483–511. DOI: 10.51594/gjabr.v3i2.93

Egdom, G. W. V. (2024). Bridging Linguistic Divides? A Critical Exploration of Machine Translation's Role in Fostering Cross-Cultural Accessibility in Literature. *Íkala, Revista de Lenguaje y Cultura, 29*(3).

Eguchi, M., & Kyle, K. (2024). Building custom NLP tools to annotate discourse-functional features for second language writing research: A tutorial. *Research Methods in Applied Linguistics*, *3*(3), 100153.

Eisenring, M., Jamiluddin, J., Hairul, M., & Putri, D. (2024). The use of chatbots in the English language teaching to promote modern language learning: A literature review. *International Journal of Indonesian Education and Teaching*, *8*(1), 127–139.

Elman, J. L. (1990). Finding structure in time. *Cognitive Science*, *14*(2), 179–211.

Emlen, Q. N. (2025). The Quechuan-Aymaran relationship. In *The oxford guide to the languages of the central andes* (Issue December). DOI: 10.1093/oso/9780198849926.003.0026

Erdem Coşgun, G. (2025). Artificial intelligence literacy in assessment: Empowering pre-service teachers to design effective exam questions for language learning. *British Educational Research Journal*.

Eriksson Krutrök, M., & Poromaa Isling, P. (2025). Technolinguistic suspension and the digital futures of minority languages: AI and Meänkieli in Sweden. *new media & society*, 14614448251385920. Firdaus, T., Bakari, L. Y., Octarianto, D. S., Khoirunnisa, R., Malini, D. R., & Mukarromah, M. Applying Grice's Maxims and Speech Act Theory to Analyze Pragmatics in Chatgpt and Gemini AI Dialogues. *Uluslararası Dil ve Çeviri Çalışmaları Dergisi*, *5*(1), 34–53.

Escalante, J., Pack, A., & Barrett, A. (2023). AI-generated feedback on writing: Insights into efficacy and ENL student preference. *International Journal of Educational Technology in Higher Education*, *20*(57), 1–20. DOI: 10.1186/s41239-023-00425-2

Eskander, R., Muresan, S., & Collins, M. (2020). Unsupervised cross-lingual part-of-speech tagging for truly low-resource scenarios. *EMNLP 2020 - 2020 Conference on Empirical Methods in Natural Language Processing, Proceedings of the Conference*, 4820–4831. DOI: 10.18653/v1/2020.emnlp-main.391

Essien, A., Bukoye, O. T., O'Dea, X., & Kremantzis, M. (2024). The influence of AI text generators on critical thinking skills in UK business schools. *Studies in Higher Education*, *49*(5), 865–882. DOI: 10.1080/03075079.2024.2316881

Fadaee, M., Bisazza, A., & Monz, C. (2017). Data augmentation for low-Resource neural machine translation. *ACL 2017 - 55th Annual Meeting of the Association for Computational Linguistics, Proceedings of the Conference (Long Papers), 2*, 567–573. DOI: 10.18653/v1/P17-2090

Fadhil, M., Septiana, E., Sawitri, O., Abdillah, R., Syahfutra, W., & Nurafrezah, N. (2025). The Role of English Proficiency in Supporting Academic Achievement and Career Opportunities of University Students. *Widya Accarya*, *16*(1), 36–41. DOI: 10.46650/wa.16.1.1645.36-41

Farooqui, Y., Ali, S. I., Shekokar, K., Patidar, P. K., Bhushanwar, K., & Nandi, P. K. (2025). A secure framework for decentralized digital lockers using blockchain technology. *Journal of Computational Science*, *21*(9), 2105–2112. DOI: 10.3844/jcssp.2025.2105.2112

Fathi, J., & Rahimi, M. (2024). Utilising artificial intelligence-enhanced writing mediation to develop academic writing skills in EFL learners: A qualitative study. *Computer Assisted Language Learning*, •••, 1–40. DOI: 10.1080/09588221.2024.2374772

Feng, D. (2025). A Critical Bibliometric Review of Technology-Enhanced Vocabulary Learning: Uncovering a Decade of Tensions and Trends. *SAGE Open*, *15*(4), 21582440251393196.

Feng, L. (2025). Investigating the effects of artificial intelligence-assisted language learning strategies on cognitive load and learning outcomes: A comparative study. *Journal of Educational Computing Research*, *62*(8), 1741–1774.

Fidan, Ş., & Kasımi, Y. (2023). An investigation into artificial intelligence (AI) in the English as a foreign language (EFL) context. *International Journal of Educational Spectrum*, *5*(2), 269–280.

Fitryansyah, M. A., & Fauziah, F. N. (2024). Bridging Traditions and Technology: AI in The Interpretation of Nusantara Religious Manuscripts. *Jurnal Lektur Keagamaan*, *22*(2), 317–346.

Flavell, J. H. (1979). Metacognition and cognitive monitoring: A new area of cognitive–developmental inquiry. *The American Psychologist*, *34*(10), 906–911. DOI: 10.1037/0003-066X.34.10.906

Floridi, L., & Cowls, J. (2019). A unified framework of five principles for AI in society. *Harvard Data Science Review*, *1*(1). Advance online publication. DOI: 10.1162/99608f92.8cd550d1

Floris, V., Van Den, D., & Natascha, K. (2023). *Debating Ethical Dilemmas in the Classroom*. ISVW.

Foung, D., & Kohnke, L. (2025). Generating synthetic data for CALL research with GenAI: A proof-of-concept study. *Research Methods in Applied Linguistics*, *4*(3), 100248.

Foung, D., Kohnke, L., & Chen, J. (2023). Predicting student success with self-regulated behaviors: A seven-year data analytics study on a Hong Kong University English Course. In *Data analytics and adaptive learning* (pp. 92–109). Routledge.

Friederici, A. D. (2011). The brain basis of language processing: From structure to function. *Physiological Reviews*, *91*(4), 1357–1392. PMID: 22013214

Frith, U. (2003). *Autism: Explaining the enigma*. Blackwell Publishing.

Fu, Y., Foell, S., Xu, X., & Hiniker, A. (2024). From text to self: Users' perception of AIMC tools on interpersonal communication and self. *Proceedings of the CHI Conference on Human Factors in Computing Systems, 17,* 1–17. https://doi.org/ DOI: 10.1145/3613904.3641955

Gacel-Ávila, J. (2005). The Internationalisation of Higher Education: A Paradigm for Global Citizenry. *Journal of Studies in International Education*, *9*(2), 121–136. DOI: 10.1177/1028315304263795

Galloway, N., & Mckinley, J. (2021). *Englishization of Higher Education. Research Questions in Language Education and Applied Linguistics*. Springer., DOI: 10.1007/978-3-030-79143-8_123

Gamage, B. G. J., Paranagama, R. D., Ranaweera, R. M. S. H., Dilshan, A. V. R., Weerathunga, I., & Kasthurirathna, D. (2023, December). Sinhala sign language translation through immersive 3D avatars and adaptive learning. In *2023 5th International Conference on Advancements in Computing (ICAC)* (pp. 1-6). IEEE.

Gaol, E. L., Noerjanto, F., Wolomasi, A. K., & Wea, D. (2024). Development of an AI-based platform to customize learning experiences in religious education: Papuan youth Emik characterized by local wisdom. *Journal of Social Science Utilizing Technology*, *2*(4), 565–573. DOI: 10.70177/jssut.v2i4.1503

Ghazanfari, M., Kumar, S., & Chen, L. (2025). Multilayered semantic complexity in non-Western proverbs: Challenges for artificial intelligence interpretation and cultural understanding. *Computational Linguistics and Digital Humanities*, *9*(1), 45–63.

Gillo, M. D. (2021). Fundamentals of hermeneutics as a qualitative research theoretical framework. *European Journal of Education and Pedagogy*, *2*(3), 42–45. DOI: 10.24018/ejedu.2021.2.3.43

Gligorea, I., Cioca, M., Oancea, R., Gorski, A., Gorski, H., & Tudorache, P. (2023). Adaptive Learning Using Artificial Intelligence in e-Learning: A Literature Review. *Education Sciences*, *13*(12). Advance online publication. DOI: 10.3390/educsci13121216

Goatly, A. (2022). *Two dimensions of meaning: Similarity and contiguity in metaphor and metonymy, language, culture, and ecology*. Routledge.

Göksu, İ., & Atici, B. (2013, November). Need for Mobile Learning: Technologies and Opportunities. *Procedia: Social and Behavioral Sciences*, *103*, 685–694. DOI: 10.1016/j.sbspro.2013.10.388

Gomashie, G. A., & Terborg, R. (2021). Nahuatl, selected vitality indicators and scales of vitality in an Indigenous language community in Mexico. *Open Linguistics*, *7*(1), 166–180. DOI: 10.1515/opli-2021-0010

Good, J. (2014). *Niger-Congo languages Niger-Congo*. The Cambridge handbook of areal linguistics.

Goyal, N., Gao, C., Chaudhary, V., Chen, P. J., Wenzek, G., Ju, D., Krishnan, S., Ranzato, M., Guzmán, F., & Fan, A. (2022). The FLORES-101 Evaluation Benchmark for Low-Resource and Multilingual Machine Translation. *Transactions of the Association for Computational Linguistics*, *10*, 522–538. DOI: 10.1162/tacl_a_00474

Graesser, A. C., Chipman, P., Haynes, B. C., & Olney, A. (2005). AutoTutor: An intelligent tutoring system with mixed-initiative dialogue. *IEEE Transactions on Education*, *48*(4), 612–618.

Greenhill, S. J., Drummond, A. J., & Gray, R. D. (2010). How accurate and robust are the phylogenetic estimates of austronesian language relationships? *PLoS One*, *5*(3), 1–6. DOI: 10.1371/journal.pone.0009573 PMID: 20224774

Gunkel, D. J. (2021). *Deconstruction*. The MIT Press., DOI: 10.7551/mit-press/13637.001.0001

Guo, S., & Xia, M. (2026). Exploring Learners' Perceptions of AI-mediated informal digital Multilingual Learning: A Q-Methodology Approach. *International Journal of Applied Linguistics*. Advance online publication. DOI: 10.1111/ijal.12815

Habibi, A., Attar, R. W., Atan, N. A., Said, M. N. H. B. M., Sofwan, M., Yaqin, L. N., & Al-Adwan, A. S. (2025). Technology use during teaching practicum: Beliefs/knowledge in urban and suburban schools. *Asia Pacific Journal of Education*, ●●●, 1–26. DOI: 10.1080/02188791.2025.2521112

Hadijah Haji Hassan. (1987). *Perambahan Brunei*. Bandar Seri Begawan: Dewan Bahasa dan Pustaka Brunei.

Hadi, M. U., Qureshi, R., Shah, A., Irfan, M., Zafar, A., Shaikh, M. B., & Mirjalili, S. (2023). Large language models: a comprehensive survey of its applications, challenges, limitations, and future prospects. *Authorea preprints, 1*(3), 1–26.

Haenlein, M., & Kaplan, A. (2019). A Brief History of Artificial Intelligence: On the Past, Present, and Future of Artificial Intelligence. *California Management Review*, *61*(4), 5–14. DOI: 10.1177/0008125619864925

Hagendorff, T., Bossert, L. N., Tse, Y. F., & Singer, P. (2023). Speciesist bias in AI: How AI applications perpetuate discrimination and unfair outcomes against animals. *AI and Ethics*, *3*(3), 717–734.

Haji-Othman, N. A., McLellan, J., & Deterding, D. (2016). *The Use and Status of Language in Brunei Darussalam*. Springer.

Hakkarainen, S., & Engelbrecht, K. (2024). *Automatic Idiomatic Expression Detection. Comparison Between GPT-4 and Gemini Pro Prompt Engineering & LSTM-RNN Construction*. [Master's Thesis, University of Gothenburg]

Hamdan, M. (2017). Transformational Change in Brunei Higher Education: A Case Study of Universiti Brunei Darussalam. *Higher Education Research*, *2*(3), 76–85. DOI: 10.11648/j.her.20170203.11

Hamdi, S. A., Hashem, R. A., Holbah, W. A., Azi, Y., & Mohammed, S. Y. (2023). Proverbs Translation for Intercultural Interaction: A Comparative Study between Arabic and English Using Artificial Intelligence. *World Journal of English Language, 13*(7), 282.

Hancock, J. T., & Levy, K. (2020). AI-mediated communication: Definition, research agenda, and ethical considerations. *Journal of Computer-Mediated Communication, 25*(1), 89–100. DOI: 10.1093/jcmc/zmz022

Haozous, E. A., Faan, J. L., & Soto, C. (2021). Urban American Indian and Alaska Native Data Sovereignty: Ethical Issues. *American Indian and Alaska Native Mental Health Research, 28*(2), 77–99. DOI: 10.5820/AIAN.2802.2021.77 PMID: 34586627

Haraway, D. J. (2016). *Staying with the trouble: Making kin in the Chthulucene.* Duke University Press.

Harisha, N., Suminih, S., & Oktaviana, F. (2024). the Power of Chatbots in English Language Learning: A New Age of Learning. *Lingua, 20*(1), 19–31. DOI: 10.34005/lingua.v20i1.3492

Haristiani, N. (2019). Artificial Intelligence (AI) Chatbot as Language Learning Medium: An inquiry. *Journal of Physics: Conference Series, 1387*(1), 1–6. DOI: 10.1088/1742-6596/1387/1/012020

Harnad, S. (2024). Language writ large: LLMS, ChatGPT, grounding, meaning and understanding. *arXiv preprint arXiv:2402.02243.*

Hasan, M. A., Tarannum, P., Dey, K., Razzak, I., & Naseem, U. (2024). Do Large Language Models Speak All Languages Equally? A Comparative Study in Low-Resource Settings. *ArXiv Preprint ArXiv:2408.02237.* https://arxiv.org/abs/2408.02237

Hayles, N. K. (2017). *Unthought: The power of the cognitive nonconscious.* University of Chicago Press., DOI: 10.7208/chicago/9780226447919.001.0001

Helm, P., Bella, G., Koch, G., & Giunchiglia, F. (2024). Diversity and language technology: How language modelling bias causes epistemic injustice. *Ethics and Information Technology, 26*(1), 8.

Heyard, R., & Hottenrott, H. (2021). The value of research funding for knowledge creation and dissemination: A study of SNSF Research Grants. *Humanities & Social Sciences Communications, 8*, 217. DOI: 10.1057/s41599-021-00891-x

Hila, A. (2025). The epistemological consequences of large language models: Rethinking collective intelligence and institutional knowledge. *AI & Society*. Advance online publication. DOI: 10.1007/s00146-025-02426-3

Hoermann, S., McCabe, K. L., Milne, D. N., & Calvo, R. A. (2017). Application of synchronous text-based dialogue systems in mental health interventions: Systematic review. *Journal of Medical Internet Research, 19*(8), e267. DOI: 10.2196/jmir.7023 PMID: 28784594

Holec, H. (1981). *Autonomy is the ability to take charge of one's own learning*. Pergamon.

Hossain, K. I. (2024). Reviewing the role of culture in English language learning: Challenges and opportunities for educators. *Social Sciences and Humanities Open, 9*(August 2023), 100781. https://doi.org/DOI: 10.1016/j.ssaho.2023.100781

Hovy, D., & Spruit, S. L. (2016). The social impact of natural language processing. *54th Annual Meeting of the Association for Computational Linguistics, ACL 2016 - Short Papers*, 591–598. DOI: 10.18653/v1/p16-2096

Huang, K., Mo, F., Li, H., Li, Y., Zhang, Y., Yi, W., Mao, Y., Liu, J., Xu, Y., Xu, J., Nie, J.-Y., & Liu, Y. (2024). A Survey on Large Language Models with Multilingualism: Recent Advances and New Frontiers. *ArXiv Preprint ArXiv:2405.10936*, 1–65. https://arxiv.org/abs/2405.10936

Huang, S., & Wang, W. (2025, March). Foreign Language Learners' Perceptions of Chatbot-Assisted Speech-Giving Tasks. In *2025 14th International Conference on Educational and Information Technology (ICEIT)* (pp. 79-83). IEEE.

Huang, G., & Wang, Z. (2022). The application of virtual reality technology in the coordination and interaction of regional economy and culture in the sustainable development of ecotourism. *Mathematical Problems in Engineering, 2022*(1), 9847749.

Huang, W., Hew, K. F., & Fryer, L. K. (2022). Chatbots for language learning—Are they really useful? A systematic review of chatbot-supported language learning. *Journal of Computer Assisted Learning, 38*(1), 237–257. DOI: 10.1111/jcal.12610

Huck, M., Dutka, D., & Fraser, A. (2019). Cross-lingual Annotation Projection Is Effective for Neural Part-of-Speech Tagging. *In Proceedings of the Sixth Workshop on NLP for Similar Languages, Varieties and Dialects, 2001*, 223–233. DOI: 10.18653/v1/w19-1425

Hu, J., Ruder, S., Siddhant, A., Neubig, G., Firat, O., & Johnson, M. (2020). XTREME: A massively multilingual multi-task benchmark for evaluating cross-lingual generalization. *37th International Conference on Machine Learning, ICML 2020, PartF16814*, 4361–4371.

Huynh, Q. Q. (2024). A literature review on EFL students' integration of Al-based tools in English learning and the perceived impact on autonomy. *International Journal of AI in Language Education, 1*(2), 30–51. DOI: 10.54855/ijaile.24123

Hwang, W. Y., Manabe, K., Cai, D. J., & Ma, Z. H. (2020). Collaborative kinesthetic English learning with recognition technology. *Journal of Educational Computing Research, 58*(5), 946–977.

Inayah, N., Afra, A., Azmar, C. N., & Yaqin, L. N. (2024). A linguistic analysis of grammatical cohesion devices in discussion sections of undergraduate theses. *International Journal of Language Studies, 18*(4), 149–166. DOI: 10.5281/zenodo.13956443

Inuwa-Dutse, I. (2025). NaijaNLP: A Survey of Nigerian Low-Resource Languages. *ArXiv Preprint ArXiv:2502.19784, 1*(1). https://arxiv.org/abs/2502.19784

Ishamina, A., & Deterding, D. (2017). English Medium Education in a University in Brunei Darussalam: Code-Switching and Intelligibility. In Fenton-Smith, B., Humphreys, P., & Walkinshaw, I. (Eds.), *English Medium Instruction in Higher Education in Asia-Pacific. Multilingual Education, 21, 281-298*. Springer., DOI: 10.1007/978-3-319-51976-0_15

Issa, F. (2024). Using Infographics as An Educational Technology Tool in EFL Writing: University of Baghdad Case Study. *Arab World English Journal, 15*(1), 166–181. DOI: 10.24093/awej/vol15no1.11

Izharuddin, M., Hadi, F., & Lestyorini, M. (2023). Evaluation of World University Rankings of Southeast Asia HEIs Image. *Wiga: Jurnal Penelitian Ilmu Ekonomi, 13*(1), 38–45. DOI: 10.30741/wiga.v13i1.941

Izzati, I. D. C., Tentama, F., & Suyono, H. (2020). Academic stress scale: A psychometric study for academic stress in senior high school. *European Journal of Education Studies, 7*(7), 153–168. DOI: 10.46827/ejes.v7i7.3161

Jeon, J. (2024). Exploring AI chatbot affordances in the EFL classroom: Young learners' experiences and perspectives. *Computer Assisted Language Learning, 37*(1-2), 1–26.

Jeon, J., Lee, S., & Choe, H. (2023). Beyond ChatGPT: A conceptual framework and systematic review of speech-recognition chatbots for language learning. *Computers & Education, 206*, 104898.

Jiang, F., Jiang, Y., Zhi, H., Dong, Y. S., Wang, Y., Dong, Q., Shen, H., & Wang, Y. (2017). Artificial intelligence in healthcare: Past, present and future. *Stroke and Vascular Neurology, 2*(4), 230–243. DOI: 10.1136/svn-2017-000101

Jiang, R. (2022). How does artificial intelligence empower EFL teaching and learning nowadays? A review on artificial intelligence in the EFL context. *Frontiers in Psychology, 13*, 1049401. DOI: 10.3389/fpsyg.2022.1049401 PMID: 36467167

Ji, H., Han, I., & Ko, Y. (2022). A systematic review of conversational AI in language education: Focusing on the collaboration with human teachers. *Journal of Research on Technology in Education, 55*(1), 1–22. DOI: 10.1080/15391523.2022.2142873

Johanson, L. (2021). *The Turkic Languages* (Johanson, L., & Csató, É. Á., Eds.). Routledge., 10.4324/9781315758115

Jolley, J., & Maimone, L. (2025). Machine translation and language teaching and learning. *Language Teaching, 58*(2), 173–191.

Jones, G. M. (2015). Bilingual and Multilingual Education in Brunei and Malaysia: Policies and Practices. In Wright, W. E., Boun, S., & Garcia, O. (Eds.), *The Handbook of Bilingual and Multilingual Education* (pp. 531–541). Wiley-Blackwell.

Joshi, B. M., Khatiwada, S. P., & Pokhrel, R. K. (2024). Influence of Socioeconomic Factors on Access to Digital Resources for Education. *Rupantaran: A Multidisciplinary Journal, 8*(01), 17–33. DOI: 10.3126/rupantaran.v8i01.65197

Joshi, P., Santy, S., Budhiraja, A., Bali, K., & Choudhury, M. (2020). The State and Fate of Linguistic Diversity and Inclusion in the NLP World. *Proceedings of the 58th Annual Meeting of the Association for Computational Linguistics*, 6282–6293. https://aclanthology.org/2020.acl-main.560/

Joshi, S. (2025). A Comprehensive Review of DeepSeek: Performance, Architecture and Capabilities [Preprint]. doi: DOI: 10.20944/preprints202503.1887.v1

Kakwani, D., Kunchukuttan, A., Golla, S., Gokul, N. C., Bhattacharyya, A., Khapra, M. M., & Kumar, P. (2020). Findings of the Association for Computational Linguistics IndicNLPSuite: Monolingual Corpora, Evaluation Benchmarks and Pre-trained Multilingual Language Models for Indian Languages. *Findings of the Association for Computational Linguistics: EMNLP 2020, 3*, 4948–4961. https://github.com/AI4Bharat/indicnlp_catalog

Kalinowski, E., Egert, F., Gronostaj, A., & Vock, M. (2020). Professional development on fostering students' academic language proficiency across the curriculum—A meta-analysis of its impact on teachers' cognition and teaching practices. *Teaching and Teacher Education*, *88*, 102971. DOI: 10.1016/j.tate.2019.102971

Kann, K., Lacroix, O., & Søgaard, A. (2020). Weakly supervised POS taggers perform poorly on truly low-resource languages. *AAAI 2020 - 34th AAAI Conference on Artificial Intelligence, 2015*, 8066–8073. DOI: 10.1609/aaai.v34i05.6317

Karjalainen, H., Puura, U., Günthal, R., & Kovaleva, S. (2013). Karelian in Russia. *Studies in European Language Diversity*, *26*, 1–269. https://phaidra.univie.ac.at/detail_object/o:80726

Karpouzis, K. (2025). AI, digital humanities, and the legacies of colonial power [Preprint]. *Preprints*.org. https://doi.org/DOI: 10.20944/preprints202502.1823.v1

Karunambiga, K., Ali, S. I., Esteban, A. P., & Pascual, M. (2023). Marketing policy in service enterprises using deep learning model. *International Journal of Intelligent Systems and Applications in Engineering*, *12*(7s), 239–243.

Kasneci, E., Seßler, K., Küchemann, S., Bannert, M., Dementieva, D., Fischer, F., & Kasneci, G. (2023). ChatGPT for good? On opportunities and challenges of large language models for education. *Learning and Individual Differences*, *103*, 102274.

Kawakibi, M. A., & Indrawan, F. (2024). The Importance of English in International Science Research Collaboration. *Teaching English as Foreign Language. Literatura y Lingüística*, *4*(1), 40–44.

Kelleher, J. D. (2023). *Deep learning*. MIT Press., DOI: 10.7551/mitpress/12689.001.0001

Kenya, P., Nugroho, R., Kuswandi, D., & Wedi, A. (2023). Is the Trend of AI Chatbot Assistants in Learning Worth Maintaining? In *EDTECH-ID* (pp. 1–3).

Khandaqji, F., Ashqar, H. I., & Atawnih, A. (2025, July). A Survey of Using Artificial Intelligence (AI) in Sign Language for Deaf and Hard-of-Hearing Students. In *2025 International Conference on Smart Learning Courses (SCME)* (pp. 172-177). IEEE.

Khoo, E., & Kang, S. (2022). Proactive learner empowerment: Towards a transformative academic integrity approach for English language learners. *International Journal for Educational Integrity*, *18*(1), 24. DOI: 10.1007/s40979-022-00111-2

Khurma, O. A., Albahti, F., Ali, N., & Bustanji, A. (2024). AI ChatGPT and student engagement: Unraveling dimensions through PRISMA analysis for enhanced learning experiences. *Contemporary Educational Technology*, *16*(2). Advance online publication. DOI: 10.30935/cedtech/14334

Kiaos, T. (2023). An interpretative framework for analysing managerial ideology, normative control, organisational culture and the self. *Cogent Business & Management*, *10*(1), 2163795.

Kim, H., Yoon, J., Lee, S. M., Cho, S. M., Kim, Y. J., Noh, G., & Lee, J. (2025). Lecture-grounded ChatGPT-4 tutor: Enhancing adult learner engagement in distance learning. *PUPIL: International Journal of Teaching, Education and Learning*. https://doi.org/DOI: 10.20319/ictel.2025.324342

Kizilcec, R. F., Huber, E., Papanastasiou, E. C., Cram, A., Makridis, C. A., Smolansky, A., Zeivots, S., & Raduescu, C. (2024). Perceived impact of generative AI on assessments: Comparing educator and student perspectives in Australia, Cyprus, and the United States. *Computers and Education: Artificial Intelligence*, *7*, 100269. DOI: 10.1016/j.caeai.2024.100269

Ko, G. (2025). Artificial Intelligence and Your Life: The Problems We Must Face. *Science Insights*, *46*(4), 1799–1803. DOI: 10.15354/si.25.pe211

Kohnke, L., Moorhouse, B. L., & Zou, D. (2023). Exploring generative artificial intelligence preparedness among university language instructors: A case study. *Computers and Education: Artificial Intelligence*, *5*, 100156.

Kolchenko, V. (2018). Can modern AI replace teachers? Not so fast! Artificial intelligence and adaptive learning: Personalized education in the AI age. *HAPS educator, 22*(3), 249-252. https://doi.org/DOI: 10.21692/haps.2018.032

Koltovskaia, S. (2020). Student engagement with automated written corrective feedback (AWCF) provided by Grammarly: A multiple case study. *Assessing Writing*, *44*, 100450.

Koriat, A. (2015). When two heads are better than one and when they can be worse: The amplification hypothesis. *Journal of Experimental Psychology. General*, *144*(5), 934–950. DOI: 10.1037/xge0000092 PMID: 26168039

Korzekwa, D., Lorenzo-Trueba, J., Zaporowski, S., Calamaro, S., Drugman, T., & Kostek, B. (2021, June). Mispronunciation detection in non-native (L2) English with uncertainty modeling. In *ICASSP 2021-2021 IEEE international* conference *on acoustics, speech and signal processing (ICASSP)* (pp. 7738-7742). IEEE.

Kovač, I., Stamenković, D., & Williams, M. (2025). Cultural bias in large language models: Mechanisms, manifestations, and mitigation strategies for reducing Western-centric epistemological dominance. *ACM Transactions on Computing and Society*, *8*(2), 1–28.

Kumar, S., & Kaur, K. (2025). Digital pedagogies in the age of AI: Leveraging emerging ICT tools for inclusive and sustainable education. *The Academic*, *3*(5), 1805–1819. DOI: 10.5281/zenodo.15659123

Kurusu, K., Oda, C., Go, M., Wu, D., Saure, K., & Narang, S. (2024). English in the internationalization of higher education and international student mobility. *AILA Review*, *37*(1), 120–136. DOI: 10.1075/aila.24002

Kwiek, M. (2020). What large-scale publication and citation data tell us about international research collaboration in Europe: Changing national patterns in global contexts. *Studies in Higher Education*, *46*(12), 2629–2649. DOI: 10.1080/03075079.2020.1749254

Labadze, L., Grigolia, M., & Machaidze, L. (2023). Role of AI chatbots in education: Systematic literature review. *International Journal of Educational Technology in Higher Education*, *20*(1), 1–17. DOI: 10.1186/s41239-023-00426-1

Lahby, M. (Ed.). (2024). *Empowering Digital Education with ChatGPT: From Theoretical to Practical Applications*. CRC Press.

Lake, B. M., Ullman, T. D., Tenenbaum, J. B., & Gershman, S. J. (2017). Building machines that learn and think like people. *Behavioral and Brain Sciences*, *40*, e253. PMID: 27881212

Lakew, S. M., Negri, M., & Turchi, M. (2020). Low Resource Neural Machine Translation: A Benchmark for Five African Languages. *ArXiv Preprint ArXiv:2003.14402*, 1–10. https://arxiv.org/abs/2003.14402

Lakoff, G. (2006). Conceptual metaphor. *Cognitive linguistics: Basic readings, 34*, 185–238.

Lample, G., Conneau, A., Ranzato, M., Denoyer, L., & Jegou, H. (2018). Word translation without parallel data. *International Conference on Learning Representations, 2015*, 1–14. https://openreview.net/pdf?id=H196sainb

Lan, G., Li, Y., Yang, J., & He, X. (2025). Investigating a customized generative AI chatbot for automated essay scoring in a disciplinary writing task. *Assessing Writing*, *66*, 100959.

Lantolf, J. P., & Thorne, S. L. (2006). *Sociocultural theory and the genesis of second language development*. Oxford University Press.

Lauscher, A., Ravishankar, V., Vulić, I., & Glavaš, G. (2020). From zero to hero: On the limitations of zero-shot language transfer with multilingual transformers. *EMNLP 2020 - 2020 Conference on Empirical Methods in Natural Language Processing, Proceedings of the Conference*, 4483–4499. DOI: 10.18653/v1/2020.emnlp-main.363

Le Scao, T., Fan, A., Akiki, C., Pavlick, E., Ilić, S., Hesslow, D., Castagné, R., Luccioni, A. S., Yvon, F., Gallé, M., Tow, J., Rush, A. M., Biderman, S., Webson, A., Ammanamanchi, P. S., Wang, T., Sagot, B., Muennighoff, N., del Moral, A. V., . . . Wolf, T. (2022). BLOOM: A 176B-Parameter Open-Access Multilingual Language Model. *ArXiv Preprint ArXiv:2211.05100*. https://arxiv.org/abs/2211.05100

Leahey, T. (2012). *Cognition and Learning*. Wiley Publisher.

Leben, W. R. (2018). Languages of the World. *Oxford Research Encyclopedia of Linguistics, January*. DOI: 10.1093/acrefore/9780199384655.013.349

Lee, H., Kim, H., & Yan, W. (2025). Promoting inclusive AI and technology in K-12: A review of context, instructional strategies, and learning outcomes. *Computers and Education: Artificial Intelligence*, 100478.

Lee, I., & Hahn, S. (2024). On the relationship between mind perception and social support of chatbots. *Frontiers in Psychology*, *15*(March), 1–10. DOI: 10.3389/fpsyg.2024.1282036 PMID: 38510306

Lee, J. H., Shin, D., & Noh, W. (2023). Artificial intelligence-based content generator technology for young English-as-a-foreign-language learners' reading enjoyment. *RELC Journal*, *54*(2), 508–516.

Lee, J., Park, H., & Kim, Y. (2025). Generative AI in language education: A systematic review. *Language Learning & Technology*, *29*(1), 45–67. DOI: 10.1016/llt.2025.45

Lee, S., Choe, H., Zou, D., & Jeon, J. (2025). Generative AI (GenAI) in the language classroom: A systematic review. *Interactive Learning Environments*. Advance online publication. DOI: 10.1080/10494820.2025.2498537

Leveraging Artificial Intelligence to Foster Tolerance and Coexistence in Multicultural Societies. A Framework for Inclusive Policy and Social Integration. (2025). *SSRN* 5318927. https://www.researchgate.net/publication/392982288_Leveraging_Artificial_Intelligence_to_Foster_Tolerance_and_Coexistence_in_Multicultural_Societies_A_Framework_for_Inclusive_Policy_and_Social_Integration

Liang, B. (2024, December). Artificial Intelligence in Language Education: CiteSpace-based Visualisation and Analysis. In *Proceedings of the 2024 2nd International Conference on Information Education and Artificial Intelligence* (pp. 721-725).

Liang, J. C., Hwang, G. J., Chen, M. R. A., & Darmawansah, D. (2023). Roles and research foci of artificial intelligence in language education: An integrated bibliographic analysis and systematic review approach. *Interactive Learning Environments, 31*(7), 4270–4296.

Liang, S. (2024). Advancing Internationalization in the Formal Curriculum: Content Integration and Inclusive Design. *Journal of Studies in International Education, 28*(5), 743–760. DOI: 10.1177/10283153241262461

Libovický, J., Rosa, R., & Fraser, A. (2019). How Language-Neutral is Multilingual BERT? *ArXiv Preprint ArXiv:1906.01502*, 4996–5001. https://arxiv.org/abs/1911.03310

Li, H., Gan, Z., Leung, S. O., & An, Z. (2022). The impact of reading strategy instruction on reading comprehension, strategy use, motivation, and self-efficacy in Chinese university EFL students. *SAGE Open, 12*(1). Advance online publication. DOI: 10.1177/21582440221086659

Li, J., & Xue, E. (2022). Exploring High-Quality Institutional Internationalization for Higher Education Sustainability in China: Evidence from Stakeholders. *Sustainability, 14*(13). Advance online publication. DOI: 10.3390/su14137572

Lincoln, Y. S., & Guba, E. G. (1985). *Naturalistic inquiry*. SAGE Publications.

Ling, Y., & Jan, J. M. (2025). Voices from the Flip: Teacher Perspectives on Integrating AI Chatbots in Flipped English Classrooms. *Education Sciences, 15*(9), 1219.

Lin, H., & Liu, X. (2025). InVideo AI: Creating instructional audiovisual input for English learning. *RELC Journal, 56*(3), 803–809.

Lin, L., Zhou, D., Wang, J., & Wang, Y. (2024). A systematic review of big data driven education evaluation. *SAGE Open, 14*(2), 21582440241242180.

Lin, S., & Crosthwaite, P. (2024). The grass is not always greener: Teacher vs. GPT-assisted written corrective feedback. *System, 127*, 103529.

Lin, Y., & Yu, Z. (2025). Learner perceptions of artificial intelligence-generated pedagogical agents in language learning videos: Embodiment effects on technology acceptance. *International Journal of Human-Computer Interaction, 41*(2), 1606–1627.

Li, P., & Xu, Q. (2023). Computational modeling of bilingual language learning: Current models and future directions. *Language Learning*, *73*(S2), 17–64.

Li, T., & Ding, Y. (2025). Exploring the GenAI Literacy of Chinese University Students in EFL Learning. [IJCALLT]. *International Journal of Computer-Assisted Language Learning and Teaching*, *15*(1), 1–24.

Litre, G., Hirsch, F., Caron, P., Andrason, A., Bonnardel, N., Fointiat, V., Nekoto, W. O., Abbott, J., Dobre, C., Dalboni, J., Luxardo, G., & Bohbot, H. (2022). Participatory Detection of Language Barriers towards Multilingual Sustainability (ies) in Africa. *Sustainability*, •••, 1–15.

Little, D. (1991). *Learner autonomy 1: Definitions, issues and problems*. Authentik.

Liu, Z., Zhang, W., & Yang, P. (2025). Can AI chatbots effectively improve EFL learners' learning effects? A meta-analysis of empirical research from 2022–2024. *Computer Assisted Language Learning*, •••, 1–27.

Li, W., & Jiang, W. (2019). Requests made by Australian learners of Chinese as a foreign language. *Journal of Language Teaching and Research, 10*(1), 23–34.

Li, Y. (2025, June). An Enhanced Efficient English Learning System based on Artificial Intelligence Driven B/S Architectures. In *2025 International Conference on Intelligent Computing and Knowledge Extraction (ICICKE)* (pp. 1-6). IEEE.

Loeffen, R. (2025). Revisiting Earth Expansion: Mass-Energy Growth in Celestial Bodies Through the Cosmic Influx Theory, in Collaboration with ChatGPT [Preprint]. Doi DOI: 10.5281/zenodo.14586147

Lomotey, E. K., Karikari-Bowen, P., & Boateng, S. (2021). Paremiology and pragmatics: Proverbs as communicative acts in a cultural context. *African Journal of Linguistics and Languages*, *7*(3), 78–96.

Luo, X. (2023). Cross-Cultural Adaptation Framework for Enhancing Large Language Model Outputs in Multilingual Contexts. *Journal of Advanced Computing Systems*, *3*(5), 48–62.

MacWhinney, B. (2025). Understanding Language Through TalkBank. *Current Directions in Psychological Science*, *34*(2), 75–81.

Mager, M., Mager, E., Medina-Urrea, A., Meza, I., & Kann, K. (2018). Lost in Translation: Analysis of Information Loss During Machine Translation Between Polysynthetic and Fusional Languages. *ArXiv Preprint ArXiv:1807.00286*, 73–83. https://arxiv.org/abs/1807.00286

Mahafdah, R., Bouallegue, S., & Bouallègue, R. (2024). Enhancing e-learning through AI: Advanced techniques for optimizing student performance. *PeerJ. Computer Science, 10*. Advance online publication. DOI: 10.7717/peerj-cs.2576

Malik, A., Pratiwi, Y., Andajani, K., Numertayasa, I., Suharti, S., & Darwis, A., &, M. (2023). Exploring Artificial Intelligence in Academic Essay: Higher Education Student's Perspective. *International Journal of Educational Research Open, 5*. Advance online publication. DOI: 10.1016/j.ijedro.2023.100296

Marshall, D. J., & Tsikouras, B. (2023). Clay-shielded estuarine gastropods are better protected against environmental acidification than unshielded individuals. *The Science of the Total Environment, 866*, 161367.

Masoud, R., Liu, Z., Ferianc, M., Treleaven, P. C., & Rodrigues, M. R. (2025). Cultural alignment in large language models: An explanatory analysis based on Hofstede's cultural dimensions. In *Proceedings of the 31st International Conference on Computational Linguistics* (pp. 8474–8503).

Maurya, K. K., & Desarkar, M. S. (2023). Towards Low-resource Language Generation with Limited Supervision. *BigPicture 2023 - Big Picture Workshop, Proceedings*, 80–92. DOI: 10.18653/v1/2023.bigpicture-1.7

Ma, Z. (2023). The Study on the Influence of Academic Pressure on Academic Performance. *Journal of Education and Educational Research, 3*(2), 106–109. DOI: 10.54097/jeer.v3i2.9045

Mittelmeier, J., Lomer, S., Al-Furqani, S., & Huang, D. (2024). Developing Meaningful Internationalisation that Impacts Students' Outcomes in Higher Education: A Scoping Review of the Literature 2011–2022. *Journal of Studies in International Education, 28*(4), 503–525. DOI: 10.1177/10283153231222278

Mittelmeier, J., Rienties, B., Gunter, A., & Raghuram, P. (2020). Conceptualizing Internationalization at a Distance: A "Third Category" of University Internationalization. *Journal of Studies in International Education, 25*(3), 266–282. DOI: 10.1177/1028315320906176

Mlambo, R., & Matfunjwa, M. (2024). The use of technology to preserve indigenous languages of South Africa. *Literator, 45*(1), 1–8. DOI: 10.4102/lit.v45i1.2007

Montanari, F. (2025). ChatGPT and the others: Artificial intelligence, social actors, and political communication. A tentative sociosemiotic glance. *Semiotica, 2025*(262), 189–212.

Moorhouse, B. L., Ho, T. Y., Wu, C., & Wan, Y. Pre-service Language Teachers' Task-specific Large Language Model Prompting Practices. *RELC Journal*. Advance online publication. DOI: 10.1177/00336882251313701

Morgado, E., Leonido, L., Pereira, A., & Gouveia, L. B. (2025). Technology-Mediated Education: Impact of AI on the main distance learning modalities. *Educational Process: International Journal, 16*, e2025211.

Moskovkin, V. (2024, July 15th). Tracing the origins of 'publish or perish'. https://blogs.lse.ac.uk/impactofsocialsciences/2024/07/15/tracing-the-origins-of-publish-or-perish/

Muhammad, S. H., Abdulmumin, I., Ayele, A. A., Ousidhoum, N., Adelani, D. I., Yimam, S. M., Ahmad, I. S., Beloucif, M., Mohammad, S. M., Ruder, S., Hourrane, O., Brazdil, P., Jorge, A., Ali, F. D. M. A., David, D., Osei, S., Bello, B. S., Ibrahim, F., Gwadabe, T., . . . Arthur, S. (2023). AfriSenti: A Twitter Sentiment Analysis Benchmark for African Languages. *EMNLP 2023 - 2023 Conference on Empirical Methods in Natural Language Processing, Proceedings*, 13968–13981. DOI: 10.18653/v1/2023.emnlp-main.862

Muhammad, N., & Petra, M. (2021). *History and Development of the Brunei Education.*

Muhammad, S. H., Ahmad, I. S., Abdulmumin, I., Lawan, F. I., Sani, B., Imam, S. H., Aliyu, Y., Sani, S. A., Umar, A. U., Gwadabe, T., Church, K., & Marivate, V. (2025). HausaNLP: Current Status, Challenges and Future Directions for Hausa Natural Language Processing. *ArXiv Preprint*. https://arxiv.org/abs/2505.14311

Munoto, M., Sumbawati, M. S., & Sari, S. F. M. (2020). The Use of Mobile Technology in Learning With Online and Offline Systems. *International Journal of Information and Communication Technology Education, 17*(2), 54–67. DOI: 10.4018/ijicte.2021040104

Namaziandost, E. (2025). Integrating flipped learning in AI-enhanced language learning: Mapping the effects on metacognitive awareness, writing development, and foreign language learning boredom. *Computers and Education: Artificial Intelligence*, 100446.

Nanduri, D. K. (2024). *Exploring the Role of Generative Artificial Intelligence in Culturally Relevant Storytelling for Native Language Learning Among Children* (Master's thesis, University of Maryland, College Park).

Naseer, F., Khan, M., Tahir, M., Addas, A., & Aejaz, S. (2024). Integrating deep learning techniques for personalized learning pathways in higher education. *Heliyon, 10*(11). Advance online publication. DOI: 10.1016/j.heliyon.2024.e32628

Nast, C., Llopis, O., Yankova, D., & D'Este, P. (2025). Sourcing insights elsewhere: The positive influence of academic engagement on scientific impact. *Technovation, 139*. Advance online publication. DOI: 10.1016/j.technovation.2024.103112

Nekoto, W., Marivate, V., Matsila, T., Fasubaa, T., Kolawole, T., Fagbohungbe, T., Oluwole Akinola, S., Hassan Muhammad, S., Kabongo, S., Osei, S., Freshia, S., Andre Niyongabo, R., Macharm, R., Ogayo, P., Ahia, O., Meressa, M., Adeyemi, M., Mokgesi-Selinga, M., Okegbemi, L., & Bashir, A. (2020). Participatory Research for Low-resourced Machine Translation: A Case Study in African Languages. In Trevor Cohn, Y. L., & He, Y. (Eds.), *Association for Computational Linguistics* (Vol. 9). Association for Computational Linguistics., DOI: 10.18653/v1/2020.findings-emnlp.195

Neuman, Y. (2024). *AI for Understanding Context.* Springer.

Ni, X., & Yin, H. (2025, May). The Impact of Mobile-Assisted Pronunciation Training on Intelligibility and Comprehensibility in EFL Learners. In *2025 5th International Conference on Artificial Intelligence and Education (ICAIE)* (pp. 839-844). IEEE.

Ning, Z., Gu, T., Song, J., Hong, S., Li, L., Liu, H., Li, J., Wang, Y., Lingyu, M., Teng, Y., & Wang, Y. (2025). LinguaSafe: A Comprehensive Multilingual Safety Benchmark for Large Language Models. *ArXiv Preprint ArXiv:2508.12733*. https://arxiv.org/abs/2508.12733

Obeidat, M. M., Haider, A. S., Tair, S. A., & Sahari, Y. (2024). Analysing the Performance of Gemini, ChatGPT, and Google Translate in Rendering English Idioms into Arabic. *FWU Journal of Social Sciences, 18*(4).

Ofosu-Asare, Y. (2025). Cognitive imperialism in artificial intelligence: Counteracting bias with indigenous epistemologies. *AI & Society*, *40*(4), 3045–3061.

Ogueji, K., Zhu, Y., & Lin, J. (2021). Small data? no problem! exploring the viability of pretrained multilingual language models for low-resourced languages. *MRL 2021 - 1st Workshop on Multilingual Representation Learning, Proceedings of the Conference*, 116–126.

Onyenwe, I. E. (2017). *Developing Methods and Resources for Automated Processing of the African Language Igbo* (Issue April). The University of Sheffield.

Owan, V., Abang, K., Idika, D., Etta, E., & Bassey, B. (2023). Exploring the potential of artificial intelligence tools in educational measurement and assessment. *Eurasia Journal of Mathematics, Science and Technology Education, 19*(8). Advance online publication. DOI: 10.29333/ejmste/13428

Oxford, R. (2017). *Teaching and researching language learning strategies: Self-regulation in context.* Routledge.

Oyetade, K., & Zuva, T. (2025). Advancing equitable education with inclusive AI to mitigate bias and enhance teacher literacy. *Educational Process: International Journal, 14*, e2025087. Advance online publication. DOI: 10.22521/edupij.2025.14.87

Pakray, P., Gelbukh, A., & Bandyopadhyay, S. (2025). Natural language processing applications for low-resource languages. *Natural Language Processing, 31*(2), 183–197. DOI: 10.1017/nlp.2024.33

Pan, Y., Zhang, L., & Chen, H. (2025). Self-regulated reading with ChatGPT: Exploring learner strategies in AI-supported contexts. *Language Learning & Technology, 29*(2), 55–73. DOI: 10.1016/llt.2025.29.2

Papakyriakopoulos, O., & Xiang, A. (2023). Considerations for Ethical Speech Recognition Datasets. *WSDM 2023 - Proceedings of the 16th ACM International Conference on Web Search and Data Mining, 1*(1), 1287–1288. DOI: 10.1145/3539597.3575793

Papouli, E. (2016). Teaching and learning for ethical practice in social work education. In *Routledge International Handbook of Social Work Education* (pp. 157–170). Issue June., DOI: 10.4324/9781315712536

Pattanaik, B., Kumari, S. S., Kumar, K. C., Pattnaik, M., Ali, S. I., & Kumarasamy, M. (2023). An innovation of algebraic mathematical–based statistical analysis model for complex number theory. In *Proceedings of the 2023 International Conference on Disruptive Technologies* (pp. 94–99). IEEE. https://doi.org/DOI: 10.1109/ICDT57929.2023.10151169

Pattnayak, P., Patel, H. L., & Agarwal, A. (2025). Tokenization Matters: Improving Zero-Shot NER for Indic Languages. *In 2025 IEEE International Conference on Electro Information Technology (EIT).* DOI: 10.1109/eIT64391.2025.11103625

Pegrum, M. (2025). From revolution to evolution: What generative AI really means for language learning. *Language Teaching*, •••, 1–17. DOI: 10.1017/S0261444825000151

Pennycook, A. (2010). Language as a local practice. In *Language as a Local Practice* (first, Issue April 2010). DOI: 10.4324/9780203846223

Perani, D., & Abutalebi, J. (2005). The neural basis of first and second language processing. *Current Opinion in Neurobiology, 15*(2), 202–206. PMID: 15831403

Pillai, M., Griffin, A. C., Kronk, C. A., & McCall, T. (2023). Toward Community-Based Natural Language Processing (CBNLP): Cocreating With Communities. *Journal of Medical Internet Research, 25*. Advance online publication. DOI: 10.2196/48498 PMID: 37540551

Politov, A., Shkalikov, O., Jäkel, R., & Färber, M. (2025). Revisiting Projection-based Data Transfer for Cross-Lingual Named Entity Recognition in Low-Resource Languages. *ArXiv Preprint ArXiv:2501.18750*. https://arxiv.org/abs/2501.18750

Ponti, E. M., O'Horán, H., Berzak, Y., Vulić, I., Reichart, R., Poibeau, T., & Korhonen, A. (2019). Modelling language variation and universals: A survey on typological linguistics for natural language processing. *Computational Linguistics, 45*(3), 559–601.

Pratap, V., Tjandra, A., Shi, B., Tomasello, P., Babu, A., Kundu, S., Elkahky, A., Ni, Z., Vyas, A., Fazel-Zarandi, M., Baevski, A., Adi, Y., Zhang, X., Hsu, W.-N., Conneau, A., & Auli, M. (2024). Scaling Speech Technology to 1,000+ Languages. *Journal of Machine Learning Research, 25*, 1–52. https://arxiv.org/abs/2305.13516

Pratiwi, A. (2021). Communication and International Language. *KOMUNIKASIA: Journal of Islamic Communication and Broadcasting, 1*(1), 69–84. DOI: 10.32923/kpi.v1i1.1909

Premkumar, P., Yatigammana, M., & Kannangara, S. (2024). Impact of generative AI on critical thinking skills in undergraduates: A systematic review. *The Journal of Desk Research Review and Analysis, 2*(1), 199–215.

Prensky, M. (2001). Digital natives, digital immigrants. *On the Horizon. The International Journal of Learning Futures, 9*(5), 1–6. DOI: 10.1108/10748120110424816

Pulvermüller, F. (2018). Neural reuse of action perception circuits for language, concepts and communication. *Progress in Neurobiology, 160*, 1–44. PMID: 28734837

Puppel, M. (2024). *A comparative study of technical and creative text translation: evaluating the performance of ChatGPT* (Master's thesis, University of Malta).

Puspitasari, F. D., Zhang, C., Dam, S. K., Zhang, M., Kim, T. H., Hong, C. S., & Yang, Y. (2025). Deepseek models: A comprehensive survey of methods and applications. *Authorea Preprints*.

Qin, Y. (2025, April). Key Factors Influencing the Deployment of Generative Large Models in Language Teaching. In *2025 7th International Conference on Computer Science and Technologies in Education (CSTE)* (pp. 592-595). IEEE.

Qomariyah, L., Ekowati, D., & Mudzakkir, M. F. (2023). The Contexts of Academic Pressure: Narrative Review. *RSF Conference Series: Business, Management and Social Sciences, 3*(3), 502–514. https://doi.org/DOI: 10.31098/bmss.v3i3.715

Queirós, A., Carvalho, T., Manatos, M., & Diogo, S. (2023). Internationalization of Portuguese Academia: The impact on academic engagement and collaboration with society. *Higher Education, 87*, 1769–1789. DOI: 10.1007/s10734-023-01090-0

Qu, X., Gu, Y., Xia, Q., Li, Z., Wang, Z., & Huai, B. (2024). A Survey on Arabic Named Entity Recognition: Past, Recent Advances, and Future Trends. *IEEE Transactions on Knowledge and Data Engineering, 36*(3), 943–959. DOI: 10.1109/TKDE.2023.3303136

Radford, A., Wu, J., Child, R., Luan, D., Amodei, D., & Sutskever, I. (2019). Language models are unsupervised multitask learners. *OpenAI Blog, 1*(8), 1–24.

Rahman, M. M., Sakib, F. A., Faisal, F., & Anastasopoulos, A. (2023). To token or not to token: A Comparative Study of Text Representations for Cross-Lingual Transfer. *MRL 2023 - 3rd Workshop on Multi-Lingual Representation Learning, Proceedings of the Workshop*, 173–183. DOI: 10.18653/v1/2023.mrl-1.6

Rahmasari, B. S., Munir, A., & Nugroho, H. M. A. (2025). Fostering learner autonomy in EFL reading: A study of strategy-based instruction at two Indonesian universities. *Cogent Education, 12*(1), 2477367. DOI: 10.1080/2331186X.2025.2477367

Raihan, M. M. H., Subroto, S., Chowdhury, N., Koch, K., Ruttan, E., & Turin, T. C. (2025). Dimensions and barriers for digital (in)equity and digital divide: A systematic integrative review. *Digital Transformation and Society, 4*(2), 111–127. DOI: 10.1108/DTS-04-2024-0054

Raimundo, A. C., & Terra, D. V. (2021). Education for ethnic-racial relations in early childhood education: The story of sophia +. *Movimento, 27*. Advance online publication. DOI: 10.22456/1982-8918.108168

Raji, I. D., Bender, E. M., Paullada, A., Denton, E., & Hanna, A. (2021). AI and the Everything in the Whole Wide World Benchmark. *Advances in Neural Information Processing Systems, NeurIPS*.

Ranalli, J. (2018). Automated written corrective feedback: How well can students make use of it? *Computer Assisted Language Learning, 31*(7), 653–674. DOI: 10.1080/09588221.2018.1428994

Rane, N., Choudhary, S., & Rane, J. (2024). Gemini versus ChatGPT: Applications, performance, architecture, capabilities, and implementation. *Journal of Applied Artificial Intelligence*, *5*(1), 69–93.

Rashid, A., & Kausik, M. (2024). AI revolutionizing industries worldwide: A comprehensive overview of its diverse applications. *Hybrid Advances*, *7*. Advance online publication. DOI: 10.1016/j.hybadv.2024.100277

Rathore, R., Nayeem, T., Agarwal, A., Kumar, S. P., & Bhardwaj, A. (2024). AI System for Autonomous Vehicles. *International Journal For Multidisciplinary Research*, *6*(6), 1–9. DOI: 10.36948/ijfmr.2024.v06i06.28785

Rianto, A. (2022). Exploring correlation between metacognitive online reading strategy use and online reading comprehension of EFL students. *Turkish Online Journal of Distance Education*, *23*(2), 223–235.

Richardson, S., Lawrence, K., Schoenthaler, A. M., & Mann, D. (2022). A framework for digital health equity. *NPJ Digital Medicine*, *5*(1). Advance online publication. DOI: 10.1038/s41746-022-00663-0 PMID: 35982146

Romero, M., Gómez-Canaval, S., & Torre, I. G. (2024). Automatic Speech Recognition Advancements for Indigenous Languages of the Americas. *Applied Sciences (Switzerland)*, *14*(15). Advance online publication. DOI: 10.3390/app14156497

Rosé, C. P., Wang, Y.-C., Cui, Y., Arguello, E., Stegmann, K., Weinberger, A., & Fischer, F. (2008). Analyzing collaborative learning processes automatically: Exploiting the advances of computational linguistics in computer-supported collaborative learning. *International Journal of Computer-Supported Collaborative Learning*, *3*(1), 237–271.

Rosyidah, N., Matin & Rosyidi, U. (2020). Internationalization in Higher Education: University's Effective Promotion Strategies in Building International Trust. *European Journal of Educational Research*, *9*(1), 351–361. DOI: 10.12973/eu-jer.9.1.351

Ruder, S., Constant, N., Botha, J., Siddhant, A., Firat, O., Fu, J., Liu, P., Hu, J., Garrette, D., Neubig, G., & Johnson, M. (2021). XTREME-R: Towards More Challenging and Nuanced Multilingual Evaluation. *EMNLP 2021 - 2021 Conference on Empirical Methods in Natural Language Processing, Proceedings*, 10215–10245. DOI: 10.18653/v1/2021.emnlp-main.802

Ruder, S., Peters, M., Swayamdipta, S., & Wolf, T. (2019). Transfer learning in natural language processing tutorial. *NAACL HLT 2019 - 2019 Conference of the North American Chapter of the Association for Computational Linguistics: Human Language Technologies - Tutorial Abstracts, 2010*, 15–18.

Ruder, S., Clark, J. H., Gutkin, A., Kale, M., Ma, M., Nicosia, M., Rijhwani, S., Riley, P., Sarr, J. M. A., Wang, X., Wieting, J., Gupta, N., Katanova, A., Kirov, C., Dickinson, D. L., Roark, B., Samanta, B., Tao, C., Adelani, D. I., & Talukdar, P. (2023). XTREME-UP: A User-Centric Scarce-Data Benchmark for Under-Represented Languages. *Findings of the Association for Computational Linguistics: EMNLP, 2023*, 1856–1884. DOI: 10.18653/v1/2023.findings-emnlp.125

Rulyansah, A., Hidayat, M. T., Rihlah, J., Shari, D., & Mariati, P. (2023). Digital Play for Enhancing Language Learning in Early Grades. *Egitim ve Ögretim, 13*(2), 182–190. DOI: 10.47750/pegegog.13.02.22

Saeed Ahmed Mohamed, S., & Mahmoud Ibrahim Alian, E. (2023). *Students' attitudes toward using chatbot in EFL learning. Arab World English Journal* (Vol. 14). AWEJ.

Safar, M., & Anggraheni, D. (2024). Language Learning through AI Chatbots : Effectiveness and Cognitive Load Analysis. *Journal of Social Science Utilizing Technology, 2*(3), 430–445.

Sallabank, J. (2011). Encyclopedia of the world's endangered languages. In *Language Learning Journal* (Vol. 39, Issue 1, pp. 122–123). DOI: 10.1080/09571736.2011.553475

Sambasivan, N., Arnesen, E., Hutchinson, B., Doshi, T., & Prabhakaran, V. (2021). Re-imagining algorithmic fairness in India and beyond. *FAccT 2021 - Proceedings of the 2021 ACM Conference on Fairness, Accountability, and Transparency*, 315–328. DOI: 10.1145/3442188.3445896

Sani, H. M. (2025). ChatGPT and the Future of Language Learning and Linguistic Diversity in Africa. In *Sociocultural and Multicultural Meanings in Online Communication* (pp. 437–464). IGI Global Scientific Publishing.

Sarı, M. H., & Yüce, E. (2020). Problems experienced in classrooms with students from different cultures. *Journal on Efficiency and Responsibility in Education and Science, 13*(2), 90–100. DOI: 10.7160/eriesj.2020.130204

Sarker, R. (2024). Academic pressure and depression. *International Journal of Multidisciplinary Research and Development, 11*(5), 46–52.

Segerer, R. (2025). Cultural Value Alignment in Large Language Models: A Prompt-based Analysis of Schwartz Values in Gemini, ChatGPT, and DeepSeek. *arXiv preprint arXiv:2505.17112*.

Sellami, A. (2025). AI-infused Immersion: Cultivating EFL Learners' Intercultural Sensitivity through Google's Gemini AI Chatbot. *Journal of Studies in Language* [JSLCS]. *Cultura e Scuola*, *8*(2), 250–270.

Seyoum, B. E., Miyao, Y., & Mekonnen, B. Y. (2019). Universal dependencies for Amharic. *LREC 2018 - 11th International Conference on Language Resources and Evaluation*, 2216–2222.

Shaikh, M. S., & Ali, S. I. (2025a). An integrated analysis of the effects of learning analytics dashboards on learner perspective, inspiration, engagement, and accomplishment. In Ponnusamy, S. (Eds.), *Revolutionizing education with remote experimentation and learning analytics* (pp. 19–36). IGI Global Scientific Publishing., DOI: 10.4018/979-8-3693-8593-7.ch002

Shaikh, M. S., Ali, S. I., & Siddiqui, U. F. (2016). Li-Fi: An emerging wireless communication technology. *International Journal of Advanced Electronics & Communication Systems*, *5*(1), 10758.

Shaikh, M. S., Baig, I., & Ali, S. I. (2022). Quantum computing for optimization problems: A review and future directions. *COMPUSOFT: An International Journal of Advanced Computer Technology*, *11*, 3995.

Shaikh, M. S., Chauhan, P. S., Baig, I., & Ali, S. I. (2023). Decentralized autonomous networks for secure data sharing: A theoretical framework and conceptual analysis. *COMPUSOFT: An International Journal of Advanced Computer Technology*, *12*, 4007.

Shaikh, M. S., Jain, N. K., Biswal, A., Patidar, P. K., Panwar, I., & Ali, S. I. (2025a). Sensors to insights revolution: Leveraging IoT, AI, and big data for next-gen patient monitoring. In *Proceedings of the IEEE International Conference on Emerging Technologies in Engineering, Technology, and Scientific Innovation*. IEEE. https://doi.org/DOI: 10.1109/ICETETSIP64213.2025

Shakirova, J., & Smolnikova, E. (2017). Influence of university internationalization on their positions in world rankings. *QUID Investigacion. Ciencia y Tecnologia*, *1*, 859–864.

Shanmuga, S. M., Jupally, S. H., & Vempalli, N. R. (2025). Models and Frameworks for Digital Preservation of Indigenous Knowledge Systems. In *Revaluation and Preservation of Indigenous Knowledge Systems in Modern Society* (pp. 511–534). IGI Global Scientific Publishing.

Sharma, K., Bajpai, B., Vaghela, H., Patidar, P. K., Ali, S. I., & Shaikh, M. S. (2025). Blockchain for intelligent transportation systems: Purposes, obstacles, and possibilities. In Ponnusamy, S. (Eds.), *Leveraging VANETs and blockchain technology for urban mobility* (pp. 45–58). IGI Global Scientific Publishing., DOI: 10.4018/979-8-3373-0265-2.ch003

Shen, M., Peng, C. F., Teng, Y. T., & Hua, F. (2022). Chinese ESL Instructors' Beliefs and Practices of Intercultural Education in a Ubiquitous Environment. In Intercultural Communication and Ubiquitous Learning in Multimodal English Language Education (pp. 68-88). IGI Global Scientific Publishing. https://doi.org/DOI: 10.4018/978-1-7998-8852-9.ch004

Shi, W., Li, R., Zhang, Y., Ziems, C., Yu, C., Horesh, R., de Paula, R. A., & Yang, D. (2024). CultureBank: An Online Community-Driven Knowledge Base Towards Culturally Aware Language Technologies. *EMNLP 2024 - 2024 Conference on Empirical Methods in Natural Language Processing, Findings of EMNLP 2024*, 4996–5025. DOI: 10.18653/v1/2024.findings-emnlp.288

Shi, L. (2023). Enriching early years English language education with translanguaging. In *Innovative practices in early English language education* (pp. 81–99). Springer International Publishing.

Shrivastava, A. (2024). Artificial Intelligence (AI): Evolution, Methodologies, and Applications. *International Journal for Research in Applied Science and Engineering Technology*, *12*(IV), 5501–5505. DOI: 10.22214/ijraset.2024.61241

Siregar, I. (2022). The pressure of academic stress and self-efficacy among student. *Jurnal Konseling Dan Pendidikan*, *10*(3), 394. DOI: 10.29210/156700

Slamet, J., Basthomi, Y., Ivone, F. M., & Eliyanah, E. (2024). Unlocking the potential in a gamification-based MOOC: Assessing autonomous learning and self-directed learning behaviors. *Teaching & Learning Inquiry*, *12*, 1–20. DOI: 10.20343/teachlearninqu.12.19

Sofyan, N. (2021). The role of English as global language. *Edukasi*, *19*(1). Advance online publication. DOI: 10.33387/J.EDU.V19I1.3200

Speer, A. (2018). Contemplation and Philosophy: A Historical and Systematic Approach. In *Contemplation and Philosophy: Scholastic and Mystical Modes of Medieval Philosophical Thought* (pp. 79–107). Brill.

Spence, P. J., & Brandao, R. (2021). Towards language sensitivity and diversity in the digital humanities. *Digital Studies/Le champ numérique, 11*(1), 9. https://doi.org/DOI: 10.16995/dscn.8098

Spencer-Oatey, H. (2008). *Culturally speaking: Culture, communication and politeness theory* (2nd ed.). Continuum.

Spennemann, D. H. (2023). ChatGPT and the generation of digitally born "knowledge": How does a generative AI language model interpret cultural heritage values? *Knowledge (Beverly Hills, Calif.)*, *3*(3), 480–512.

Srivastava, S. S., Ali, S. I., Srivastava, A. K., & Goswami, C. (2024). Improving 5G network performance with MIMO technology using beamforming algorithm. *ICTACT Journal on Communication Technology*, *15*(2), 3173–3180. DOI: 10.21917/ijct.2024.0472

Stockemer, D., & Wigginton, M. (2019). Publishing in English or another language: An inclusive study of scholar's language publication preferences in the natural, social and interdisciplinary sciences. *Scientometrics*, *118*, 645–652. DOI: 10.1007/s11192-018-2987-0

Stuckey, H. (2015). Three types of interviews: Qualitative research methods in social health. *Journal of Social Health and Diabetes, 01*(02), 056–059. https://doi.org/DOI: 10.4103/2321-0656.115294

Sumakul, D. T. Y., Hamied, F. A., & Sukyadi, D. (2022). Artificial intelligence in EFL classrooms: Friend or foe? *LEARN Journal: Language Education and Acquisition Research Network*, *15*(1), 232–256.

Sun, C., Wei, L., & Young, R. F. (2022). Measuring teacher cognition: Comparing Chinese EFL teachers' implicit and explicit attitudes toward English language teaching methods. *Language Teaching Research*, *26*(3), 382–410. DOI: 10.1177/1362168820903010

Sun, Y., & Gao, F. (2023). Exploring the risks of generative AI in intercultural education. *Computers & Education*, *200*, 104792. DOI: 10.1016/j.compedu.2023.104792

Suthar, B. S., Ali, S. I., & Patel, K. (2025). Hand gesture sign language recognition embedded with Arduino Uno. In *IET Conference Proceedings* (Vol. 2025, Issue 7). https://doi.org/DOI: 10.1049/icp.2025.1565

Swiecki, Z., Khosravi, H., Chen, G., Martinez-Maldonado, R., Lodge, J. M., Milligan, S., Selwyn, N., & Gašević, D. (2022). Assessment in the age of artificial intelligence. *Computers and Education: Artificial Intelligence*, *3*, 100075. DOI: 10.1016/j.caeai.2022.100075

Syafiee, A. S. E. E. A., & Yaqin, L. N. (2025). A comparative analysis of Google Translate and ChatGPT in translating Borneo Bulletin News into standard Malay. *Journal of Research on English and Language Learning (J-REaLL)*, *6*(1), 115–126.

Tabiati, S. E. (2016). Factors contributing to learners' autonomy in EFL reading. *Jurnal Pendidikan Humaniora*, *4*(2), 82–94.

Tacheva, Z., & Ramasubramanian, S. (2023). *Challenging AI empire: Toward a decolonial and queer framework of data resurgence* [Preprint]. Advance., DOI: 10.31124/advance.22012724.v1

Tahir, B., & Mehmood, M. A. (2021). Corpulyzer: A Novel Framework for Building Low Resource Language Corpora. *IEEE Access : Practical Innovations, Open Solutions*, *9*, 8546–8563. DOI: 10.1109/ACCESS.2021.3049793

Tambets, K., Yunusbayev, B., Hudjashov, G., Ilumäe, A. M., Rootsi, S., Honkola, T., Vesakoski, O., Atkinson, Q., Skoglund, P., Kushniarevich, A., Litvinov, S., Reidla, M., Metspalu, E., Saag, L., Rantanen, T., Karmin, M., Parik, J., Zhadanov, S. I., Gubina, M., & Metspalu, M. (2018). Genes reveal traces of common recent demographic history for most of the Uralic-speaking populations. *Genome Biology*, *19*(1), 1–20. DOI: 10.1186/s13059-018-1522-1 PMID: 30241495

Tan, J. C. A., Chan, W. P., Robinson, N. L., Kulić, D., & Croft, E. A. (2022, August). Impacts of Teaching towards Training Gesture Recognizers for Human-Robot Interaction. In *2022 31st IEEE International Conference on Robot and Human Interactive Communication (RO-MAN)* (pp. 618-623). IEEE.

Tang, K., Li, Y., & Qin, Y. (2025). *DIY-MKG: An LLM-based polyglot language learning system* [Preprint]. arXiv. https://doi.org//arXiv.2507.01872DOI: 10.48550

Tanverdi, B. (2024). *Accumulating Architectural Knowledge Through Practice-Based Engagement: Experiential Learning Modules* (Doctoral dissertation, Middle East Technical University (Turkey)).

Taşkın, M. (2025). Artificial Intelligence in Personalized Education: Enhancing Learning Outcomes Through Adaptive Technologies and Data-Driven Insights. *Human-Computer Interaction*, *8*(1). Advance online publication. DOI: 10.62802/ygye0506

Team, G., Anil, R., Borgeaud, S., Alayrac, J. B., Yu, J., Soricut, R., . . . Blanco, L. (2023). Gemini: a family of highly capable multimodal models. *arXiv preprint arXiv:2312.11805*.

Teubner, T., Flath, C. M., Weinhardt, C., Van Der Aalst, W., & Hinz, O. (2023). Welcome to the era of ChatGPT et al., the prospects of large language models. *Business & Information Systems Engineering, 65*(2), 95–101.

Thenmozhi, V. (2025). A Study on the Effectiveness of AI-Powered Customer Service Enhancement at Syena Logistics PVT LTD. *International Journal of Scientific Research in Engineering and Management*, *9*(4), 1–9. DOI: 10.55041/ijsrem45403

Thomas, J. (1983). Cross-cultural pragmatic failure. *Applied Linguistics*, *4*(2), 91–112. DOI: 10.1093/applin/4.2.91

Tsai, S. C. (2022). Chinese students' perceptions of using Google Translate as a translingual CALL tool in EFL writing. *Computer Assisted Language Learning*, *35*(5-6), 1250–1272.

Tulasi, S., & Ahamed, I. (2025). AI-Powered E-Commerce Personalization and Recommendation Systems. *Journal of Informatics Education and Research*, *5*(1), 2092–2099. DOI: 10.52783/jier.v5i1.2197

UBD. (2025). *UBD Moves Up in QS World University Rankings, Reinforces Global Standing*. Retrieved September 16, 2025, from https://ubd.edu.bn/2025/07/05/ubd-moves-up-in-qs-world-university-rankings-reinforces-global-standing/

Umbrello, S., & van de Poel, I. (2021). Mapping value sensitive design onto AI for social good principles. *AI and Ethics*, *1*, 283–296. DOI: 10.1007/s43681-021-00038-y PMID: 34790942

UNESCO. (2024). Indigenous languages in the digital age: Preservation, transmission, and epistemological sovereignty in technological contexts. United Nations Educational. *Scientific and Cultural Organization Reports*, *45*(1), 1–89.

UNESCO. (2025). *Languages matter: Global guidance on multilingual education*. https://doi.org/DOI: 10.54675/MLIO7101

Utku Bilici, Ö. (2024). *The multi-dimensional nature of teacher cognition: The case of EFL teachers* (Publication No. 31653733) [Doctoral dissertation, Middle East Technical University]. ProQuest. https://www.proquest.com/docview/3122678784?pq-origsite=gscholar&fromopenview=true

Utku Bilici, Ö., & Subaşı, G. (2022). EFL learners' metacognitive awareness of global reading strategies and L2 reading comprehension skills: The efficacy of reciprocal teaching. *The Reading Matrix: An International Online Journal*, *22*(2), 170–185.

Vassilakopoulou, P., & Hustad, E. (2023). Bridging Digital Divides: A Literature Review and Research Agenda for Information Systems Research. *Information Systems Frontiers*, *25*(3), 955–969. DOI: 10.1007/s10796-020-10096-3 PMID: 33424421

Vaswani, A., Shazeer, N., Parmar, N., Uszkoreit, J., Jones, L., Gomez, A. N., Kaiser, Ł., & Polosukhin, I. (2017). Attention is all you need. In *Advances in Neural Information Processing Systems* (Vol. 30).

Vibhavi, R., Swetha, A., Shruthi, B., Shravya, S., Balal, R., & Narayan, R. (2024). AI-Driven Chatbots in Mental Health : Enhancing Emotional Support and AI-Driven Chatbots in Mental Health : Enhancing Emotional Support and Therapy Through Cognitive Behavioural Approaches. *Gradiva Review Journal, 10*(9), 323–328.

Viidalepp, A. (2023). The Expected AI as a Sociocultural Construct and Its Impact on the Discourse on Technology.

Vygotsky, L. (1978). *Mind in Society: The Development of Higher Psychological Processes*. MIT Press.

Vygotsky, L. S. (1978). *Mind in society: The development of higher psychological processes*. Harvard University Press.

Walter, L., Denter, N. M., & Kebel, J. (2022). A review on digitalization trends in patent information databases and interrogation tools. *World Patent Information, 69*, 102107. Advance online publication. DOI: 10.1016/j.wpi.2022.102107

Walter, M. (2024). AI literacy in education: Preparing learners for critical engagement with AI. *Computers & Education, 195*, 104744. DOI: 10.1016/j.compedu.2024.104744

Walters, W. (2017). Citation-Based Journal Rankings: Key Questions, Metrics, and Data Sources. *IEEE Access : Practical Innovations, Open Solutions, 5*, 22036–22053. DOI: 10.1109/ACCESS.2017.2761400

Waluyo, B., & Rouaghe, F. (2025). Beyond teacher-led approaches: Student-initiated translanguaging with artificial intelligence tools in foreign language acquisition. *SAGE Open, 15*(3), 21582440251362998.

Wang, Y., & Zhao, Y. (2023). Gemini in reasoning: Unveiling commonsense in multimodal large language models. *arXiv preprint arXiv:2312.17661.*

Wang, D., & Chen, G. (2025). Evaluating the use of BERT and Llama to analyse classroom dialogue for teachers' learning of dialogic pedagogy. *British Journal of Educational Technology*. Advance online publication. DOI: 10.1111/bjet.13604

Wang, D., Yang, T., & Chen, G. (2024). Artificial intelligence in classroom discourse: A systematic review of the past decade. *International Journal of Educational Research, 123*, 102275. Advance online publication. DOI: 10.1016/j.ijer.2023.102275

Wang, H. (2024). *The design and evaluation of intelligent computer-assisted language learning tools for beginner learners of Mandarin Chinese*. Trinity College.

Wang, J., & Fan, W. (2025). The effect of ChatGPT on students' learning performance, learning perception, and higher-order thinking: Insights from a meta-analysis. *Humanities & Social Sciences Communications*, *12*(1), 1–21. DOI: 10.1057/s41599-025-04787-y

Wang, L. (2024). Self-regulated use of generative AI in language learning: Monitoring, evaluation, and adaptation. *ReCALL*, *36*(1), 22–40. DOI: 10.1017/S0958344023000219

Wang, X. (2025, June). Design and Development of Smart Learning System for College English based on Intelligent Cloud Platform. In *2025 International Conference on Intelligent Computing and Knowledge Extraction (ICICKE)* (pp. 1-8). IEEE.

Wang, Y., Zhang, T., Yao, L., & Seedhouse, P. (2025). A scoping review of empirical studies on generative artificial intelligence in language education. *Innovation in Language Learning and Teaching*. Advance online publication. DOI: 10.1080/17501229.2025.2509759

Wei, L. Z., Vijayasingam, V., & Kaur, P. (2023, December). Empowering Local Communities through an Intelligent Digital Platform for Rural Tourism in Malaysia. In *2023 IEEE 21st Student Conference on Research and* Development *(SCOReD)* (pp. 455-462). IEEE.

Weigang, L., & Brom, P. C. (2025). The paradox of poetic intent in back-translation: Evaluating the quality of large language models in Chinese translation. *arXiv preprint arXiv:2504.16286.*

Wellen, K. (2019). Families in austronesian expansion. *Journal of Family History*, *44*(3), 239–253. DOI: 10.1177/0363199019839330

Wenden, A. (1998). Metacognitive knowledge and language learning. *Applied Linguistics*, *19*(4), 515–537. DOI: 10.1093/applin/19.4.515

Wen, Y., Zhao, X., Li, X., & Zang, Y. (2025). Attitude Mining Toward Generative Artificial Intelligence in Education: The Challenges and Responses for Sustainable Development in Education. *Sustainability*, *17*(3), 1127.

Wilcox, E. G., Hu, M. Y., Mueller, A., Warstadt, A., Choshen, L., Zhuang, C., & Linzen, T. (2025). Bigger is not always better: The importance of human-scale language modeling for psycholinguistics. *Journal of Memory and Language*, *144*, 104650.

Winne, P. H., & Azevedo, R. (2014). Metacognition. In Sawyer, R. K. (Ed.), *The Cambridge handbook of the learning sciences* (2nd ed., pp. 63–87). Cambridge University Press.

Woo, J. H., & Choi, H. (2021). Systematic review for AI-based language learning tools. *Journal of Digital Contents Society*, *22*(11), 1783–1792. DOI: 10.48550/arXiv.2111.04455

Woolf, B. P. (2021). AI in education. In Shapiro, S. (Ed.), *Encyclopedia of artificial intelligence* (2nd ed.). John Wiley & Sons.

Wu, M., Wang, W., Liu, S., Yin, H., Wang, X., Zhao, Y., Lyu, C., Wang, L., Luo, W., & Zhang, K. (2025). The Bitter Lesson Learned from 2,000+ Multilingual Benchmarks. *ArXiv Preprint ArXiv:2504.15521*. https://arxiv.org/abs/2504.15521

Wu, J., & Cai, Y. (2025). The Paradox of AI Knowledge: A Blockchain-Based Approach to Decentralised Governance in the Chinese New Media Industry. *Future Internet*, *17*(10), 479.

Wu, M.-J., Zhao, K., & Fils-Aime, F. (2022). Response rates of online surveys in published research: A meta-analysis. *Computers in Human Behavior Reports*, *7*. Advance online publication. DOI: 10.1016/j.chbr.2022.100206

Wunder, I. (2017). *The influence of cultural background on teaching and learning in synchronous online sessions*. Lancaster University (United Kingdom).

Wu, T., & Yu, Z. (2024). Bibliometric and systematic analysis of artificial intelligence chatbots' use for language education. *Journal of University Teaching & Learning Practice*, *21*(6), 174–198.

Wy, I., & Kurniawan, A. (2024). *English Language and Its Importance as Global Communication. Samā Jiva Jnānam*. International Journal of Social Studies., DOI: 10.25078/ijoss.v2i1.3920

Xiao, F., Zhu, S., & Xin, W. (2025). Exploring the landscape of generative AI (ChatGPT)-powered writing instruction in English as a foreign language education: A scoping review. *ECNU Review of Education*, 20965311241310881.

Xia, Y., Shin, S. Y., & Kim, J. C. (2024). Cross-cultural intelligent language learning system (CILS): Leveraging AI to facilitate language learning strategies in cross-cultural communication. *Applied Sciences (Basel, Switzerland)*, *14*(13), 5651.

Xue, Y. (2024). Towards automated writing evaluation: A comprehensive review with bibliometric, scientometric, and meta-analytic approaches. *Education and Information Technologies*, *29*(15), 19553–19594.

Xu, N., Gui, T., Ma, R., Zhang, Q., Ye, J., Zhang, M., & Huang, X. (2022). Cross-Linguistic Syntactic Difference in Multilingual BERT: How Good is It and How Does It Affect Transfer? *Proceedings of the 2022 Conference on Empirical Methods in Natural Language Processing, EMNLP 2022, Section 2*, 8073–8092. DOI: 10.18653/v1/2022.emnlp-main.552

Yakar, U., Sülü, A., Porgali, M., & Çaliş, N. (2020). From Constructivist Educational Technology to Mobile Constructivism: How mobile learning serves constructivism? *International Journal of Academic Research in Education*, *6*(1), 56–75. DOI: 10.17985/ijare.818487

Yan, D. (2023). Impact of ChatGPT on learners in a L2 writing practicum: An exploratory investigation. *Education and Information Technologies*, *28*(11), 13943–13967.

Yang, E., Jänich, T., Mayfield, J., & Lawrie, D. (2024). Language Fairness in Multilingual Information Retrieval. *SIGIR 2024 - Proceedings of the 47th International ACM SIGIR Conference on Research and Development in Information Retrieval*, 2487–2491. DOI: 10.1145/3626772.3657943

Yang, X. (2025, May). Generative AI Driven College English Multimodal Teaching Based on BERT Optimization: An Empirical Study on Interactive Participation and Learning Outcomes in Digital Twin Classrooms. In *2025 International Conference on Artificial Intelligence and Digital Ethics (ICAIDE)* (pp. 513-516). IEEE.

Yaqin, L. N. (2025). Emerging research in AI-assisted language learning: A systematic literature review. *Multi-Industry* Digitalization *and Technological Governance in the AI Era*, 29–50.

Yaqin, L. N., Hassan, H., & Yusof, B. (2025b). Performance And Accuracy Of Chatgpt In Generating Malay Academic Texts: A Comparative Study With Expert Corrections. *LLT Journal: A Journal on Language and Language* Teaching, *28*(1), 495–517.

Yaqin, L. N., Bilad, M. R., Yusof, B., Hassan, H., Damit, A. R., & Chuchu, I. F. (2025a). Mapping research evolution in higher education: A scientometric analysis of Brunei Darussalam (1986–2024). *Discover Sustainability*, *6*(1). Advance online publication. DOI: 10.1007/s43621-025-00917-3

Yaqin, L. N., McLellan, J., & Najib, N. (2025c). Shifting paradigms in translanguaging: A scientometric analysis of research trends and trajectories. *Studies in English Language and Education*, *12*(2), 684–702. https://doi.org/https://doi.org/10.24815/siele.v12i2.40243

Ye, Y., Zhang, Z., Ma, T., Wang, Z., Li, Y., Hou, S., ... & Chawla, N. (2025). LLM-s4All: A Review of Large Language Models Across Academic Disciplines.

Yenduri, G., Ramalingam, M., Selvi, G. C., Supriya, Y., Srivastava, G., Maddikunta, P. K. R., & Gadekallu, T. R. (2024). GPT (generative pre-trained transformer)—A comprehensive review on enabling technologies, potential applications, emerging challenges, and future directions. *IEEE Access : Practical Innovations, Open Solutions*, *12*, 54608–54649.

Yildiz, T. (2021). Degree Mobility in the Higher Education Context: Could the English Language be a Key Driver of Internationalisation? *Journal of University Research*, *4*(3), 220–227. DOI: 10.32329/uad.876946

Yuan, H., Che, Z., Li, S., Zhang, Y., Hu, X., & Luo, S. (2024). The high-dimensional psychological profile and cultural bias of ChatGPT. *arXiv preprint arXiv:2405.03387.*

Yuan, R., Lee, I., De Costa, P. I., Yang, M., & Liu, S. (2022). TESOL teacher educators in higher education: A review of studies from 2010 to 2020. *Language Teaching*, *55*(4), 434–469. DOI: 10.1017/S0261444822000209

Yunos, R. (2013). *The Golden Warisan Brunei Darussalam (volume 1).* Bandar Seri Begawan: Dewan Bahasa dan Pustaka.

Zampatti, S., Peconi, C., Megalizzi, D., Calvino, G., Trastulli, G., Cascella, R., & Giardina, E. (2024). Innovations in medicine: Exploring ChatGPT's impact on rare disorder management. *Genes*, *15*(4), 421. PMID: 38674356

Zawacki-Richter, O., Marín, V. I., Bond, M., & Gouverneur, F. (2019). Systematic review of research on artificial intelligence applications in higher education–where are the educators? *International Journal of Educational Technology in Higher Education*, *16*(1), 1–27.

Zhai, X. (2025). DeepSeek: Transforming the foundations of education [Preprint]. doi: DOI: 10.20944/preprints202503.1776.v1

Zhang, C., Song, J., Zhu, X., Zhu, L., & Zhang, S. (2021). Hcmsl: Hybrid cross-modal similarity learning for cross-modal retrieval. [TOMM]. *ACM Transactions on Multimedia Computing Communications and Applications*, *17*(1s), 1–22.

Zhang, R. W., Liang, X., & Wu, S. H. (2024). When chatbots fail: Exploring user coping following a chatbots-induced service failure. *Information Technology & People*, *37*(8), 175–195. DOI: 10.1108/ITP-08-2023-0745

Zhang, T., Erlam, R., & de Magalhães, M. B. (2025). Exploring the dual impact of AI in post-entry language assessment: Potentials and pitfalls. *Annual Review of Applied Linguistics*. Advance online publication. DOI: 10.1017/S0267190525000030

Zhao, J., Huang, C., & Li, X. (2024). A comparative study of cultural hallucination in large language models on culturally specific ethical questions [Preprint]. https://doi.org/DOI: 10.21203/rs.3.rs-4566507/v1

Zhong, T., Yang, Z., Liu, Z., Zhang, R., Liu, Y., Sun, H., Pan, Y., Li, Y., Zhou, Y., Jiang, H., Chen, J., & Liu, T. (2024). Opportunities and Challenges of Large Language Models for Low-Resource Languages in Humanities Research. *ArXiv Preprint*, 1–31. https://arxiv.org/abs/2412.04497

Zhu, M., & Wang, C. (2025). A systematic review of artificial intelligence in language education: Current status and future implications. *Language Learning & Technology*, *29*(1), 1–29. https://hdl.handle.net/10125/73606

Ziems, C., Held, W. B., Yu, J., Goldberg, A., Grusky, D., & Yang, D. (2025, November). Culture Cartography: Mapping the Landscape of Cultural Knowledge. In *Proceedings of the 2025 Conference on Empirical Methods in Natural Language Processing* (pp. 1739-1757).

Zohuri, B., & Behgounia, F. (2023). Application of artificial intelligence driving nano-based drug delivery system. In Philip, A., Shahiwala, A., Rashid, M., & Faiyazuddin, M. (Eds.), *A handbook of artificial intelligence in drug delivery* (pp. 145–212). Academic Press., DOI: 10.1016/B978-0-323-89925-3.00007-1

Zou, B., Guan, X., Shao, Y., & Chen, P. (2023). Supporting speaking practice by social network-based interaction in artificial intelligence (AI)-assisted language learning. *Sustainability*, *15*(4), 2872.

Zuboff, S. (2019). *The age of surveillance capitalism: The fight for a human future at the new frontier of power*. PublicAffairs.

About the Contributors

Lalu Nurul Yaqin is an Assistant Professor of Malay Language, Linguistics and Literature at Universiti Brunei Darussalam. His research focuses on cultural linguistics, comparative politeness, sociolinguistics, and AI in language studies. He serves as Editor-in-Chief and reviewer for international journals, lectures as a visiting academic across Southeast Asia, and examines Master's and PhD theses internationally. He has secured national and international research grants and delivers workshops on systematic reviews, scientometrics, and research software/AI.

Yabit Alas serves as the Director of the Continuing Education Centre. He held the position of Head of Malay Language and Linguistics from 2006 to 2008 and was subsequently appointed as the Director of the Language Centre from 2009 to 2010. In 2011, he spearheaded the establishment of the newly formed Continuing Education Centre before being entrusted with the leadership of the Faculty of Arts and Social Sciences from January 2012 until June 2013, after which he resumed his role at the Continuing Education Centre in July 2013. His area of expertise is Comparative Linguistics, particularly Austronesian Languages. However, his research predominantly focuses on the Dusunic Languages spoken in northern Borneo, with particular emphasis on their origin, dispersal, and classification. He is also actively engaged in research on Malay languages, mainly Standard Malay, and its relevance to globalisation.

Akhmad Habibi (Assoc. Prof. at Universitas Jambi and Adjunct Prof. at Korea University), a graduate of the University of Malaya, has expertise in educational technology, curriculum and instruction, English education, and statistical research. His scholarly contributions are reflected in 119 articles published in Scopus-indexed journals (h-index 23) and 64 articles in Web of Science (h-index 15). Beyond journal publications, he has authored five books and co-edited a book chapter for Springer, published in the prestigious Lecture Notes in Educational Technology series. He serves on the editorial boards of high-impact journals, including International

Studies in Sociology of Education, Discover Sustainability, the International Journal of Information and Education Technology, and Sustainability. As a trusted reviewer, he has consistently evaluated research for leading journals, including Computers and Education, Technology in Society, Technological Forecasting and Social Change, International Journal of Educational Development, and Education & Information Technologies. In recognition of his exceptional research contributions, he was honoured as the Best Researcher in both 2022 and 2023 by Universitas Jambi.

***

Syed Ibad Ali holds a distinguished academic background with a Bachelor of Engineering in Electronics and Communication Engineering (ECE), a Master of Technology in VLSI Design, and a PhD in Electronics and Communication Engineering. With over 13 years of rich academic and research experience, Dr Ali has been at the forefront of interdisciplinary innovation, merging traditional ECE principles with cutting-edge advancements in artificial intelligence and data science. Currently serving as an associate professor in the Department of Artificial Intelligence and Data Science at Parul Institute of Engineering and Technology, Parul University, Gujarat, he is recognised for his commitment to fostering analytical thinking and research-led pedagogy. His work often bridges theory with impactful applications, inspiring the next generation of engineers and AI specialists. Dr Ali's passion lies in advancing transparent and ethical AI systems, a vision that is deeply embedded in the themes of this book.

Constantine Andoniou is an Associate Professor of Education at Abu Dhabi University. With an extensive academic background, he holds a Bachelor of Education in Teacher Education, a Diploma in Education in Educational Psychology and Sociology of Education, a Master of Education in Philosophy of Education and Educational Sociology, and a Doctorate of Philosophy in Education. With teaching and leadership experience at higher education institutions across Europe, Australia, and the Middle East, Dr Andoniou actively collaborates with international researchers and industry leaders to bridge the gap between theoretical advances and the practical implementation of AI and digital technologies in education. Dr Andoniou's research spans information theory, postmodernism, digital learning theories, digital transformation, and the integration of educational technology in teaching and learning environments. His research focuses on AI integration in education, particularly on how artificial intelligence reshapes human cognition, transforms pedagogical practices, and redefines educators' roles.

Siti Ifwah Fauzani binti Awang Haji Chuchu is a Tutor in the Malay Language and Linguistics programme at Universiti Brunei Darussalam (UBD). Her teaching encompasses Malay syntax and generative grammar, grounded in morpho-syntactic analysis, and she draws on this expertise to engage with Brunei Darussalam's minority and endangered-language contexts. Her research interests include morphosyntax, broader work on language attitudes and language maintenance, and the documentation of Brunei Darussalam's minority and endangered languages, including Belait.

Ashrol Rahimy Damit is a Lecturer at Universiti Brunei Darussalam, where he teaches sociolinguistics, language and communication, and research methodology in linguistics in the Malay Language and Linguistics Programme. His research interests include language maintenance and language shift, the endangered languages of Brunei Darussalam, Malay language studies, and language use in cyberspace.

Rabi'atul Adawiyyah binti Haji Hassan is an education officer at the Language Centre, Universiti Brunei Darussalam (UBD), with ten years of teaching experience. She teaches Malay Language and Communication to both local and international students, with a particular focus on the fundamentals of the Malay language and its cultural contexts. She holds a Master's degree in Malay Language and Linguistics from Universiti Brunei Darussalam and is currently pursuing a PhD in Linguistics at Universiti Malaysia Sabah, with completion expected in 2025. Beyond her formal teaching and research, she pursues an active interest in foreign languages and cultures, particularly Korean and Filipino. Her research interests include multilingualism and translanguaging, language acquisition, and cross-cultural communication.

Vijayaprabakaran K is an Assistant Professor at Vellore Institute of Technology University. His research focuses on medical imaging, IoT, and UAV applications, with innovations in smart agriculture and healthcare AI. He is also an educator and frequent FDP speaker, delivering training and lectures on emerging AI technologies and applied intelligent systems.

Riya Kesharwani is an emerging researcher and a final-year Computer Science and Engineering undergraduate at Chandigarh University, with a keen interest in Artificial Intelligence, Data Analytics, and IoT systems. She has worked on impactful projects, including a hybrid machine learning model for solar power forecasting, and completed a specialisation in Industrial IoT, deepening her understanding of intelligent systems and edge computing. In addition to her technical endeavours, Riya has gained practical experience through internships focused on digital communication and brand strategy. She continues to explore how technology can drive innovation in sustainability, healthcare, and real-world problem-solving.

Min Shen is an educator and researcher specialising in foreign language education and communication studies. With over 15 years of experience in higher education and cross-cultural engagement, she has contributed significantly to language education, intercultural competence, and technology-mediated communication. She has served as chief editor and editor for several academic journals and translated works, with expertise spanning scholarly, literary, and interdisciplinary publications. Her recent research explores how artificial intelligence shapes communication, multilingual education, and digital transformation in language and communication practices across Southeast Asia.

Utkarsh Mishra is pursuing an Integrated MTech in Software Engineering at VIT Chennai, ranking among the top students with a 9.26 CGPA and a Meritorious Award (7th of 200+). He developed the government-funded, patented VSMART Trolley and has interned in Android development and cybersecurity. A researcher and inventor, he holds multiple patents, publishes in IEEE and the Alexandria Engineering Journal, and holds certifications from Google, the University of Helsinki, and IIT Bombay.

Nanlir Sallau Mullah holds a PhD in Computer Science from Universiti Sains Malaysia, an MSc from Coventry University, UK, and a B.Tech. (Hons) from Abubakar Tafawa Balewa University, Nigeria. His research interests include Big Data Mining, Agentic AI, and Natural Language Processing.

Özlem Utku-Bilici is a lecturer at Recep Tayyip Erdoğan University, with research interests in the social sciences and humanities, education, and philology. She received her PhD in Foreign Language Education from Middle East Technical University (2024) and completed a postdoctoral fellowship at Université de Strasbourg (2025). Her work focuses on teacher cognition, EFL teacher education, and L2 reading comprehension, including eye-tracking methods. She previously served at Trabzon and Bayburt Universities and has completed professional training in teacher education and R&D project management.

Wan Mohd Nazmee Wan Zainon received the PhD degree in computer science from Universiti Sains Malaysia. He is currently working as an Associate Professor at the School of Computer Sciences, Universiti Sains Malaysia. His research interests include visual computing, data mining and software engineering with a focus on software reuse, requirement engineering practices, visual data mining, and big data analytics.

Zulfadzlee Zulkiflee, a PhD candidate in historical linguistics in the Malay Language, Linguistics and Literature programme at the Faculty of Arts and Social Sciences, Universiti Brunei Darussalam. His research focuses on the classification of language variants within the Baram–Tinjar subgroup, the reconstruction of proto-forms, and the tracing of the phonological, lexical, and migration histories of indigenous communities in Borneo. Combining comparative linguistics with immersive fieldwork in Brunei and Sarawak, his work advances Austronesian subgrouping theory while prioritising community-engaged and open-access linguistic documentation.

Index

www.ingramcontent.com/pod-product-compliance
Lightning Source LLC
LaVergne TN
LVHW080907300926
849162LV00003B/585